5th EDITION

STUDENT
SOLUTIONS
MANUAL

for Smith's
PRECALCULUS
WITH GRAPHING AND PROBLEM SOLVING

MW00723632

Mary Ellen White
Portland City College

Jean Woody
Tulsa Junior College

Brooks/Cole Publishing Company
Pacific Grove, California

Brooks/Cole Publishing Company
A Division of Wadsworth, Inc.

Printed in the United States of America

10 9 8 7 6 5 4 3 2 1

ISBN 0-534-16784-5

Sponsoring Editor: Audra Silverie
Editorial Assistant: Carol Ann Benedict
Production Coordinator: Dorothy Bell
Cover Design: Kelly Shoemaker
Printing and Binding: Malloy Lithographing, Inc.

CONTENTS

CHAPTER 1

PROBLEM SET 1.1

1. a. Z, Q, R   b. Q, R   c. N, W, Z, Q, R   d. Q′, R   e. Q′, R   f. Q′, R

3. a. Q, R   b. Q, R   c. Q′, R   d. Q, R   e. Q′, R

5. $\pi \approx 3.141592654$ since π is non-repeating, non terminating so $\pi > 3.141592654$

7.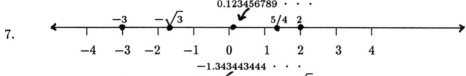

9.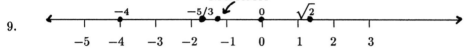

11.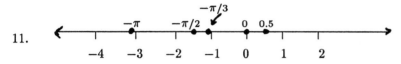

13. a. $<$ b. $-3 < 3$ 15. a. $-2.\overline{6} > -2.66$ b. $=$ 17. a. $=$ b. $<$

19. a. $\pi - 2$ b. $5 - \pi$ c. $2\pi - 6$ d. $7 - 2\pi$

21. a. $\sqrt{2} - 1$ b. $2 - \sqrt{2}$ c. $1 - \frac{\pi}{6}$ d. $\frac{2\pi}{3} - 1$

23. $-3 - (-\pi) = \pi - 3$ b. $-96 - (23) = |-119| = |119|$

   c. $-3 - (-\sqrt{5}) = |-3 + \sqrt{5}| = |3 - \sqrt{5}| = 3 - \sqrt{5}$

25. reflexive   27. distributive   29. distributive and closure   31. transitive   33. substitution

35. closure   37. multiplicative inverse (reciprocal)   39. distributive   41. multiplicative identity

43. No, since $1 + 1 = 2$ which is not an element of the original set.

45. No, since $-1 + -1 = -2$ which is not an element of the original set.

47. Yes, $0 + 3 = 3$; $3 + 3 = 6$, $3 + 6 = 9$, etc. You are working with an infinite set of multiples of 3 and you will always get an answer that is a multiple of 3.

49. No, $\frac{3}{6} = \frac{1}{2}$ which is not an element of the set.

51. $\frac{15}{3} \neq \frac{3}{15}$ or $5 \neq \frac{1}{5}$ and there are many other examples.

53. No, for example $(50 \div 10) \div 5 = 50 \div (10 \div 5)$ says that $1 = 25$, which is not true.

55. Additive inverse of $\frac{\pi}{3} + 1$ is $\frac{-\pi}{3} - 1$ since when added together these will equal 0.

57. The multiplicative inverse is $\frac{3}{\pi + 3}$ since when multiplied the answer is equal to 1.

1

59. For addition: closure, commutative and associative. For multiplication: closure, commutative, associative, identity and distributive.

61. For addition: closure, commutative, associative, identity and inverse. For multiplication: closure, commutative, associative, identity and distributive.

63. For addition: not closed, commutative and associative. For multplication: not closed, commutative, associative, and distributive.

65. For addition: none. For multiplication: closure, commutative, associative, identity, and inverse.

PROBLEM SET 1.2

1. a. $(x + 2)(x + 1) = x^2 + 2x + x + 2 = x^2 + 3x + 2$

   b. $(y - 2)(y + 3) = y^2 + 3y - 2y - 6 = y^2 + y - 6$

   c. $(x + 1)(x - 2) = x^2 - 2x + x - 2 = x^2 - x - 2$

   d. $(y - 3)(y + 2) = y^2 + 2y - 3y - 6 = y^2 - y - 6$

3. a. $(2x + 1)(x - 1) = 2x^2 - 2x + x - 1 = 2x^2 - x - 1$

   b. $(2x - 3)(x - 1) = 2x^2 - 2x - 3x + 3 = 2x^2 - 5x + 3$

   c. $(x + 1)(3x + 1) = 3x^2 + x + 3x + 1 = 3x^2 + 4x + 1$

   d. $(x + 1)(3x + 2) = 3x^2 + 2x + 3x + 2 = 3x^2 + 5x + 2$

5. a. $(a + 2)^2 = (a + 2)(a + 2) = a^2 + 2a + 2a + 4 = a^2 + 4a + 4$

   b. $(b - 2)^2 = (b - 2)(b - 2) = b^2 - 2b - 2b + 4 = b^2 - 4b + 4$

   c. $(x + 4)^2 = (x + 4)(x + 4) = x^2 + 4x + 4x + 16 = x^2 + 8x + 16$

   d. $(y - 3)^2 = (y - 3)(y - 3) = y^2 - 3y - 3y + 9 = y^2 - 6y + 9$

7. a. $me + mi + my = m(e + i + y)$ b. $a^2 - b^2 = (a - b)(a + b)$

   c. $a^2 + b^2$ is not factorable   d. $a^3 - b^3 = (a - b)(a^2 + ab + b^2)$

9. a. $a^3 + 3a^2b + 3ab^2 + b^3 = (a + b)^3$ b. $p^3 - 3p^2q + 3pq^2 - q^3 = (p - q)^3$

   c. $-c^3 + 3cd - 3cd + d^3 = (d - c)^3$ d. $x^2y + xy^2 = xy(x + y)$

11. a. $3x^2 - 5x - 2 = (3x + 1)(x - 2)$ b. $6y^2 - 7y + 2 = (3y - 2)(2y - 1)$

    c. $8a^2b + 10ab + 3b = b(8a^2 + 10a + 3) = b(4a - 1)(2a + 3)$

    d. $2s^2 - 10s - 48 = 2(s^2 - 5s - 24) = 2(s - 8)(s + 3)$

13. a. $(3x - 1)(x^2 + 3x - 2) = 3x^3 + 9x^2 - 6x - x^2 - 3x + 2 = 3x^3 + 8x - 9x + 2$

    b. $(2x + 1)(x^2 + 2x - 5) = 2x^3 + 4x^2 - 20x + x^2 + 2x - 5 = 2x^3 + 5x^2 - 20x - 5$

15. a. $(x + 1)(x - 3)(2x + 1) = (x + 1)(2x^2 - 5x - 3) = 2x^3 - 3x^2 - 8x - 3$

    b. $(2x - 1)(x + 3)(3x + 1) = (2x - 1)(3x^2 + 10x + 3) = 6x^3 + 17x^2 - 4x - 3$

17. a. $(x - 2)^2(x + 1) = (x^2 - 4x - 4)(x + 1) = x^3 - 3x^2 + 4$

    b. $(x - 2)(x + 1)^2 = (x - 2)(x^2 + 2x + 1) = x^3 - 4x^2 - 3x - 2$

19. $(x - y)^2 - 1 = [(x - y) + 1][(x - y) - 1] = (x - y + 1)(x - y - 1)$

21. $(5a - 2)^2 - 9 = [(5a - 2) + 3][(5a - 2) - 3] = (5a - 2 + 3)(5a - 2 - 3)$

    $(5a + 1)(5a - 5) = 5(5a + 1)(a - 1)$

23. $\frac{4}{25}x^2 - (x + 2)^2 = \frac{1}{25}[4x^2 - 25(x + 2)^2] = \frac{1}{25}[2x + 5(x + 2)][2x - 5(x + 2)]$

    $\frac{1}{25}[2x + 5x + 10][2x - 5x - 10] = \frac{1}{25}(7x + 10)(-3x - 10) = \frac{1}{25}(7x + 10)(3x + 10)$

25. $\frac{x^6}{y^8} - 169 = \frac{1}{y^8}[x^6 - 169y^8] = \frac{1}{y^8}(x^3 - 13y^4)(x^3 + 13y^4)$

27. $(a + b)^2 - (x + y)^2 = [(a + b) + (x + y)][(a + b) - (x + y)] = (a + b + x + y)(a + b - x - y)$

29. $2x^2 + x - 6 = (2x - 3)(x + 2)$ 31. $6x^2 + 47x - 8 = (6x - 1)(x + 8)$

33. $6x^2 + 49x + 8 = (6x + 1)(x + 8)$ 35. $4x^2 + 13x - 12 = (4x - 3)(x + 4)$

37. $9x^2 - 56x + 12 = 9x^2 - 2x - 54x + 12 = x(9x - 2) - 6(9x - 2) = (9x - 2)(x - 6)$

39. $4x^4 - 17x^2 + 4 = (4x^2 - 1)(x^2 - 4) = (2x + 1)(2x - 1)(x + 2)(x - 2)$

41. $x^6 + 9x^3 + 8 = (x^3 + 8)(x^3 + 1) = (x + 2)(x^2 - 2x + 4)(x + 1)(x^2 - x + 1)$ OR

    $(x + 1)(x + 2)(x^2 - x + 1)(x^2 - 2x + 4)$

43. $(x^2 - \frac{1}{4})(x^2 - \frac{1}{9}) = \frac{1}{4}(4x^2 - 1)\frac{1}{9}(9x^2 - 1) = \frac{1}{36}(2x + 1)(2x - 1)(3x + 1)(3x - 1)$

45. $(x + y)^3 = (x + y)(x + y)^2 = (x + y)(x^2 + 2xy + y^2) = x^3 + 3x^2y + 3xy^2 + y^3 \neq x^3 + y^3$

47. $\sqrt{x^2 + y^2}$ is not factorable but $\sqrt{(x + y)^2}$ is and equals $x + y$.

49. $\sqrt{x^3} = \sqrt{x^2\, x} = \sqrt{x^2}\sqrt{x} = |x|\sqrt{x}$ or $x\sqrt{x}$

51. $-2(2x^2 - 5x)^{-3}(4x - 5) = \dfrac{-2(4x - 5)}{(2x^2 - 5x)^3} = \dfrac{-2(4x - 5)}{x^3(2x - 5)^3} = -2x^{-3}(4x - 5)(2x - 5)^{-3}$

53. $(2x + 1)^2[3(3x + 2)^2(3)] + (3x + 2)^3[2(2x + 1)(2)]$

    $= (2x + 1)(3x + 2)^2[(2x + 1)(9) + (3x + 2)(4)] = (2x + 1)(3x + 2)^2[18x + 9 + 12x + 8]$

    $= (2x + 1)(3x + 2)^2(30x + 17)$

55. $\dfrac{[(x^2 + 3)(3)(7x + 11)^2(7)] - [(7x + 11)^3(2x)]}{(x^2 + 3)^2} = \dfrac{(7x + 11)^2}{(x^2 + 3)^2}\left([21(x^2 + 3)] - [(7x + 11)2x]\right)$

    $= \dfrac{(7x + 11)^2}{(x^2 + 3)^2}\left(21x^2 + 63 - 14x^2 + 22x\right) = \dfrac{(7x + 11)^2}{(x^2 + 3)^2}(7x^2 + 22x + 63)$

57. $(2x - 1)^2(3x^4 - 2x^3 + 3x^2 - 5x + 12) = (4x^2 - 4x + 1)(3x^4 - 2x^3 + 3x^2 - 5x + 12) =$

$$
\begin{array}{r}
3x^4 - 2x^3 + 3x^2 - 5x + 12 \\
4x^2 - 4x + 1 \\
\hline
12x^6 - 18x^5 + 12x^4 - 20x^3 + 48x^2 \\
12x^5 + 8x^4 - 12x^3 + 20x^2 - 48x \\
3x^4 - 2x^3 + 3x^2 - 5x + 12 \\
\hline
12x^6 - 20x^5 + 23x^4 - 34x^3 + 71x^2 - 53x + 12
\end{array}
$$

59. $(2x + 1)^2(5x^4 - 6x^3 - 3x^2 + 4x - 5)$
$= (4x^2 + 4x + 1)(5x^4 - 6x^3 - 3x^2 + 4x - 5)$

$$
\begin{array}{r}
5x^4 - 6x^3 - 3x^2 + 4x - 5 \\
4x^2 + 4x + 1 \\
\hline
20x^6 - 24x^5 - 12x^4 + 16x^3 - 20x^2 \\
20x^5 - 24x^4 - 12x^3 + 16x^2 - 20x \\
5x^4 - 6x^3 - 3x^2 + 4x - 5 \\
\hline
20x^6 - 4x^5 - 31x^4 - 2x^3 - 7x^2 - 16x - 5
\end{array}
$$

61. $(2x^3 + 3x^2 - 2x + 4)^3 = (2x^3 + 3x^2 - 2x + 4)(2x^3 + 3x^2 - 2x + 4)^2$ Use the vetical process as shown in problem #59 square the last two and then multiply by the first one and you will get the following answer: $8x^9 + 36x^8 + 30x^7 + 3x^6 + 114x^5 + 48x^4 + 56x^3 + 192x^2 - 96x + 64$.

63. $x^{2n} - y^{2n} = (x^n + y^n)(x^n - y^n)$ 65. $x^{3n} + y^{3n} = (x^n + y^n)(x^{2n} - x^n y^n + y^{2n})$

67. $(x - 2)^2 - (x - 2)^1 - 6 = [(x - 2) - 3][(x - 2) + 2] = (x - 2 - 3)(x - 2 + 2) = x(x - 5)$

69. $z^5 - 8z^2 - 4z^3 + 32 = z^2(z^3 - 8) - 4(z^3 - 8) = (z^3 - 8)(z^2 - 4)$

$= (z + 2)(z - 2)(z - 2)(z^2 + 2z + 4) = (z + 2)(z - 2)^2(z^2 + 2z + 4)$

71. $x^2 - 2xy + y^2 - a^2 - 2ab - b^2 = (x - y)^2 - (a + b)^2 = [(x - y) + (a + b)][(x - y) - (a + b)]$

$= (x - y + a + b)(x - y - a - b)$

73. $x^2 + y^2 - a^2 - b^2 - 2xy + 2ab = (x^2 - 2xy + y^2) - (a^2 - 2ab + b^2) = (x - y)^2 - (a - b)^2$

$= [(x - y) + (a - b)][(x - y) - (a - b)] = (x - y + a - b)(x - y - a + b)$

75. $(x + y + 2z)^2 - (x - y + 2z)^2 = [(x + y + 2z) - (x - y + 2z)][(x + y + 2z) + (x - y + 2z)]$

$= (x + y + 2z - x + y - 2z)(x + y + 2z + x - y + 2z) = 2y(2x + 4z)$ or $4xy + 8yz$

77. $2(x + y)^2 - 5(x + y)(a + b) - 3(a + b)^2$

$= [2(x + y) + (a + b)][(x + y) - 3(a + b)] = [2x + 2y + a + b][x + y - 3a - 3b\}$

79. $3(x + 1)^2(x - 2)^4 + 4(x + 1)^3(x - 2)^3 = (x + 1)^2(x - 2)^3[3(x - 2) + 4(x + 1)]$

$= (x + 1)^2(x - 2)^3[3x - 6 + 4x + 4] = (x + 1)^2(x - 2)^3(7x - 2)$

81. $3(2x - 1)^2(2)(3x + 2)^2 + 2(2x - 1)^3(3x + 2)^2(3)$

$= 6(2x - 1)^2(3x + 2)^2[1 + (2x - 1)] = 6(2x - 1)^2(3x + 2)^2(2x) = 12x(2x - 1)^2(3x + 2)^2$

83. $4(x+5)^3(x^2-2)^3 + (x+5)^4(3)(x^2-2)^2(2x) = 2(x+5)^3(x^2-2)^2[2(x^2-2) + 3x(x+5)]$

$= 2(x+5)^3(x^2-2)^2[2x^2 - 4 + 3x^2 + 15] = 2(x+5)^3(x^2-2)^2(5x^2+11)$

PROBLEM SET 1.3

1. a. $d = \sqrt{(5-8)^2 + (1-5)^2} = \sqrt{(-3)^2 + (-4)^2} = \sqrt{9+16} = \sqrt{25} = 5$

   b. $d = \sqrt{(13-1)^2 + (9-5)^2} = \sqrt{(12)^2 + 4^2} = \sqrt{144+16} = \sqrt{160} = 4\sqrt{10}$

3. a. $d = \sqrt{(4-3)^2 + (5+1)^2} = \sqrt{1+36} = \sqrt{37}$

   b. $d = \sqrt{[(-1-(-2)]^2 + (-5-1)^2} = \sqrt{1^2 + (-6)^2} = \sqrt{1+36} = \sqrt{37}$

5. $d = \sqrt{(-3x-x)^2 + (2x-5x)^2} = \sqrt{(-4x)^2 + (-3x)^2} = \sqrt{16x^2 + 9x^2}$

   $= \sqrt{25x^2} = 5x$ with $x < 0$

7. a. $m = \left(\dfrac{5+8}{2}, \dfrac{1+5}{2}\right) = \left(\dfrac{13}{2}, \dfrac{6}{2}\right) = \left(\dfrac{13}{2}, 3\right)$

   b. $m = \left(\dfrac{1+13}{2}, \dfrac{4+9}{2}\right) = \left(7, \dfrac{13}{2}\right)$

9. a. $m = \left(\dfrac{4+3}{2}, \dfrac{5+-1}{2}\right) = \left(\dfrac{7}{2}, \dfrac{4}{2}\right) = \left(\dfrac{7}{2}, 2\right)$

   b. $m = \left(\dfrac{-2+-1}{2}, \dfrac{1-5}{2}\right) = \left(\dfrac{-3}{2}, \dfrac{-4}{2}\right) = \left(-\dfrac{3}{2}, -2\right)$

11. $m = \left(\dfrac{x+-3x}{2}, \dfrac{5x+2x}{2}\right) = \left(\dfrac{-2x}{2}, \dfrac{7x}{2}\right) = \left(-x, \dfrac{7x}{2}\right)$

13.

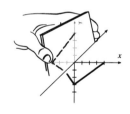

15.

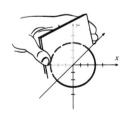

5

17. $x + y + 3 = 0$
$x + y = -3$

x	0	-3	1
y	-3	0	-4

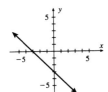

19. $3x + 2y = -6$

x	0	-2
y	-3	0

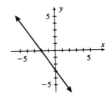

21. $y = -x^2$

x	0	1	-1	2	-2
y	0	-1	-1	-4	-4

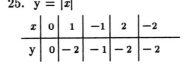

23. $x^2 + 3y = 0$
$y = -\frac{1}{3} x^2$

x	0	1	-1	2	-2	3	-3
y	0	$-\frac{1}{3}$	$-\frac{1}{3}$	$-\frac{4}{3}$	$-\frac{4}{3}$	-3	-3

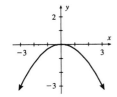

25. $y = |x|$

x	0	1	-1	2	-2
y	0	-2	-1	-2	-2

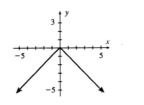

27. $y = 2 |x|$

x	0	1	-1	2	-2
y	0	2	2	4	4

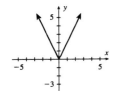

6

29.

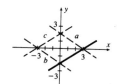

31.

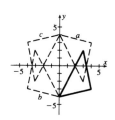

33.

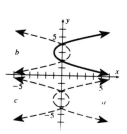

35. $y = x^2$;   a. y-axis: $y = (-x)^2 = x^2$ so y-axis symmetry

b. x-axis: $-y = x^2$; $y = -x^2$ so no x-axis symmetry

c. origin: $-y = (-x)^2$; $-y = x^2$; $y = -x^2$ so no origin symmetry

37. $y = \dfrac{x^3 - 1}{x - 1}$;

a. y-axis: $y = \dfrac{(-x)^3 - 1}{-x - 1}$; $y = \dfrac{-x^3 - 1}{-x - 1}$; $y = \dfrac{-(x^3 + 1)}{-(x + 1)}$; $y = \dfrac{x^3 + 1}{x + 1}$, no y-axis symmetry.

b. x-axis: $-y = \dfrac{x^3 - 1}{x - 1}$; $y = \dfrac{-(x^3 - 1)}{x - 1}$, no x-axis symmetry

c. origin: $-y = \dfrac{(-x)^3 - 1}{-x - 1}$; $-y = \dfrac{-x^3 - 1}{-x - 1}$; $-y = \dfrac{-(x^3 - 1)}{-(x + 1)}$;

$-y = \dfrac{x^3 - 1}{x - 1}$, no origin symmetry

39. $y = -\dfrac{\sqrt{x + 3}}{x - 1}$;

a. y-axis: $y = -\dfrac{\sqrt{-x + 3}}{-x - 1}$; $y = -\dfrac{\sqrt{3 - x}}{-(x + 1)}$; $y = \dfrac{\sqrt{3 - x}}{x + 1}$, no y-axis symmetry.

b. x-axis: $-y = -\dfrac{\sqrt{-x + 3}}{-x - 1}$; $y = \dfrac{\sqrt{-x + 3}}{-x - 1}$, no x-axis symmetry.

c. origin: $-y = -\dfrac{\sqrt{-x + 3}}{-x - 1}$; $y = -\dfrac{\sqrt{3 - x}}{-(x + 1)}$; $y = \dfrac{\sqrt{3 - x)}}{x + 1}$, no origin symmetry.

41. $2|x| - |y| = 5$;   a. y-axis: $2|-x| - |y| = 5$; $2|x| - |y| = 5$, y-axis symmetry.

b. x-axis: $2|x| - |-y| = 5$; $2|x| - |y| = 5$, x-axis symmetry.

c. origin: $2|-x| - |-y| = 5$; $2|x| - |y| = 5$, origin symmetry. Remenber that if two of the symmetries are true then all three are true.

43. $xy = 1$;   a. y-axis: $-xy = 1$, no y-axis symmetry.
b. x-axis: $x(-y) = 1$; $-xy = 1$, no x-axis symmetry.
c. origin: $-x(-y) = 1$; $xy = 1$, origin symmetry.

45. $x^2 + 2xy + y^2 = 4$;
a. y-axis: $(-x)^2 + 2(-x)y + y^2 = 4$; $x^2 - 2xy + y^2 = 4$, no y-axis symmetry.
b. x-axis: $x^2 + 2x(-y) + (-y)^2 = 4$; $x^2 - 2xy + y^2 = 4$, no x-axis symmetry.
c. origin: $(-x)^2 + 2(-x)(-y) + (-y)^2 = 4$; $x^2 + 2xy + y^2 = 4$, origin symmetry.

47. $C(5, -1)$, $r = 4$
$(x - h)^2 + (y - k)^2 = r^2$; $(x - 5)^2 + (y - (-1))^2 = 4^2$; $(x - 5)^2 + (y + 1)^2 = 16$

49. endpoints: $(1, -4)$, $(3, 6)$;
midpoint: $\left(\frac{1 + 3}{2}, \frac{-4 + 6}{2}\right)$; $\left(\frac{4}{2}, \frac{2}{2}\right)$; $(2, 1)$ is the center of the circle.

The distance between the two points is the diameter and one-half the distance is the radius.

$d = \sqrt{(1 - 3)^2 + (-4 - 6)^2} = \sqrt{(-2)^2 + (-10)^2} = \sqrt{4 + 100} = \sqrt{104} = 2\sqrt{26}$;

$\frac{1}{2}(2\sqrt{26}) = \sqrt{26} = r$ so $(x - 2)^2 + (y - 1)^2 = 26$

51. $C(a, b)$ and $P(x, 0)$; radious $= \sqrt{(x - a)^2 + (0 - b)^2} = \sqrt{(x - a)^2 + b^2} = r$; $r^2 = (x - a)^2 + b^2$;
The equation is: $(x - a)^2 + (y - b)^2 = (x - a)^2 + b^2$

53. $A(1, 3)$, $B(7, 1)$, $C(7, 10)$; $d_{ab} = \sqrt{(1 - 7)^2 + (3 - 1)^2} = \sqrt{36} = \sqrt{40}$;
$d_{ac} = \sqrt{(1 - 7)^2 + (3 - 10)^2} = \sqrt{36 + 49} = \sqrt{85}$
$d_{bc} = \sqrt{(7 - 7)^2 + (1 - 10)^2} = \sqrt{0 + 81} = \sqrt{81}$; Not a right triangle.

55. $A(0, 0)$, $B(4, 3)$, $C(-3, 8)$; $d_{ab} = \sqrt{(0 - 4)^2 + (0 - 3)^2} = \sqrt{16 + 9} = \sqrt{25}$
$d_{ac} = \sqrt{(0 + 3)^2 + (0 - 8)^2} = \sqrt{9 + 64} = \sqrt{73}$
$d_{bc} = \sqrt{(4 + 3)^2 + (3 - 8)^2} = \sqrt{49 + 25} = \sqrt{74}$; Not a right triangle.

57. $A(3, 2)$, $B(7, -4)$, $C(4, -6)$; $d_{ab} = \sqrt{(3 - 7)^2 + (2 + 4)^2} = \sqrt{16 + 36} = \sqrt{52}$
$d_{ac} = \sqrt{(3 - 4)^2 + (2 + 6)^2} = \sqrt{1 + 64} = \sqrt{65}$
$d_{bc} = \sqrt{(7 - 4)^2 + (-4 + 6)^2} = \sqrt{9 + 4} = \sqrt{13}$
Yes it is a right triangle since $(\sqrt{52})^2 + (\sqrt{13})^2 = (\sqrt{65})^2$.

59. $\sqrt{(x - 2)^2 + (y - 3)^2} = 7$; $(x - 2)^2 + (y - 3)^2 = 49$

61. $\sqrt{(x + 4)^2 + (y - 1)^2} = 3$; $(x + 4)^2 + (y - 1)^2 = 9$

63. Find the points that are 8 units from $(2, 4)$ on the y-axis, $x = 0$, $d = 8$

$(0 - 2)^2 + (y - 4)^2 = 8$; $4 + (y - 4)^2 = 8$; $4 + (y - 4)^2 = 64$;

$y^2 - 8y - 44 = 0$ and $y = 4 \pm 2\sqrt{15}$; $\left\{(0, 4 \pm 2\sqrt{15})\right\}$

65. $A = \pi r^2$

r	A
1	3.14
2	12.6
3	28.3
.5	0.8
.25	0.2

r	A
−1	3.14
−2	12.6
.5	0.8
.25	0.2
0	0

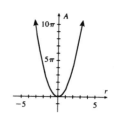

67. $h = \frac{a}{2}\sqrt{3}$

a	h
0	0
1	0.9
2	1.7
3	2.6

a	h
−1	−0.9
−2	−1.7
−3	−2.6

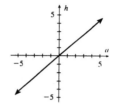

69. $P_1(x_1, y_1)$ and $P_2(x_2, y_2)$; $x_1 = x_2$; show that $P_1 \rightarrow P_2$ distance is $|y_2 - y_1|$.

$$d = \sqrt{(x_2 - x_1)^2 - (y_2 - y_1)^2} = \sqrt{(x_2 - x_2)^2 + (y_2 - y_1)^2} = \sqrt{0^2 + (y_2 - y_1)^2} = |y_2 - y_1|.$$

71. 1) $(-x, y)$ If symmetry works or 1 but not for 2, then $-y$ is not true so 3 cannot
    2) $(x, -y)$ true. If the summetry works for 2 but not for 1, then $-x$ is not true so
    3) $(-x, -y)$ 3 cannot be true.

73. $\{ (x, y) \text{ to } (4, 0) \} + \{ (x, y) \text{ to } (-4, 0) \} = 10$

$$\sqrt{(x-4)^2 + (y-0)^2} + \sqrt{(x+4)^2 + (y-0)^2} = 10;$$
$$\sqrt{(x-4)^2 + y^2} + \sqrt{(x+4)^2 + y^2} = 10; \quad \sqrt{(x-4)^2 + y^2} = 10 - \sqrt{(x+4)^2 + y^2};$$
$$(x-4)^2 + y^2 = 100 - 20\sqrt{(x+4)^2 + y^2} + (x+4)^2 + y^2;$$
$$x^2 - 8x + 16 = 100 - 20\sqrt{(x+4)^2 + y^2} + x^2 + 8x + 16;$$

Continued on next page.

9

$-16x - 100 = -20\sqrt{(x+4)^2 + y^2}$; $\quad 4x + 25 = 5\sqrt{(x+4)^2 + y^2}$;

$16x^2 + 200x + 625 = 25[(x+4)^2 + y^2]$; $\quad 16x^2 + 200x + 625 = 25x^2 + 200x + 400 + 25y^2$;

$0 = 9x^2 - 225 + y^2$; $\quad 9x^2 + y^2 = 225$

75. $\left\{(x, y) \text{ to } (0, 5)\right\} - \left\{(x, y) \text{ to } (0, -5)\right\} = 8$

$\sqrt{x^2 + (y-5)^2} - \sqrt{(x-0)^2 + (y+5)^2} = 8$; $\quad \sqrt{x^2 + (y-5)^2} = 8 + \sqrt{x^2 + (y+5)^2}$;

$x^2 + (y-5)^2 = 64 + 16\sqrt{x^2 + (y+5)^2} + x^2 + (y+5)^2$;

$y^2 - 10y + 25 = 64 + 16\sqrt{x^2 + (y+5)^2} + y^2 + 10y + 25$;

$-20y - 64 = 16\sqrt{x^2 + (y+5)^2}$; $\quad 5y + 16 = -4\sqrt{x^2 + (y+5)^2}$;

$25y^2 + 160y + 256 = 16\left[x^2 + (y+5)^2\right]$; $\quad 25y^2 + 160y + 256 = 16x^2 + 16y^2 + 160y + 400$;

$0 = 16x^2 - 9y^2 + 144$; $\quad 9y^2 - 16x^2 = 144$

PROBLEM SET 1.4

1. a. $(3, 7)$,

   c. $[-2, 6]$,

   b. $(-4, -1)$,

   d. $(-3, 0)$,

3. a. $(-\infty, -3]$,

   c. $(-\infty, 0)$,

   b. $[-2, \infty)$,

   d. $(2, \infty)$;

5. a. $[-3, 2]$,

   c. $(-\infty, 3]$,

   b. $(-2, 2)$,

   d. $[-2, \infty)$,

7. a.

   b.

   c.

   d.

9. a. $-4 \leq x \leq 2$ b. $-1 \leq x \leq 2$ c. $0 < x < 8$ d. $-5 < x \leq 3$

11. a. $x < 2$ b. $x > 6$ c. $x > -1$ d. $x \leq 3$

13. False; if written with the largest number on the right you have $1 < x < 5$ which says that x ranges between 1 and 5 and does not include either end point.

15. True; multiplying both sides of inequality by the same positive value keeps the sense of the inequality.

17. False; $|x - 3| > 4$ says that the solutions four units farther away from three in each direction.

19. $|x| = 5$ so $x = \pm 5$

21. $|x| = -1$ There is no solution since the absolute value is equal to a positive number so { }.

23. $|x - 3| = 4$ means that $x - 3 = 4$ so $x = 7$ or $x - 3 = -4$ and $x = -1$ and the solution set is $\{-1, 7\}$.

25. $|x - 9| = 15$ which means that $x - 9 = 15$ and $x = 24$ or $x - 9 = -15$ and $x = -6$, $\{-6, 24\}$

27. $|2x + 4| = -12$, no solution since the answer has to be a positive 12, { }

29. $|5x + 4| = 6$, $5x + 4 = 6$ and $x = \frac{2}{5}$, $5x + 4 = -6$ and $x = -2$, $\{-2, \frac{2}{5}\}$

31. $3x - 9 \geq 12$
$3x \geq 21$
$x \geq 7$
$[\, 7, \infty)$

33. $-3x \geq 123$
$x \leq -\dfrac{123}{3}$
$x \leq -41$
$(-\infty, -41]$

35. $5(3 - x) > 3x - 1$
$15 - 5x > 3x - 1$
$15 > 8x - 1$
$x < 2$, $(-\infty, 2)$

37. $-8 < 5x < 0$
$-\frac{8}{5} < x < 0$
$\left(-\frac{8}{5}, 0\right)$

39. $-3 \leq -x < -1$
$3 \geq x > 1$
$1 < x \leq 3$
$(1, \ 3]$

41. $-7 \leq 2x + 1 < 5$
$-8 \leq 2x < 4$
$-4 \leq x < 2$
$[-4, 2)$

43. $-5 \leq 3 - 2x < 18$
$-8 \leq -2x < 15$
$4 \geq x > -\frac{15}{2}$
$\left(-\frac{15}{2}, 4\right]$

45. $|x - 3| \leq 5$
$-5 \leq x - 3 \leq 5$
$-2 \leq x \leq 8$
$[-2, 8]$

47. $|x - 3| \leq 0.001$
$-0.001 \leq x - 3 \leq 0.001$
$-2.999 \leq x \leq 3.001$
$[-2.999, 3.001]$

49. $|2x + 1| < -1$, Not possible so { }

51. $|4 - 3x| < 3$
$-3 < 4 - 3x < 3$
$-7 < -3x < -1$
$\frac{1}{3} < x < \frac{7}{3}$
$\left(\frac{1}{3}, \frac{7}{3}\right)$

53. $|2x + 7| > 5$
$2x + 7 > 5$ and $2x + 7 < -5$
$2x > -2$ $2x < -12$
$x > -1$ $x < -6$
$(-\infty, -6) \cup (-1, \infty)$

55. $35 + 0.45x < 125$

$0.45x < 90$
$x < 200$
Drive less than 200 miles/day.

57. $20 < \frac{5}{9}(F - 32) < 30$

$20 < \frac{5}{9}F - \frac{160}{9} < 30$
$180 < 5F - 160 < 270$
$340 < 5F < 430$
$68° < F < 86°$

59. $s < 10,000$ and $s > 20,000$, $|x - 15,000| > 5,000$; answers in thousands would be
$s > 10$ or $s < 20$ and $|x - 15| > 5$

61. $\left|\dfrac{3 - x}{2}\right| > 3$

$\dfrac{3 - x}{2} > 3$ and $\dfrac{3 - x}{2} < -3$

$3 - x > 6$ $3 - x < -6$
$-x > 3$ $-x < -9$
$x < -3$ $x > 9$
$(-\infty, -3) \cup (9, \infty)$

63. $\dfrac{3}{|x + 1|} \geq 3$

$3 \geq 3|x + 1|$

$1 \geq |x + 1|$
$-1 \leq x + 1 \leq 1$
$-2 \leq x \leq 0$
$[-2, 0]$

11

65. $|3x + 1| + |x| - 5 = 0$, $|3x + 1| = 5 - |x|$,
    $3x + 1 = 5 - x$ and $-(3x + 1) = 5 + x$
    $3x + 1 = 5 - x$ $-3x - 1 = 5 + x$
        $4x = 4$ $-4x = 6$
        $x = 1$ $x = -\frac{3}{2}$ $\{-\frac{3}{2}, 1\}$

67. If $|a| \leq b$, show that $-b \leq a \leq b$.
    By definition $|a| = a$ if $x \geq 0$ and $|a| = -a$ if $a < 0$.
Case 1:   If $a \geq 0$
           $|a| \leq b$ given
           $a \leq b$ substitution of a for $|a|$
Case 2:   If $a < 0$
           $|a| \leq b$ given
           $-a \leq b$ substitution of $-a$ for $|a|$
           $a - a \leq a + b$ definition of equality
           $0 \leq a + b$ additive inverse
           $-b + 0 \leq a + b - b$ definition of equality
           $-b \leq a$ additive inverse, and definition of zero
So $-b \leq a$ and $a \leq b$. Combine in one statement by the transitive property: $-b \leq a \leq b$

69. If $|a| \geq b$, show that either $a \geq b$ or $a \leq -b$
Case 1:   If $a \geq 0$
           $|a| \geq b$ given
           $a \geq b$ substitution of a for $|a|$
Case 2:   If $|a| \geq b$
           $|a| \geq b$ given
           $-a \geq b$ substitution of $-a$ for $|a|$
           $(-1) - a \geq (-1)b$ definition of equality
           $a \geq -b$
So $a \geq b$ or $a \leq -b$

PROBLEM SET 1.5

1. $\sqrt{-36} = \sqrt{-1}\sqrt{36} = 6i$ 3. $\sqrt{-49} = 7i$ 5. $\sqrt{-20} = 2i\sqrt{5}$

7. $(3 + 3i) + (5 + 4i) = (3 + 5) + (3i + 4i) = 8 + 7i$

9. $(5 - 3i) - (5 + 2i) = (5 - 5) + (-3i - 2i) = 0 - 5i = 5i$

11. $(4 - 2i) - (3 + 4i) = 1 - 6i$ 13. $5 - (2 - 3i) = 3 + 3i$

15. $6(3 + 2i) + 4(-2 - 3i) = 18 + 12i - 8 - 12i = 10 + 0i = 10$

17. $i(5 - 2i) = 5i - 2i^2 = 2 + 5i$ 19. $(3 - i)(2 + i) = 6 + 3i - 2i - i^2 = 7 + i$

21. $(5 - 2i)(5 + 2i) = 25 + 10i - 10i - 4i^2 = 29 + 0i = 29$

23. $(3 - 5i)(3 + 5i) = 9 + 15i - 15i - 25i^2 = 34 + 0i = 34$

25. $-i^2 = -(-1) = 1$ 27. $i^3 = i(i^2) = i(-1) = -i$ 29. $-i^4 = -(i^2)(i^2) = -(-1)(-1) = -1$

31. $-i^6 = -(i^2)(i^2)(i^2) = -(-1)(-1)(-1) = 1$ 33. $i^{11} = (i^4)^2 (i^3) = (1)^2(i^2)(i) = 1(-1)(i) = -i$

35. $-i^{1980} = -[(i^4)^{495}] = -(1)^{495} = -1$ 37. $(6 - 2i)^2 = 36 - 24i + 4i^2 = 36 - 4 - 24i = 32 - 24i$

39. $(4 + 5i)^2 = 16 + 40i + 25i^2 = -9 + 40i$

41. $(3 - 5i)^3 = (3 - 5i)(3 - 5i)^2 = (3 - 5i)(9 - 30i + 25i^2) = (3 - 5i)(-16 - 30i) =$
$= -48 - 90i + 80i + 150i^2 = -198 - 10i$

43. $\left(\dfrac{-3}{1+i}\right)\left(\dfrac{1-i}{1-i}\right) = \dfrac{-3+3i}{1-i^2} = \dfrac{-3+3i}{2} = \dfrac{-3}{2} + \dfrac{3}{2}i$

45. $\left(\dfrac{2}{1-i}\right)\left(\dfrac{1+i}{1+i}\right) = \dfrac{2+2i}{1-i^2} = \dfrac{2(1+i)}{2} = 1 + i$ 47. $\left(\dfrac{2}{i}\right)\left(\dfrac{i}{i}\right) = \dfrac{2i}{i^2} = -2i$

49. $\left(\dfrac{-2i}{3+i}\right)\left(\dfrac{3-i}{3-i}\right) = \dfrac{-2i(3-i)}{9-i^2} = \dfrac{-6i+2i^2}{10} = \dfrac{-2-6i}{10} = -\dfrac{1}{5} - \dfrac{3}{5}i$

51. $\left(\dfrac{-i}{2-i}\right)\left(\dfrac{2+i}{2+i}\right) = \dfrac{-2i-i^2}{4-i^2} = \dfrac{1-2i}{5} = \dfrac{1}{5} - \dfrac{2}{5}i$

53. $\left(\dfrac{4-2i}{3+i}\right)\left(\dfrac{3-i}{3-i}\right) = \dfrac{12-4i-6i+2i^2}{9-i^2} = \dfrac{10-10i}{10} = 1 - i$

55. $\left(\dfrac{1+3i}{1-2i}\right)\left(\dfrac{1+2i}{1+2i}\right) = \dfrac{1+2i+3i+6i^2}{1-4i^2} = \dfrac{-5+5i}{5} = -1 + i$

57. $\left(\dfrac{2+7i}{2-7i}\right)\left(\dfrac{2+7i}{2+7i}\right) = \dfrac{4+28i+49i^2}{4-49i^2} = \dfrac{-45+28i}{53} = -\dfrac{45}{53} + \dfrac{28}{53}i$

59. $\left(\dfrac{-3.2253+8.4022\,i}{3.4985+1.9392\,i}\right)\left(\dfrac{3.4985-1.9392\,i}{3.4985-1.9392\,i}\right) \approx \dfrac{5.00983419+35.64959846\,i}{15.99999889}$

$\approx \dfrac{5.00983419}{15.99999889} + \dfrac{35.64959846}{15.99999889}i \approx 0.313114659 + 2.228100058\,i \approx 0.3131 + 2.2281i$

61. $\left\{(1 + \sqrt{3}) + (2 + \sqrt{3})\,i\right\} - \left\{(2 + 2\sqrt{3}) + \sqrt{3}i\right\}$
$= \left\{(1 + \sqrt{3}) - (2 + 2\sqrt{3})\right\} + \left\{(2i + \sqrt{3}i) - \sqrt{3}i\right\}$
$= 1 + \sqrt{3} - 2 - 2\sqrt{3} + 2i = -1 - \sqrt{3} + 2i$ or $-(1 + \sqrt{3}) + 2i$

63. $\dfrac{(1 + \sqrt{3}) + (2 + \sqrt{3})\,i}{(2 + 2\sqrt{12}) + \sqrt{3}i} = \dfrac{(1 + \sqrt{3}) + (2 + \sqrt{3})\,i}{(2 + 2\sqrt{3}) + \sqrt{3}i}$

$= \dfrac{(1 + \sqrt{3}) + (2 + \sqrt{3})i}{(2 - 2\sqrt{3}) + \sqrt{3}i} \cdot \dfrac{(2 + 2\sqrt{3}) - \sqrt{3}i}{(2 + 2\sqrt{3}) - \sqrt{3}i} = \dfrac{11 + 6\sqrt{3} - (7 + 5\sqrt{3})i}{19 + 8\sqrt{3}} \cdot \dfrac{19 - 8\sqrt{3}}{19 - 8\sqrt{3}}$

$= \dfrac{(209 - 88\sqrt{3} + 114\sqrt{3} - 144) + (-133 + 56\sqrt{3} - 95\sqrt{3} + 120)i}{361 - 192}$

$= \dfrac{65 + 26\sqrt{3}}{169} - \dfrac{13 + 39\sqrt{3}}{169}i = \dfrac{5 + 2\sqrt{3}}{13} - \dfrac{1 + 3\sqrt{3}}{13}i$

65. $\left[(2 + \sqrt{12}) + \sqrt{3}i\right]^2 = (2 + 2\sqrt{3})^2 + 2(2 + 2\sqrt{3})(\sqrt{3}i) + (\sqrt{3}i)^2$

$= 4 + 8\sqrt{3} + 12 + 4\sqrt{3}i + 12i + 3i^2 = 16 - 3 + 8\sqrt{3} + 4\sqrt{3}i + 12i$
$= (13 + 8\sqrt{3}) + (12 + 4\sqrt{3})i$

67. Prove that addition is associative for complex numbers.

$$(a + bi) + \left[(c + di) + (e + fi)\right] = a + bi + \left[(c + e) + (d + f)i\right]$$

$$= \underline{(a + c + e) + (b + d + f)i} \quad \text{and} \quad \left[(a + bi) + (c + di)\right] + (e + fi)$$

$$= \left[(a + c) + (b + d)i\right] + e + fi = \underline{(a + c + e) + (b + d + f)i}$$

So it is proven that addition is associative.

Prove that multiplication is associative for complex numbers.

$$(a + bi) \left[(c + di) \cdot (e + fi)\right] = (a + bi) + \left[ce + cfi + dei + dfi^2\right]$$

$$= (a + bi) \left[(ce - df) + (cf + de)i\right] = a(ce - df) + a(cf + de)i + bi(ce - df) + bi(cf + de)i$$

$$= ace - adf + (acf + ade)i + (bce - bdf)i + (bcf + bde)i^2$$

$$= \underline{\left[ace - (adf + bcf + bde)\right] + \left[acf + ade + bce - bdf\right] i} \quad \text{and}$$

$$\left[(a + bi)(c + di)\right](e + fi) = \left[ac + adi + bci + bdi^2\right](e + fi)$$

$$= \left[(ac - bd) + (ad + bc)i\right](e + fi) = ace - bde + acfi - bdfi + (ade + bce)i + (adf + bcf)i^2$$

$$= ace - bde - adf - bcf + acfi - bdfi + adei + bcei =$$

$$= \underline{\left[ace - (adf + bcf + bde)\right] + \left[acf + ade + bce - bdf\right] i}$$

So it is proven that multiplication is associative.

PROBLEM SET 1.6

1. $x^2 + 2x - 15 = (x + 5)(x - 3) = 0$ so $x = -5, 3$ and $\{-5, 3\}$

3. $x^2 + 7x - 18 = (x + 9)(x - 2) = 0$ so $x = -9, 2$ and $\{-9, 2\}$

5. $6x^2 = 5x; \ 6x^2 - 5x = x(6x - 5) = 0$ so $x = 0, \frac{6}{5}$ and $\{0, \frac{6}{5}\}$

7. $10x^2 - 3x - 4 = (2x + 1)(5x - 4) = 0$ so $x = -\frac{1}{2}, \frac{4}{5}$ and $\{-\frac{1}{2}, \frac{4}{5}\}$

9. $9x^2 - 34x - 8 = (9x + 2)(x - 4) = 0$ so $x = -\frac{2}{9}, 4$ and $\{-\frac{2}{9}, 4\}$

11. $x^3 + 2x^2 + x = x(x^2 + 2x + 1) = x(x + 1)^2$ so $x = -1, 0$ and $\{-1. 0\}$

13. $x^2 + 4x - 5 = 0; \ x^2 + 4x + 4 = 5 + 4; \ (x + 2)^2 = 9, \ x + 2 = \pm 3$
    $x = -2 \pm 3$ so $x = -5, 1$ and $\{-5, 1\}$

15. $x^2 + 2x - 8 = 0; \ x^2 + 2 + 1 = 8 + 1, \ (x + 1)^2 = 9, \ x + 1 = \pm 3, \ x = -4, 2; \ \{-4, 2\}$

17. $x^2 + 7x + 12 = 0; \ x^2 + 7x + \frac{49}{4} = -12 + \frac{49}{4}, \ (x + \frac{7}{2})^2 = \frac{1}{4},$
    $x + \frac{7}{2} = \pm\frac{1}{2}; \ x = -\frac{7}{2} \pm \frac{1}{2},$ so $x = -4, -3$ and $\{-4, -3\}$

19. $x^2 - 10x - 2 = 0; \ x^2 - 10x + 25 = 2 + 25, \ (x - 5)^2 = 27,$
    $x - 5 = \pm \sqrt{27},$ so $x = 5 \pm 3\sqrt{3}$ and $\{5 \pm 3\sqrt{3}\}$

21. $x^2 - 3x = 1$; $x^2 - 3x + \frac{9}{4} = 1 + \frac{9}{4}$, $(x - \frac{3}{2})^2 = \frac{13}{4}$, $x - \frac{3}{2} = \pm \frac{\sqrt{13}}{2}$;

$x = \frac{3}{2} \pm \frac{\sqrt{13}}{2}$ and $\left\{ \frac{3 \pm \sqrt{13}}{2} \right\}$

23. $6x^2 = x + 2$; $6x^2 - x = 2$, $x^2 - \frac{1}{6}x = \frac{1}{3}$, $x^2 - \frac{1}{6}x + \frac{1}{144} = \frac{1}{3} + \frac{1}{144}$

$(x - \frac{1}{12})^2 = \frac{49}{144}$, $x - \frac{1}{12} = \pm \frac{7}{12}$, $x = \frac{1}{12} \pm \frac{7}{12}$, so $x = -\frac{1}{2}, \frac{3}{2}$ and $\left\{ -\frac{1}{2}, \frac{3}{2} \right\}$

25. $x^2 + 5x - 6 = (x + 6)(x - 1) = 0$ so $x = -6, 1$ and $\{-6, 1\}$

27. $x^2 - 10x + 25 = (x - 5)^2 = 0$ so $x = 5$ and $\{5\}$

29. $12x^2 + 5x - 2 = (3x + 2)(4x - 1) = 0$, so $x = -\frac{2}{3}, \frac{1}{4}$ and $\left\{ -\frac{2}{3}, \frac{1}{4} \right\}$

31. $5x^2 - 4x + 1 = 0$; $x = \dfrac{-(-4) \pm \sqrt{(-4)^2 - 4(5)(1)}}{2(5)} = \dfrac{4 \pm \sqrt{16 - 20}}{10} = \dfrac{4 \pm \sqrt{-4}}{10} =$

$\frac{4 \pm 2i}{10} = \frac{2}{5} \pm \frac{1}{5}i$ not real numbers so $\{\ \}$

33. $4x^2 - 5 = 0$; $4x^2 = 5$, $x^2 = \frac{5}{4}$, so $x = \pm \frac{\sqrt{5}}{2}$ and $\left\{ \pm \frac{\sqrt{5}}{2} \right\}$

35. $3x^2 = 7x$; $3x^2 - 7x = 0$, $x(3x - 7) = 0$, so $x = 0, \frac{7}{3}$ and $\{0, \frac{7}{3}\}$

37. $3x^2 = 5x + 2$; $3x^2 - 5x + 2 = 0$, $(3x + 1)(x - 2) = 0$, so $x = -\frac{1}{3}, 2$ and $\left\{ -\frac{1}{3}, 2 \right\}$

39. $5x = 3 - 4x^2$; $4x^2 + 5x - 3 = 0$, $x = \dfrac{-5 \pm \sqrt{5^2 - 4(4)(-3)}}{2(4)}$

$= \dfrac{-5 \pm \sqrt{25 + 48}}{8} = \dfrac{-5 \pm \sqrt{73}}{8}$ and $\left\{ \dfrac{-5 \pm \sqrt{73}}{8} \right\}$

41. $3x = 1 - 2x^2$; $2x^2 + 3x - 1 = 0$, $x = \dfrac{-3 \pm \sqrt{3^2 - 4(2)(-1)}}{2(2)}$

$= \dfrac{-3 \pm \sqrt{9 + 8}}{4} = \dfrac{-3 \pm \sqrt{17}}{4}$ and $\left\{ \dfrac{-3 \pm \sqrt{17}}{4} \right\}$

43. $\sqrt{5} - 4x^2 = 3x$; $4x^2 + 3x - \sqrt{5} = 0$, $x = \dfrac{-3 \pm \sqrt{(-3)^2 - 4(4)(-\sqrt{5})}}{2(4)}$

$= \dfrac{-3 \pm \sqrt{9 + 16\sqrt{5}}}{8}$ and $\left\{ \dfrac{-3 \pm \sqrt{9 + 16\sqrt{5}}}{8} \right\}$

45. $3x^2 - 4x = \sqrt{5}$; $3x^2 - 4x - \sqrt{5} = 0$, $x = \dfrac{4 \pm \sqrt{(-4)^2 - 4(3)(-\sqrt{5})}}{2(3)}$

$= \dfrac{4 \pm \sqrt{16 + 12\sqrt{5}}}{6} = \dfrac{4 \pm \sqrt{4(4 + 3\sqrt{5})}}{6} = \dfrac{4 \pm 2\sqrt{4 + 3\sqrt{5}}}{6}$

$= \dfrac{2 \pm \sqrt{4 + 3\sqrt{5}}}{3}$ and $\left\{ \dfrac{2 \pm \sqrt{4 + 3\sqrt{5}}}{3} \right\}$

47. $x^4 + 6x^2 = 25$; $x^4 + 6x^2 - 25 = 0$, $x^2 = \dfrac{-6 \pm \sqrt{(-6)^2 - 4(1)(-25)}}{2(1)}$

$= \dfrac{-6 \pm \sqrt{36 + 100}}{2} = \dfrac{-6 \pm \sqrt{136}}{2} = \dfrac{-6 \pm \sqrt{4(34)}}{2} = \dfrac{-6 \pm 2\sqrt{34}}{2} = -3 \pm \sqrt{34} = x^2$;

$x = \pm \sqrt{-3 \pm \sqrt{34}}$, $x \approx \pm 1.6825$

49. $2x^2 + x - w = 0$; $x = \dfrac{-1 \pm \sqrt{(1)^2 - 4(2)(w)}}{2(2)} = \dfrac{-1 \pm \sqrt{1 + 8w}}{4}$, $\left\{\dfrac{-1 \pm \sqrt{1 + 8w}}{4}\right\}$, $w > -\dfrac{1}{8}$

51. $3x^2 + 2x + (y + 2) = 0$; $x = \dfrac{-2 \pm \sqrt{2^2 - 4(3)(y + 2)}}{2(3)} = \dfrac{-2 \pm \sqrt{4 - 12(y + 2)}}{6}$

$= \dfrac{-2 \pm \sqrt{4 - 12y - 24}}{6} = \dfrac{-2 \pm \sqrt{-20 - 12y}}{6} = \dfrac{-2 \pm \sqrt{-4(3y + 5)}}{6} = \dfrac{-2 \pm 2i\sqrt{3y + 5}}{6}$

$= \dfrac{-1 \pm \sqrt{3y + 5}}{3}$, $\left\{\dfrac{-1 \pm \sqrt{3y + 5}}{6}\right\}$

53. $4x^2 - 4x + (1 - t^2) = 0$; $x = \dfrac{4 \pm \sqrt{(4)^2 - 4(4)(1 - t^2)}}{2(4)} = \dfrac{4 \pm \sqrt{16 - 16(1 - t^2)}}{8}$

$= \dfrac{4 \pm \sqrt{16[1 - (1 - t^2)]}}{8} = \dfrac{4 \pm 4\sqrt{1 - 1 + t^2}}{8} = \dfrac{4 \pm 4\sqrt{t^2}}{8} = \dfrac{4 \pm 4t}{8} = \dfrac{1 \pm t}{2}$, $\left\{\dfrac{1 \pm t}{2}\right\}$

55. $2x^2 + 3x + 4 - y = 0$; $x = \dfrac{-3 \pm \sqrt{3^2 - 4(2)(4 - y)}}{2(2)} = \dfrac{-3 \pm \sqrt{3^2 - 4(2)(4 - y)}}{4} =$

$\dfrac{-3 \pm \sqrt{9 - 8(4 - y)}}{4} = \dfrac{-3 \pm \sqrt{9 - 32 + 8y}}{4} = \dfrac{-3 \pm \sqrt{-23 + 8y}}{4} = \left\{\dfrac{-3 \pm \sqrt{8y - 23}}{4}\right\}$, $y > \dfrac{23}{8}$

57. $x^2 - (3t + 10)x + (6t + 4) = 0$

$x = \dfrac{(3t + 10) \pm \sqrt{(3t + 10)^2 - 4(4)(6t + 4)}}{2(4)} = \dfrac{(3t + 10) \pm \sqrt{9t^2 + 60t + 100 - 96t - 64}}{8}$

$= \dfrac{(3 + 10) \pm \sqrt{9t^2 - 36t + 36}}{8} = \dfrac{(3t + 10) \pm \sqrt{9(t^2 - 4t + 4)}}{8} = \dfrac{(3t + 10) \pm \sqrt{9(t - 2)^2}}{8}$

$= \dfrac{(3t + 10) \pm 3(t - 2)}{8} = \dfrac{(3t + 10) \pm (3t - 6)}{8}$

$\dfrac{3t + 10 + 3t - 6}{8}$ $\dfrac{3t + 10 - (3t - 6)}{8}$

$= \dfrac{6t + 4}{8} = \dfrac{3t + 2}{4}$ $= \dfrac{16}{8} = 2$ $\left\{\dfrac{3t + 2}{4}, 2\right\}$

59. $(x + 2)^2 - (y + 1)^2 = 9$; $(x + 2)^2 = (y + 1)^2 + 9$, $x + 2 = \pm\sqrt{(y + 1)^2 + 9}$,

$x = -2 \pm \sqrt{(y + 1)^2 + 9}$, $x = -2 \pm \sqrt{y^2 + 2y + 1 + 9}$, $-2 \pm \sqrt{y^2 + 2y + 10}$

16

61. $r_1 = \dfrac{2c}{-b + \sqrt{b^2 - 4ac}}$, $\quad x - \dfrac{2c}{-b + \sqrt{b^2 - 4ac}} = 0;$ and

$r_2 = \dfrac{2c}{-b - \sqrt{b^2 - 4ac}}$, $\quad x - \dfrac{2c}{-b - \sqrt{b^2 - 4ac}} = 0,$ multiply the two binomials;

$\left(x - \dfrac{2c}{-b + \sqrt{b^2 - 4ac}} \right)\left(x - \dfrac{2c}{-b - \sqrt{b^2 - 4ac}} \right) = 0,$

$x^2 - \dfrac{2cx}{-b + \sqrt{b^2 - 4ac}} - \dfrac{2cx}{-b - \sqrt{b^2 - 4ac}} + \dfrac{4c^2}{(-b + \sqrt{b^2 - 4ac})(-b - \sqrt{b^2 - 4ac})} = 0,$

Find the least common denominator: $b^2 - b\sqrt{b^2 - 4ac} + b\sqrt{b^2 - 4ac} - (\sqrt{b^2 - 4ac})^2$

$= b^2 - b^2 + 4ac = 4ac,$ insert this value into your equation and you have:

$x^2 + \dfrac{-2cx(-b + \sqrt{b^2 - 4ac})}{4ac} + \dfrac{-2cx(-b - \sqrt{b^2 - 4ac})}{4ac} + \dfrac{4c^2}{4ac},$ multiply by 4ac and

simplify: $4acx^2 + 2bcx + 2bcx + 4c^2 = 0,$ $\quad 4acx^2 + 4bcx + 4c^2 = 0,$ $\quad 4c(ax^2 + bx + c) = 0,$

$ax^2 + bx + c = 0,$ therefore r_1 and r_2 are the roots of $ax^2 + bx + c = 0.$

PROBLEM SET 1.7

1. The solutions for this problem are only in the interval $(-5, 0)$.

3. $5x^2 = 125x$ becomes $5x^2 - 125x = 0$ and $5x(x - 125) = 0$ gives you that $x = 0, 125$.

5. The standard quadratic equation that generated the quadratic formula is $ax^2 + bx + c = 0$.

7. $x(x + 3) < 0$

$(-3, 0)$

8. $x + 3 \quad \; - \; - \; | \; + \; + \; | \; + \; + \; +$

$x \quad \; - \; - \; | \; - \; - \; | \; + \; + \; +$

$\qquad\qquad -3 \qquad 0$

9. $(x - 6)(x - 2) \geq 0$

$(-\infty, 2] \cup [6, \infty)$

$x - 6 \quad \; - \; - \; | \; - \; - \; - \; | \; + \; +$

$x - 2 \quad \; - \; - \; | \; + \; + \; + \; | \; + \; +$

$\qquad\qquad 2 \qquad\qquad 6$

11. $(x - 8)(x + 7) < 0$

$(-7, 8)$

$x - 8 \quad \; - \; - \; | \; - \; - \; - \; | \; + \; +$

$x + 7 \quad \; - \; - \; | \; + \; + \; + \; | \; + \; +$

$\qquad\qquad -7 \qquad\qquad 8$

13. $(x + 2)(2x - 1) \le 0$

$[-2, \frac{1}{2}]$

$x + 2$	$- -$	$+ +$	$+ +$
$2x - 1$	$- -$	$- -$	$+ +$

$\qquad -2 \qquad \frac{1}{2}$

15. $(3x + 2)(x - 3) > 0$

$(-\infty, -\frac{2}{3}) \cup (3, \infty)$

$3x + 2$	$- -$	$+ + +$	$+ +$
$x - 3$	$- -$	$- - -$	$+ +$

$\qquad -\frac{1}{2} \qquad 3$

17. $(x + 2)(8 - x) > 0$

$(-\infty, -2) \cup [8, \infty)$

$x + 2$	$- -$	$+ +$	$+ +$
$8 - x$	$+ +$	$+ +$	$- -$

$\qquad -2 \qquad 8$

19. $(1 - 3x)(x - 4) < 0$

$(-\infty, \frac{1}{3}) \cup (4, \infty)$

$1 - 3x$	$+ +$	$- - -$	$- -$
$x - 4$	$- -$	$- - -$	$+ +$

$\qquad \frac{1}{3} \qquad 4$

21. $x(x - 3)(x + 4) \le 0$

$(-\infty, -4] \cup [0, 3]$

x	$- -$	$- -$	$+ +$	$+ +$
$x - 3$	$- -$	$- -$	$- -$	$+ +$
$x + 4$	$- -$	$+ +$	$+ +$	$+ +$

$\qquad -4 \qquad 0 \qquad 3$

23. $(x - 2)(x + 3)(x - 4) \ge 0$

$[-3, 2] \cup [4, \infty)$

$x - 2$	$- -$	$- -$	$+ +$	$+ +$
$x + 3$	$- -$	$+ +$	$+ +$	$+ +$
$x - 4$	$- -$	$- -$	$- -$	$+ +$

$\qquad -3 \qquad 2 \qquad 4$

25. $(x + 1)(2x + 5)(7 - 3x) > 0$

$(-\infty, -\frac{5}{2}) \cup (-1, \frac{7}{3})$

$x + 1$	$- -$	$- -$	$+ +$	$+ +$
$2x + 5$	$- -$	$+ +$	$+ +$	$+ +$
$7 - 3x$	$+ +$	$+ +$	$+ +$	$- -$

$\qquad -\frac{5}{2} \qquad -1 \qquad \frac{7}{3}$

27. $\frac{x + 2}{x} < 0$

$(-2, 0)$

$x + 2$	$- -$	$+ +$	$+ +$
x	$- -$	$- -$	$+ +$

$\qquad -2 \qquad 0$

29. $\dfrac{x}{x-8} > 0$

$(-\infty, 0) \cup (8, \infty)$

x	$--$	$++$	$++$	
$x-8$	$--$	$--$	$++$	

$\qquad\qquad 0 \qquad 8$

31. $\dfrac{x-2}{x+5} \le 0$

$(-5, 2]$

$x-2$	$--$	$--$	$++$
$x+5$	$--$	$++$	$++$

$\qquad -5 \qquad 2$

33. $\dfrac{x(2x-1)}{5-x} > 0$

$(-\infty, 0) \cup (\frac{1}{2}, 5)$

x	$--$	$++$	$++$	$++$
$2x-1$	$--$	$--$	$++$	$++$
$5-x$	$++$	$++$	$++$	$--$

$\qquad\quad 0 \qquad \frac{1}{2} \qquad 5$

35. $\dfrac{1}{x(x-3)(x+2)} \le 0$

$(-\infty, -2) \cup (0, 3)$

x	$--$	$--$	$++$	$++$
$x-3$	$--$	$--$	$--$	$++$
$x+2$	$--$	$++$	$++$	$++$

$\qquad\quad -2 \qquad 0 \qquad 3$

37. $x^2 \ge 9$

$x^2 - 9 \ge 0$

$(x+3)(x-3) \ge 0$

$(-\infty, -3] \cup [3, \infty)$

$x-3$	$--$	$--$	$++$
$x+3$	$--$	$++$	$++$

$\qquad -3 \qquad 3$

39. $x^2 + 9 \ge 0$ is going to always be positive so $(-\infty, \infty)$.

41. $x^2 - x - 6 > 0$

$(x-3)(x+2)$

$(-\infty, -2) \cup (3, \infty)$

$x-3$	$--$	$--$	$++$
$x+2$	$--$	$++$	$++$

$\qquad -2 \qquad 3$

43. $5x - 6 \ge x^2$

$x^2 - 5x + 6 \le 0$

$(x-3)(x-2) \le 0$

$[2, 3]$

$x-3$	$--$	$--$	$++$
$x-2$	$--$	$++$	$++$

$\qquad 2 \qquad 3$

45. $5 - 4x \geq x^2$

$x^2 + 4x - 5 \leq 0$

$(x-1)(x+5) \leq 0$

$x + 5$	$--$	$+\,+$	$+\,+$
$x - 1$	$--$	$--$	$+\,+$

-51

$[-5, 1]$

47. $x^2 - 2x - 2 < 0$

$\left(1 - \sqrt{3},\, 1 + \sqrt{3}\right)$

$x - 1 + \sqrt{3}$	$--$	$+\,+$	$+\,+$
$x - 1 - \sqrt{3}$	$--$	$--$	$+\,+$

$\approx -0.7 \quad \approx 2.7$

49. $2x^2 + 4x + 5 \geq 0$ answer will always be positive so $(-\infty, \infty)$.

51. $x^2 + 3x - 7 \geq 0$

$x = \dfrac{-3 \pm \sqrt{37}}{2} \geq 0$

$2x + 3 - \sqrt{37}$	$--$	$-\,-\,-$	$+\,+$
$2x + 3 + \sqrt{37}$	$--$	$+\,+\,+$	$+\,+$

$\approx -4.5 \qquad \approx 1.5$

$\left(-\infty, \dfrac{-3-\sqrt{37}}{2}\right) \cup \left(\dfrac{-3+\sqrt{37}}{2}, \infty\right)$

53. $\dfrac{x(x+5)(x-3)}{(x+3)(x-4)} \geq 0$, where $x \neq -3, 4$

x	$--$	$--$	$--$	$+\,+$	$+\,+$	$+\,+$
$x + 5$	$--$	$+\,+$	$+\,+$	$+\,+$	$+\,+$	$+\,+$
$x - 3$	$--$	$--$	$--$	$--$	$+\,+$	$+\,+$
$x + 3$	$--$	$--$	$+\,+$	$+\,+$	$+\,+$	$+\,+$
$x - 4$	$--$	$--$	$--$	$--$	$--$	$+\,+$

$-5\quad -3\quad 0\quad 3\quad 4$

$[-5, -3) \cup [0, 3] \cup (4, \infty)$

55. $\dfrac{2}{x - 2} \leq \dfrac{3}{x + 3}$

$\dfrac{2}{x - 2} - \dfrac{3}{x + 3} \leq 0$

$\dfrac{12 - x}{(x - 2)(x + 3)} \leq 0$ where $x \neq -3, 2$

$12 - x$	$+\,+$	$+\,+$	$+\,+$	$--$
$x - 2$	$--$	$--$	$+\,+$	$+\,+$
$x + 3$	$--$	$--$	$+\,+$	$+\,+$

$-3\quad 2\quad 12$

$(-3, 2) \cup [12, \infty)$

57. $\dfrac{x-3}{3x-1} \geq \dfrac{x+3}{2x+1}$

$\quad \dfrac{x-3}{3x-1} - \dfrac{x+3}{2x+1} \geq 0$

$\quad \dfrac{(x-3)(2x+1) - (3x-1)(x+3)}{(3x-1)(2x+1)} \geq 0$

$\dfrac{-x^2 - 3x}{(3x-1)(2x+1)} \geq 0$

$\dfrac{-x(x+3x)}{(3x-1)(2x+1)} \geq 0$

$x = -\dfrac{1}{2}, \dfrac{1}{3}$ so $\left[-13, -\dfrac{1}{2}\right) \cup \left[0, \dfrac{1}{3}\right)$

$-x$	$+\ +$	$+\ +$	$+\ +$	$-\ -$	$-\ -$
$x+13$	$-\ -$	$+\ +$	$+\ +$	$+\ +$	$+\ +$
$3x-1$	$-\ -$	$-\ -$	$-\ -$	$-\ -$	$+\ +$
$2x+1$	$-\ -$	$-\ -$	$+\ +$	$+\ +$	$+\ +$

$\qquad -13 \qquad -\dfrac{1}{2} \qquad 0 \qquad \dfrac{1}{3}$

59. $\dfrac{x-4}{x^2 - 5x + 6} \leq 0$

$\dfrac{x-4}{(x-2)(x-3)} \leq 0$
where $x \neq 2, 3$

$(-\infty, 2) \cup (3, 4]$

$x-4$	$-\ -$	$-\ -$	$+\ +$	$+\ +$
$x-2$	$-\ -$	$+\ +$	$+\ +$	$+\ +$
$x-3$	$-\ -$	$-\ -$	$+\ +$	$+\ +$

$\qquad 2 \qquad 3 \qquad 4$

PROBLEM SET 1.8

1. $x + y = 125$, $y = x + 1$, $2x + 1 = 125$, $x = 62$ and $y = 63$.

3. $x + y + z = 99$, $y = x + 2$, $z = x + 4$; $3x + 6 = 99$, x or the first integer) $= 31$.

5. $xy = wz - 62$; $y = x + 1$ and $w = x + 2$ and $z = x + 3$; $x(x+1) = (x+2)(x+3) - 62$,
    $x^2 + x = x^2 + 5x + 6 - 62$, $0 = 4x - 56$, $x = 14$ so the 2nd integer is 15 and the 4th is 17.

7. $lw = sw + 20$, $l = w + 6$, $s = w + 2$; $(w+6)w = (w+2)w + 20$,
    $w^2 + 6w = w^2 + 2w + 20$, $4w = 20$, $w = 5$, larger width = 5 ft., larger height = 11 ft.

9. $\dfrac{1}{2}$ bh $= \dfrac{1}{2}$ sh $+ 3$, $b = h + 3$, $s = h + 1$; $\dfrac{1}{2}(h+3)h = \dfrac{1}{2}(h+1)h + 3$,
    $\dfrac{1}{2}(h^2 + 3h) = \dfrac{1}{2}(h^2 + h) + 3$, $\dfrac{3}{2}h = \dfrac{1}{2}h + 3$, $h = 3$ cm and the smaller height = 4 cm.

11. A $= lw = mx + 3$, $l = w + 2$, $m = l + 1$ so $m = w + 3$, $x = w - 1$;
    $(w+2)w = (w+3)(w-1) + 3$, $w^2 + 2w = w^2 + 2w - 3 + 3$
    w (the width of the old figure) is any value; x (the width of the new figure) $= w - 1$.

21

13. If x is the amount invested at 10%, and $1000 - x = 12.5\%$; $0.10x + 0.125(1000 - x) = 110$, $0.10x + 125 - 0.125x = 110$, $0.025x = 15$, $x = \$600$.

15. If x is the amount invested at 9%, and $8000 - x = 8\%$; $0.09x + 0.08(8000 - x) = 665$, $0.01x = 25$, $x = \$2500$.

17. If x is the amount invested at 9.5%, and $1500 - x = 14\%$; $0.095x + 0.14(1500 - x) = 183$, $0.045x = 27$, $x = \$600$ and the amount at 14% is \$900.

19. If x is the amount invested at 10.25%, and $20,000 - x = 11.25\%$; $0.1025x + 0.1125(20,000 - x) = 2165$, $0.01x = 85$, $x = \$8500$.

21. If x is the time of train travel; $(72\text{mph})(x \text{ hrs}) + (39\text{mph})(7 - x \text{ hrs}) = 405 \text{ m}$; $72x + 39(7 - x) = 405$, $33x - 273 = 405$, $33x = 132$, $x = 4 \text{ hrs}$; The train travels 288 miles.

23. If x is the time in the auto; $(50 \text{ mph})(x \text{ hrs}) + 600 \text{ mph}(5.5 \text{ hrs} - x \text{ hrs}) = 1100 \text{ m}$, $50x + 600(5.5 - x) = 1100$, $550x = 2200$, $x = 4 \text{ hrs}$.

25. If x is the time on the bus; $(25 \text{ mph})(x \text{ hrs}) + (65 \text{ mph})(1 - x)(\text{hrs}) = 61 \text{ m}$, $40x = 4$, ; $x = 0.10 \text{ hr}$. OR 6 minutes.

27. If x is the speed of the slower jogger; $(1.5 \text{ hrs})(x \text{ mph}) + (1.5 \text{ hrs})(x + 2)(\text{mph}) = 21 \text{ m}$, $3x = 18$, $x = 6$, $x + 2 = 6 + 2$ so the faster jogger speed is 8 mph.

29. If x is the city speed (mph); $(2x \text{ mph})(0.25 \text{ hr}) + (x \text{ mph})(0.75 \text{ hr}) = 30 \text{ m}$, $0.5x + 0.75 x = 30$, x (the city speed) $= 24$ mph.

31. Let x be the gallons of cream needed; $0.69 x + 0.12(150 - x) = 150(0.5)$, $0.57x = 57$, x (the cream required) $= 100$ gallons.

33. Let x be the liters of 72% solution required; $(4.5)(0.60) = 0.72x + 0.45(4.5 - x)$, $0.63 = 0.27x$; $x = 2.49$ liters of 72% solution, 2.01 liters of 45% solution required.

35. Let x be the volume of 35% solution, a 35% solution is 65% water, and a 21% solution is 79% water; $0.79(100) = 0.65(x) + (100 - x)$, $0.35x = 21$, $x = 60$ ml.

37. Let x be the volume of water added, remember, 20% alcohol is 80% water, while 50% alcohol is 50% water; $0.80(8 + x) = 0.50(8) + x$, $0.20x = 2.4$, $x = 12$ oz of water.

39. Let x be the volume of 15% alcohol; $0.25(8 + x) = 0.50(8) + 0.15x$, $0.10x = 2$, $x = 20$ oz.

41. Let x be the value of the larger integer; $x(x - 3) = 340$, $x^2 - 3x - 340 = 0$,
    $x = \frac{3}{2} \pm \frac{1}{2}\sqrt{9 - (4)(1)(-340)} = \frac{1}{2}(3 \pm 37)$, x- range: $20 \to \infty$ and $-\infty \to -17$.

43. Let x be the smaller number, $\frac{x}{x + 3} > 0$, $x < x + 3$, $x \neq -3$; the possible values
    for x are: $(-\infty, -3) \cup [0, \infty)$.

45. Let a be the width of the rectangle, $6a = $ perimeter, $2a^2 = $ area; $2a^2 > 6a$, $a > 3$, if width > 3 and length > 6; area $>$ perimeter.

47. Let x be the number of radios produced, Profit $= -30x^2 + 360x - 600$; for the profit to be zero

$$30x^2 - 360x + 600 = 0, \quad x^2 - 12x + 20 = 0, \quad x = \frac{12 \pm \sqrt{(12)^2 - 4(1)(20)}}{2(1)}$$

$$= \frac{12 \pm \sqrt{144 - 80}}{2} = \frac{12 + \sqrt{64}}{2} = \frac{12 \pm 8}{2} = 10 \text{ or } 2 \quad \text{so 10 or 2 radios give a zero profit.}$$

49. Find time, t, when the distance, $d \geq 240$ ft., if the initial velocity is 256 ft/sec.;

$240 \geq 256t - 16t^2, \quad 16t^2 - 256t + 240 \geq 0, \quad t^2 - 16t + 15 \geq 0,$

$$t = \frac{16 \pm \sqrt{(16)^2 - 4(1)(15)}}{2(1)} = \frac{16 \pm \sqrt{256 - 60}}{2} = \frac{16 \pm 14}{2} = 15 \text{ or } 1; \quad t \geq 1 \text{ or } 15 \text{ sec.}$$

51. Let x be the amount of pure antifreeze added; $0.60(0.80) = 0.40(8 - x) + x, \ 4.8 = 3.2 + 0.6x,$ $1.6 = 0.6x, \ x = 2.67$ quarts.

53. Let x be the cost of the first 1000 kwh, let y be the cost of the second 1000 kwh and let z be the remaininig power. $\$119.20 = x + y + z$ when $x = \$40$ then $\$119.20 = \$40 + y + z$ and $y + z = \$79. 20;$ the next 1000 kwh is at \$.06 or \$60 then y = \$60 and z = 19.20 at the power rate of \$.08 per kwh for 240kwh. Total power $= x(\text{power}) + y(\text{power}) + z(\text{power})$ $= 1000$ kwh $+1000$ kwh $+ 240$ kwh $= 2240$ kwh power used.

55. Let x be the time the boys walked; $(3.5 \text{ mph})(t) + (4.0 \text{ mph})(t) = 15$ m, $3.5t + 4.0t = 15,$ $7.5t = 15, \ t = 2$hours, slower boy walked $(3.5)(2) = 7$ miles, faster boy walked 8 miles.

57. Let t be the time for the girl to overtake her brother;
brother' distance 25miles $+ (10\text{mph})t =$ sister' distance $(35 \text{ mph})t$
$25 + 10t = 35t, \ 25t = 25, \quad t = 1$ hour.

59. Solve for the extension x, $4x^2 - y^2 - 200y = 0, \quad 4x^2 = y^2 + 200y, \quad x = \pm \frac{1}{2}\sqrt{y^2 + 200y}$.

material	coefficient(y)	extension(x)
Brick	0.03	1.22 inches
Steel	0.06	1.73 inches
Aluminum	0.12	2.45 inches
Concrete	0.05	1.58 inches

CHAPTER 1 SUMMARY PROBLEMS

1. $\frac{14}{7}$: N, W, Z, Q, R; $\quad \sqrt{144}$: N, W, Z, Q, R; $\quad 6.\overline{2}$: Q, R $\quad \pi$: Q', R

3. $3.\overline{1}$: Q, R; $\quad \frac{5\pi}{6}$: Q', R; $\quad \frac{22}{7}$: Q, R; $\quad \sqrt{10}$: Q', R

5.

7.

$$\overset{\pi\ \ \sqrt{10}}{\underset{\underset{\underset{3.1}{\big\downarrow\big\downarrow}}{}}{}}$$

On the number line: $5\pi/6$ labeled above, with tick marks at 2.5, 2.75, 3, 3.25, 3.5; 3.1 and π, $\sqrt{10}$ marked.

9. $\dfrac{5}{8} = 0.625$ 11. $\dfrac{5}{7} < \dfrac{8}{11}$ 13. $\sqrt{11}$ 15. $\sqrt{11} - 3$ 17. $3 + |-5| = 3 + 5 = 8$

19. $|-\pi| + 2 = \pi + 2$ 21. $(b + c)a$ 23. $a(b + c) = 5$ 25. $(b + c)a$ 27. $ab + ac$

29. x^n 31. b^{m+n} 33. $(3x + 1)(3x^3 + 4x^2 - 35x - 12) = 9x^4 + 15x^3 + 101x^2 - 71x - 12$

35. $(3x + 1) - (3x^3 + 4x^2 - 35x - 12) = -3x^3 - 4x^2 + 38x + 13$

37. $\dfrac{4x^2}{y^2} - (2x + y)^2 = 4x^2 - y^2(2x + y)^2 = [2x + y(2x + y)]\,[2x - y(2x + y)]$
$= [2x + 2xy + y^2]\,[2x - 2xy - y^2]$

39. $\left(x^3 - \dfrac{1}{8}\right)\left(8x^3 + 8\right) = \dfrac{1}{8}\left(8x^3 - 1\right)(8)(x^3 + 1) = \dfrac{1}{8}(8)(8x^3 - 1)(x^3 + 1) = (8x^3 - 1)(x^3 + 1)$
$(2x - 1)(4x^2 + 2x + 1)(x + 1)(x^2 - x + 1)$ OR $(2x - 1)(x + 1)(4x^2 + 2x + 1)(x^2 - x + 1$

41. Plot: $\left(\dfrac{\pi}{2}, 0\right)$

43. Plot: $\left(\dfrac{\pi}{4}, \dfrac{\sqrt{2}}{2}\right)$

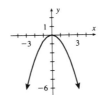

45. $d = \sqrt{(\gamma - \alpha)^2 + (\gamma - \beta)^2}$ 47. $d = \sqrt{(5x - x)^2 + (4x - x)^2} = \sqrt{16x^2 + 9x^2} = \sqrt{25x^2} = 5x$

49. $\left(\dfrac{\alpha + \gamma}{2}, \dfrac{\beta + \delta}{2}\right)$ 51. $\left(\dfrac{x + 5x}{2}, \dfrac{x + 4x}{2}\right) = \left(\dfrac{6x}{2}, \dfrac{5x}{2}\right) = \left(3x, \dfrac{5}{2}x\right)$

53. A relation is a <u>set of ordered pairs.</u>   55. $\cdots$ the ordered pair <u>satisfy</u> the equation.

57. $y = 2x + 3$ 59. $y = -\dfrac{2}{3}x^2$

24

61. Symmetric with x-axis

63. Symmetric with the origin

65. $(-4, 2)$ **67.** $(-3, \infty)$ **69.**

71.

73. $-8 \leq x < 5$ **75.** $x < 3$

77. $3x + 2 \leq 14$, $3x \leq 12$, $x \leq 4$; $(\infty, 4]$

79. $5 \leq 1 - x < 9$, $4 \leq -x < 8$, $-4 \geq x > -8$, $(-8, -4]$ **81.** $x = \pm 8$, $\{-8, 8\}$

83. $|2x + 3| = 8$; $2x + 3 = 8$, $2x = 5$, $x = \frac{5}{2}$; $2x + 3 = -8$, $2x = -11$, $x = -\frac{11}{2}$ $\left\{\frac{5}{2}, -\frac{11}{2}\right\}$

85. $|x - 4| < 5$, $x - 4 < 5$, $x < 9$; $x - 4 > -5$, $x > -1$; $(-1, 9)$

87. $|5 - 2x| \leq 25$; $-25 \leq 5 - 2x \leq 25$, $-30 \leq -2x \leq 20$, $15 \geq x \geq -10$, $-10 \leq x \leq 15$, $[-10, 15]$

89. $-i^7 = -(i^4)(i^3) = -(1)(-i) = i$ **91.** $(2 + 5i)(2 - 5i) = 4 - 25i^2 = 4 + 25 = 29$

93. $x^2 - x - 12 = 0$, $(x - 4)(x + 3) = 0$, $x = -3, 4$ **95.** $(3 - x)(5 + 2x) = 0$, $x = 3$, $-\frac{5}{2}$

97. $x^2 - 2x - 15 = 0$; $x^2 - 2x + 1 = 15 + 1$, $(x - 1)^2 = 16$, $x - 1 = \pm 4$, $x = 1 \pm 4$, $x = 5, -3$

99. $(3 - x)(5 + 2x) = 0$; $-2x^2 + x + 15 = 0$, $\left(x^2 - \frac{1}{2}x + \frac{1}{16}\right) = \frac{-15}{-2} + \frac{1}{16}$

$\left(x - \frac{1}{4}\right)^2 = \frac{121}{16}$, $x - \frac{1}{4} = \pm \frac{11}{4}$, $x = \frac{1}{4} \pm \frac{11}{4}$, $x = -\frac{5}{2}, 3$

101. $x^2 - 5x + 3 = 0$; $x = \dfrac{5 \pm \sqrt{(-5)^2 - 4(1)(3)}}{2(1)} = \dfrac{5 \pm \sqrt{25 - 12}}{2} = \dfrac{5 \pm \sqrt{13}}{2}$

103. $x^2 + 2x - 5 = 0$; $x = \dfrac{-2 \pm \sqrt{(2)^2 - 4(1)(-5)}}{2(1)} = \dfrac{-2 \pm \sqrt{24}}{2} = -1 \pm \sqrt{6}$

105. $3x^2 - 2x - 1 < 0$

$(3x+1)(x-1) < 0$

$(-\frac{1}{3}, 1)$

$3x - 1$	$-\ -$	$+\ +$	$+\ +$
$x - 1$	$-\ -$	$-\ -$	$+\ +$

$\frac{1}{3}$ 1

107. $x^2 + 2x + 1 \geq 0$

$(x+1)(x+1) \geq 0$

$(-\infty, \infty)$

$x + 1$	$-\ -$	$+\ +$
$x + 1$	$-\ -$	$+\ +$

-1

109. $x(3-x)(x+1) < 0$

$(-1, 0) \cup (3, \infty)$

x	$-\ -$	$-\ -$	$+\ +$	$+\ +$
$x + 1$	$-\ -$	$+\ +$	$+\ +$	$+\ +$
$3 - x$	$+\ +$	$+\ +$	$+\ +$	$-\ -$

-1 0 3

111. $x(x-2)^2(x+1) > 0$

$(-\infty, -1) \cup (0, 2) \cup (2, \infty)$

x	$-\ -$	$-\ -$	$+\ +$	$+\ +$
$x - 2$	$-\ -$	$-\ -$	$-\ -$	$+\ +$
$x - 2$	$-\ -$	$-\ -$	$-\ -$	$+\ +$
$x + 1$	$-\ -$	$+\ +$	$+\ +$	$+\ +$

-1 0 2

113. Let w = width, larger is $l = w + 8$, smaller is $s = w + 4$;
$(l)(w) = (s)(w) + 48$, $(w+8)(w) = (w+4)(w) + 48$, $w^2 + 8w = w^2 + 4w + 48$,
$8w = 4w + 48$, $4w = 48$, $w = 12$ and $l = w + 8 = 12 + 8 = 20$; $w = 12$, $l = 20$.

115. Let x = gallons of milk; $0.10x + 0.70(15 - x) = 0.50(15)$, $0.10x + 10.5 - 0.70x = 7.5$,
$-0.60x = -3$, $x = \frac{3}{0.6}$, $xy = 5$; 5 gallons of 10% milk and 10 gallons of 70% cream.

CHAPTER 2

PROBLEM SET 2.1

1. function, one-to-one   3. not a function   5. function, one-to-one   7. function one-to-one

9. not a function, x values repeated at $x = -1$ 11. function, one-to-one   13. $f(x) = y = 2x + 3$

15. $\dfrac{f(x+h) - f(x)}{h} = \dfrac{(x+h)^2 - (x^2)}{h} = \dfrac{x^2 + 2hx + h^2 - x^2}{h} = \dfrac{2hx + h^2}{h} = 2x + h$ since
$(x+h)^2 \neq x^2 + h^2$

17. a. $f(0) = 2(0) + 1 = 1$ b. $f(2) = 2(2) + 1 = 5$ c. $f(-3) = 2(-3) + 1 = -6 + 1 = -5$
    d. $f(\sqrt{5}) = 2(\sqrt{5}) + 1 = 2\sqrt{5} + 1$ e. $f(\pi) = 2(\pi) + 1 = 2\pi + 1$

19. a. $f(w) = 2w + 1$ b. $g(w) = 2(w)^2 - 1 = 2w^2 - 1$ c. $g(t) = 2(t)^2 - 1 = 2t^2 - 1$
    d. $g(v) = 2v^2 - 1$ e. $f(m) = 2m + 1$

21. a. $f(1 + \sqrt{2}) = 2(1 + \sqrt{2}) + 1 = 2 + 2\sqrt{2} + 1 = 3 + 2\sqrt{2}$
    b. $g(1 + \sqrt{2}) = 2(1 + \sqrt{2})^2 - 1 = 2(1 + 2\sqrt{2} + 2) - 1 = 2 + 4\sqrt{2} + 4 - 1 = 5 + 4\sqrt{2}$
    c. $f(t^2 + 2t + 1) = 2(t^2 + 2t + 1) + 1 = 2t^2 + 4t + 2 + 1 = 2t^2 + 4t + 3$
    d. $g(m - 1) = 2(m-1)^2 - 1 = 2(m^2 - 2m + 1) - 1 = 2m^2 - 4m + 2 - 1 = 2m^2 - 4m + 1$

23. $\dfrac{f(t+3) - f(t)}{3} = \dfrac{[2(t+3) + 1] - (2t+1)}{3} = \dfrac{2t + 6 + 1 - 2t - 1}{3} = \dfrac{6}{3} = 2$

25. $\dfrac{f(x+h) - f(x)}{h} = \dfrac{[2(x+h) + 1] - (2x+1)}{h} = \dfrac{2x + 2h + 1 - 2x - 1}{h} = \dfrac{2h}{h} = 2$

27. $\dfrac{g(t+h) - g(t)}{h} = \dfrac{[2(t+h)^2 - 1] - (2t^2 - 1)}{h} = \dfrac{[2(t^2 + 2ht + h^2) - 1] - 2t^2 + 1)}{h}$
$= \dfrac{2t^2 + 4ht + 2h^2 - 1 - 2t^2 + 1}{h} = \dfrac{4ht + 2h^2}{h} = 4t + 2h$

29. a. $f(w) = w^2 - 1$ b. $f(h) = h^2 - 1$ c. $f(w + h) = (w+h)^2 - 1 = w^2 + 2hw + h^2 - 1$
    d. $f(w) - f(h) = (w^2 - 1) - (h^2 - 1) = w^2 - h^2$

31. a. $f(x^2) = (x^2)^2 - 1 = x^4 - 1$ b. $f(\sqrt{x}) = (\sqrt{x})^2 - 1 = x - 1$
    c. $f(x + h) = (x+h)^2 - 1 = x^2 + 2hx + h^2 - 1$ d. $f(-x) = (-x)^2 - 1 = x^2 - 1$

33. $\dfrac{g(x+h) - g(x)}{h} = \dfrac{[2(x+h) + 5] - (2x+5)}{h} = \dfrac{2x + 2h + 5 - 2x - 5}{h} = \dfrac{2h}{h} = 2$

35. $\dfrac{[9(x+h) + 3] - (9x + 3)}{h} = \dfrac{9x + 9h + 3 - 9x - 3}{h} = \dfrac{9h}{h} = 9$

37. If x is positive: $\dfrac{|x+h| - |x|}{h} = \dfrac{(x+h) - x}{h} = \dfrac{h}{h} = 1.$
    If x is negative: $\dfrac{|-x+h| - |-x|}{h} = \dfrac{|h-x| - |-x|}{h} = \dfrac{|h-x| - x}{h}.$

39. $\dfrac{[5(x+h)^2 - (5x^2)]}{h} = \dfrac{5(x^2 + 2hx + h^2) - 5x^2}{h} = \dfrac{5x^2 + 10hx + 5h^2 - 5^3}{h} = \dfrac{10hx + 5h^2}{h}$
$= 10x + 5h$

41. $\dfrac{[2(x+h)^2 + 3(x+h) - 4] - (2x^2 + 3x - 4)}{h}$

$= \dfrac{2(x^2 + 2hx + h^2 + 3x + 3h - 4 - 2x^2 - 3x + 4)}{h} = \dfrac{2x^2 + 4hx + 2h^2 + 3h - 2x^2}{h}$

$= \dfrac{4hx + 2h^2 + 3h}{h} = 4x + 2h + 3$

43. $\dfrac{\left(\dfrac{(x+h)+1}{(x+h)-1}\right) - \left(\dfrac{x+1}{x-1}\right)}{h} = \dfrac{\left(\dfrac{x+h+1}{x+h-1} - \dfrac{x+1}{x-1}\right)}{h}$

$= \dfrac{\dfrac{(x+h+1)(x-1) - (x+1)(x+h-1)}{(x+h-1)(x-1)}}{h} = \dfrac{(x^2 + hx - h - 1) - (x^2 + hx + h - 1)}{h(x+h-1)(x-1)}$

$= \dfrac{-2h}{h(x+h-1)(x-1)} = \dfrac{-2}{(x+h-1)x-1)} = \dfrac{-2}{x^2 + hx - 2x - h - 1}$

45. a. $r(1954) = \$0.92$ b. $m(1954) = \$0.45$

47. $s(1984) - s(1944) = \$1.49 - \$0.34 = \$1.15$

49. $e(1984) - e(1044) = \$1.15 - \$.64 = \$.51$

51. a. $\dfrac{g(1944 + 40) - g(1944)}{40} = \dfrac{g(1984) - g(1944)}{40} = \dfrac{1.10 - 0.21}{40} = \dfrac{.89}{40} = \$0.02225$

b. average change in gasoline prices from 1944 to 1984

53. a. $\dfrac{s(1944 + 10) - s(1944)}{10} = \dfrac{s(1954) - s(1944)}{10} = \dfrac{0.52 - 0.34}{10} = \dfrac{.18}{10} = \$0.018$

b. $\dfrac{s(1944 + 20) - s(1944)}{20} = \dfrac{s(1964) - s(1944)}{20} = \dfrac{0.59 - 0.34}{20} = \dfrac{0.25}{20} = \$0.125$

c. $\dfrac{s(1974) - s(1944)}{30} = \dfrac{2.08 - 0.34}{30} = \dfrac{1.74}{30} = \$0.058$

d. $\dfrac{s(1984) - s(1944)}{40} = \dfrac{1.49 - 0.34}{40} = \dfrac{1.15}{40} = \$0.02875$ e. $\dfrac{s(1944 + h) - s(1944)}{h}$

55. a. $\dfrac{m(1982) - m(1987)}{5} = \dfrac{2,421,000 - 2,495,000}{5} = \dfrac{-74,000}{2} = -14,800$ average change or

14,800 fewer marriages per year.

b. $\dfrac{m(1982 + h) - m(1982)}{h}$ is the average change of marriages between 1982 and 1982 + h years

where h = 5.

57. $c(x) = -0.02x^2 + 4x + 500 \ (0 \le x \le 150)$

a. $c(50) = -0.02(50)^2 + 4(50) + 500 = -0.02(2500) + 200 + 500 = -50 + 700 = \$650.$

b) $c(100) = -0.02(100)^2 + 4(100) + 500 = -200 + 400 + 500 = \$700.$

28

59. $c(51) = -0.02(51)^2 + 4(51) + 500 = -0.02(2601) + 204 + 500 = \$\,651.98$; per unit cost for $50 = \$\,13$ and per unit cost for $51 = \$\,12.78$.

61. $d(t) = 16t^2$ when t=2, t=6

a. $\dfrac{t(6) - t(2)}{4} = \dfrac{16(6)^2 - 16(2)^2}{4} = \dfrac{576 - 64}{4} = \dfrac{512}{4} = 128$ units/sec.

b. $\dfrac{t(4) - t(2)}{2} = \dfrac{16(4)^2 - 16(2)^2}{2} = \dfrac{256 - 64}{2} = \dfrac{192}{2} = 96$ units/sec.

c. $\dfrac{t(3) - t(2)}{1} = 16(3)^2 - 16(2)^2 = 144 - 64 = 80$ units/sec.

d. $\dfrac{t(2 + h) - t(2)}{h} = \dfrac{[16(2 + h)] - [16(2)^2]}{h} = \dfrac{16(h^2 + 4h + 4) - 16(4)}{h}$

$= \dfrac{16h^2 + 64h + 64 - 64}{h} = \dfrac{16h^2 + 64h}{h} = 16h + 64$

e. $\dfrac{t(x + h) - t(x)}{h} = \dfrac{[16(x + h)^2] - [16(x^2)]}{h} = \dfrac{16(x^2 + 2hx + h^2) - 16x^2}{h}$

$= \dfrac{16x^2 + 32hx + 16h^2 - 16x^2}{h} = \dfrac{32hx + 16h^2}{h} = 32x + 16h$

63. Example: $f(x) = 2x^2$ then $\dfrac{1}{f(x)} = \dfrac{1}{2x^2}$ and $f\left(\dfrac{1}{x}\right) = 2\left(\dfrac{1}{x}\right)^2 = \dfrac{2}{x^2}$ then $f(x) \neq f\left(\dfrac{1}{x}\right)$.

PROBLEM SET 2.2

1. $f(x) = 3x + 1$ for all reals; domain: $(-\infty, \infty)$

3. $h(x) = \dfrac{(3x + 1)(x + 2)}{x + 2}$; $x \neq -2$, domain: $(-\infty, -2) \cup (-2, \infty)$

5. $G(x) = \sqrt{2x + 1}$; $x > -\frac{1}{2}$, domain: $[-\frac{1}{2}, \infty)$

7. $f(x) = \sqrt{2 - x - x^2}$; $2 - x - x^2 \geq 0$

$2 + x$	$--$	$+\ +$	$+\ +$
$1 - x$	$+\ +$	$+\ +$	$+\ +$
	-2		1

9. $f \neq g$, $f(x) = g(x)$ so not equal   11. $f = g$, $f(x) \neq g(x)$ so not equal

13. $f = g$, $f(x) \neq g(x)$ so not equal

15. $f(x) = x^2 + 1$, $f(-x) = (-x)^2 + 1 = x^2 + 1$ since $f(x) = f(-x)$, even

17. $f(x) = \dfrac{1}{3x^3 - 4}$; $f_3(-x) = \dfrac{1}{-3x^3 - 4}$ since $f(x) \neq f(-x)$ and $f(x) \neq -f(x)$ then neither.

19. $f(x) = |x|$; $f(-x) = |-x|$ since $f(x) = f(-x)$, even

21. $R = (x_0, g(x_0))$ and $S = (x_0 + h, g(x_0 + h))$

23. D: $[-4, 7]$, R: $[-4, 5]$; intercepts: $(3, 0)$, $(0, 5)$; turning points: $(1, 5)$ and $(5, 5)$; constant: $[-4, 1]$, increasing: $[0, 5]$, decreasing: $[1, 0]$ and $[5, -4]$.

25. D: $[-5, \infty)$, R: $[-3, \infty)$; intercepts: $(-1.75, 0)$, $(1.75, 0)$, $(0, -3)$; turning points: $(-2, 0)$ and $(0, -3)$; constant: $[-5, -2]$; increasing: $[0, \infty)$; decreasing: $[-2, 0]$.

27. D: $[-6, 6]$, R: $[-5, 5]$; intercepts: $(-3, 0)$, $(3, 0)$, $(0, 5)$; turning point: $(0, -6)$; increasing: $[-6, 0]$, decreasing: $[0, -6]$

29. $(-5, 0)$ and $(0, 5)$ 31. $(\pm 2, 0)$ and $(0, -4)$ 33. $(0, 0)$

35. $(0, 3)$ and no x-intercept   37. $y = \dfrac{x^2 - 9}{x - 2}$; $x = 0$, $y = \dfrac{9}{2}$ and when $y = 0$,

$0 = x^2 - 9$ so $x = \pm 3$; $(0, \frac{9}{2})$ and $(\pm 3, 0)$.

39. $y = \dfrac{x^3 - 8}{x^2 + 2x + 4}$; $x = 0$ then $y = \dfrac{0 - 8}{0 + 0 + 4} = -2$ and when $y = 0$ then $0 = x^3 - 8$

$= (x - 2)(x^2 + 2x + 4)$ and $x = 2, -1 \pm i\sqrt{3}$ not a plotting point so $x = 2$; $(0, -2)$, $(2, 0)$

41. $y = \dfrac{x^3 - 1}{x - 1}$; $x = 0$ then $y = \dfrac{0 - 1}{0 - 1} = 1$ and $y = 0$ then $0 = x^3 - 1 = (x - 1)(x^2 + x + 1)$

and $x = 1$, since $x^2 + x + 1$ will give you complex answers; $(1, 0)$, $(0, 1)$

43. $y = -\dfrac{\sqrt{x + 3}}{x - 1}$; $x = 0$ then $y = -\dfrac{\sqrt{0 + 3}}{0 - 1} = \dfrac{-\sqrt{3}}{-1} = \sqrt{3} \approx 1.7$ and $y = 0$ then

$0 = -\dfrac{\sqrt{x + 3}}{x - 1}$; $0 = -\sqrt{0 + 3}$; $0 = -\sqrt{3}$ and $x = -3$ so $(-3, 0)$, $(0, \sqrt{3})$

45. $2|x| - |y| = 5$; $x = 0$ then $|y| = -5$ which is not possible so no y-intercept;

$y = 0$ then $2|x| = 5$; $|x| = \dfrac{5}{2}$ and $x = \pm \dfrac{5}{2}$ so $\left(\dfrac{5}{2}, 0\right)$, $\left(-\dfrac{5}{2}, 0\right)$

47. $xy = 1$; $x = 0$ then $0 = 1$ so no y-intercept and $y = 0$ then $0 = 1$ so no x-intercept

49. $x^2 + 2xy + y^2 = 4$; $x = 0$ then $0 + 0 + y^2 = 4$, $y = \pm 2$; $y = 0$ then $x^2 + 0 + 0 = 4$,

$x^2 = 4$, $x = \pm 2$ so $(2, 0)$, $(-2, 0)$, $(0, 2)$, $(0, -2)$

51. $y = \sqrt{x}$
D $= [0, \infty)$
R $= [0, \infty)$

53. $y = \sqrt{x - 2}$
D $= [2, \infty)$
R $= [0, \infty)$

55. $y = \sqrt{2x - 1}$
D $= [\frac{1}{2}, \infty)$
R $= [0, \infty)$

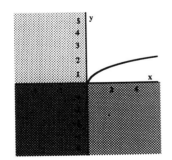
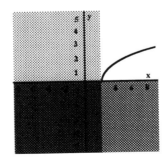
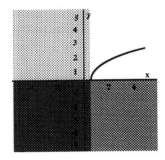

57. $y = x\sqrt{x}$
    $D = [0, \infty)$
    $R = [0, \infty)$

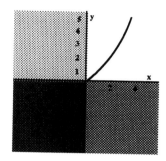

59. $y = \sqrt{x^3}$
    $D = [0, \infty)$
    $R = [0, \infty)$

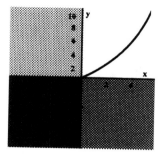

61. $y^2 = x^3$
    $D = [0, \infty)$
    $R = (-\infty, \infty)$

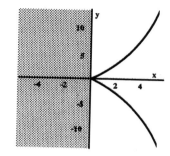

63. $y^2 = x^2 - 9$ or $x^2 = y^2 + 9$
    $D = (-\infty, -3] \cup [3, \infty)$ by
        solving for y
    $R = (-\infty, \infty)$ by solving for x

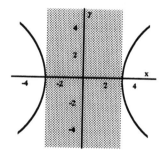

65. $y = \sqrt{x^2 - 4}$, $x^2 - 4 \geq 0$

$x - 2$	$- -$	$- -$	$+ +$
$x + 2$	$- -$	$+ +$	$+ +$
	-2	2	

    $D = (-\infty, -2] \cup [2, \infty)$
    $R = [0, \infty)$

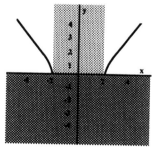

67. $y = \sqrt{x^2 + x - 12}$

$x^2 + x - 12 \geq 0$

$$
\begin{array}{c|c|c|c}
x + 4 & -\,- & +\,+ & +\,+ \\
x - 3 & -\,- & -\,- & +\,+ \\
\hline
 & -4 & & 3
\end{array}
$$

$D = (-\infty, -4] \cup [3, \infty)$

$R = [0, \infty)$

69. $y = 3, \quad 3 = \dfrac{5x^2 - 8x}{2x + 1}; \quad 3(2x + 1) = 5x^2 - 8x; \quad 6x + 3 = 5x^2 - 8x; \quad 0 = 5x^2 - 14x + 3;$

by the quadratic formula: $(3, 3)$ and $(-0.2, 3)$ OR $\left(-\frac{1}{5}, 3\right)$

71. $y = -4, \quad -4 = \dfrac{5x^2 - 8x}{2x + 1}; \quad -4(2x + 1) = 5x^2 - 8x; \quad -8x - 4 = 5x^2 - 8x; \quad 0 = 5x^2 + 4;$

$5x^2 = 4; \quad x^2 = \frac{4}{5}; \quad x = \pm \dfrac{2}{\sqrt{5}} = \pm \dfrac{2\sqrt{5}}{5}$

73. $y = x + 1, \quad x + 1 = \dfrac{x^3 + 2x^2 - 2x}{x^2 - 2}; \quad (x + 1)(x^2 - 2) = x^3 + 2x^2 - 2x;$

$x^3 - 2x + x^2 - 2 = x^3 + 2x^2 - 2x; \quad 0 = x^2 + 2; \quad x^2 = -2$ so no intersection; this is a
rational function with an oblique (slant) asymptote at $y = x + 1$.

75. A one-to-one function with 5 elements in set x will mean that you have only 5 elements in set Y.

77. f is decreasing on its domain, $f(x_1) > f(x_2)$ provided $x_1 < x_2$; an infinite number of x-values
generate an infinite number of y-values with no repetition. Therefore f is one-to-one.

PROBLEM SET 2.3

1. a. $(h, k) = (6, 3)$ b. $(-3, 5)$ 3. a. $(0, \sqrt{2})$ b. $y - 6 = f(x)$ so $(0, 6)$

5. a. $(0, 0)$ b. $(-\sqrt{2}, 0)$

7. intercepts: $(-2, 0)$ and $(2, 0)$ and vertex: $(0, -3)$ which means we have a $y = x^2$ basic parabola
shifted down 3 units; the equation is $y = x^2 - 3$.

9. intercept: $(-2, 0)$ and $(-4, 0)$ and vertex: $(-4, -4)$ which means a basic $y = x^2$ graph
shifted down 4 units and left 4 units; the equation is $y + 4 = (x + 4)^2$.

11. An absolute value graph shifted up 4 units; the equation is $y - 4 = |x|$ OR $y = |x| + 4$.

13. A square root graph shifted up 3 units; the equation is $y - 3 = \sqrt{x}$ OR $y = \sqrt{x} + 3$.

15. $y + 3 = f(x)$
    OR $y = f(x) - 3$
    shift: down 3 units

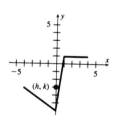

17. $y + \dfrac{1}{2} = h(x)$
    OR $y = h(x) - \dfrac{1}{2}$
    shift: down $\dfrac{1}{2}$ unit

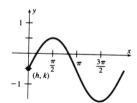

19. $y - 1 = h(x)$
    OR $y = h(x) + 1$
    up 1 unit

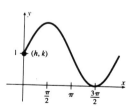

21. $y = h\left(x - \dfrac{\pi}{2}\right)$
    shift: right $\dfrac{\pi}{2}$

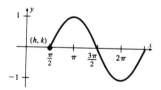

23. $y = f(x - \sqrt{5})$
    shift: right $\sqrt{5}$

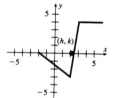

25. $y + 1 = g(x - 4)$
    shift: right 4, down 1

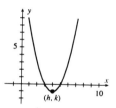

27. $y - 3 = g(x + 5)$
    shift: left 5, up 3

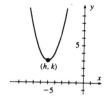

29. $y + 2 = f(x - 4)$
    shift: right 4, down 2

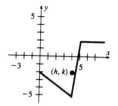

31. $y - \sqrt{3} = f(x + \sqrt{2})$
    shift: left $\sqrt{2}$, up $\sqrt{3}$

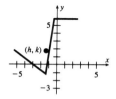

33

33. $y = 2c\,(x)$, dilation in y-direction

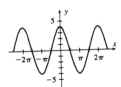

35. $y = -c\,(x)$, reflection over x-axis

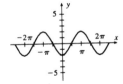

37. $y = c\,(2x)$, dilation in y-direction

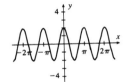

39. $y - 2 = c\,(x - 3)$, dilation in y-direction and shift: right 3, up 2

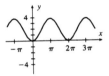

41. $y - 3 = (x - 2)^2$
shift: right 2, up 3

43. $y = x^2 - 1$
shift: down 1

45. $y = \sqrt{x} + 4$
shift: up 4

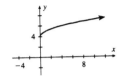

34

47. $y = |x + \pi|$
shift: left π

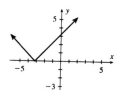

49. $y + \sqrt{3} = |x - \sqrt{2}|$
shift: right $\sqrt{2}$, down $\sqrt{3}$

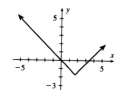

51. $y + 2 = (x + \sqrt{3})^2$
shift: left $\sqrt{3}$, down 2

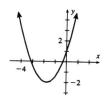

53. $y - \sqrt{2} = \sqrt{x + 5}$
shift: left 5, up $\sqrt{5}$

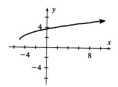

55. $y = -3 |x + 5|$
shift: left 5,
y-dilation

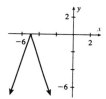

57. $y = -2 (x + 4)^2$
shift: left 4,
y-dilation

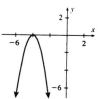

59. $y = |(x - 4)^2 - 9|$
shifts: right 4, up 9
because of absolute value

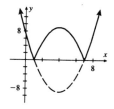

61. $y = (|x| - 2)^2 - 4$
shift: down 4, reflected graph over the
y-axis because of absolute value

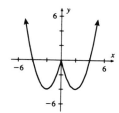

63. $y + 3 = (x + 8)^2,$
$-14 \leq x \leq -8$

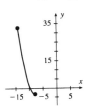

65. $y + 12 = \left(x + \dfrac{25}{2}\right)^2$
$y > -10$

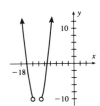

67. $y + 12 = (x - 8)^2$
$y < 4$

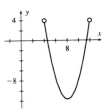

PROBLEM SET 2.4

1. $(f + g)(5) = f(5) + g(5) = [2(5) - 3] + [5^2 + 1] = 10 - 3 + 25 + 1 = 33$

3. $(f \cdot g)(2) = [f(2)][g(2)] = [2(2) - 3][2^2 + 1] = (4 - 3)(4 + 1) = 1(5) = 5$

5. $(f \circ g)(2) = f[g(2)] = f[5] = 2(5) - 3 = 7 \qquad$ since $\left\{g(2) = 2^2 + 1 = 5\right\}.$

7. $(f - g)(5) = f(5) - g(5) = \left(\dfrac{5 - 2}{5 + 1}\right) - (5^2 - 5 - 2) = \dfrac{3}{6} - (25 - 7) = \dfrac{1}{2} - 18 = -17\frac{1}{2}$

9. $\left(\dfrac{f}{g}\right)(99) = \dfrac{f(99)}{g(99)} = \dfrac{\frac{99 - 2}{99 + 1}}{99^2 - 99 - 2} = \dfrac{\frac{97}{100}}{9801 - 101} = \dfrac{.97}{9700} = 0.0001 \ \text{OR} \ 1 \ x \ 10^{-4}$

11. $(f + g)(-1) = f(-1) + g(-1) = \left\{\dfrac{2(-1)^2 - 5(-1) + 2}{-1 - 2} - 3\right\} + \left\{(-1)^2 - (-1) - 2\right\}$

$= \left\{\dfrac{2 + 5 + 2}{-3} - 3\right\} + (1 + 1 - 2) = \dfrac{9}{-3} - 3 + 0 = -3 + -3 = -6$

13. $(f \cdot g)(9) = f(9) \cdot g(9) = \left\{\dfrac{2(9^2) - 5(9) + 2}{9 - 2} - 3\right\} + \left\{9^2 - (9) - 2\right\} = \dfrac{2(81) - 45 + 2}{7} + (81 - 11)$

$\dfrac{162 - 43}{7} + 70 = \dfrac{119}{7} + 70 = 7 + 70 = 77$

15. $(f \circ g)(0) = f[g(0)] = f(-2) = \dfrac{2(-2)^2 - 5(-2) + 2}{-2 - 2} = \dfrac{8 + 10 + 2}{-4} = \dfrac{20}{-4} = -5$
$\left\{g(0) = 0^2 - 0 - 2 = -2\right\}$

17. $(f - g)(3) = f(3) - g(3) = 10 - 3 = 7 \quad$ since f has $(3, 10)$ and g has $(3, 3)$.

19. $\left(\dfrac{f}{g}\right)(0) = \dfrac{f(0)}{g(0)} = \dfrac{1}{3} \quad$ since f has $(0, 1)$ and g has $(0, 3)$

21. $(f + g)(x) = f(x) + g(x) = (2x - 3) + (x^2 + 1) = x^2 + 2x - 2$
$(f - g)(x) = f(x) - g(x) = (2x - 3) - (x^2 + 1) = 2x - 3 - x^2 - 1 = -x^2 + 2x - 4$

23. $(f + g)(x) = f(x) + g(x) = \left(\dfrac{2x^2 - x - 3}{x - 2}\right) + (x^2 - x - 2) = \dfrac{(2x - 3)(x + 1)}{x - 2} + \dfrac{(x - 2)(x + 1)}{1}$

$= \dfrac{(2x - 3)(x + 1) + (x - 2)^2(x + 1)}{x - 2} = \dfrac{2x^2 - x - 3 + x^3 - 3x^2 + 4}{x - 2} = \dfrac{x^3 - x^2 - x + 1}{x - 2}$

$(f - g)(x) = \dfrac{2x^2 - x - 3}{x - 2} - (x^2 - x - 2) = \dfrac{2x^2 - x - 3}{x - 2} - \dfrac{(x - 2)^2(x + 1)}{x - 2}$

$= \dfrac{2x^2 - x - 3 - (x^3 - 3x^2 + 4)}{x - 2} = \dfrac{2x^2 - x - 3 - x^3 + 3x^2 - 4}{x - 2} = \dfrac{-x^3 + 5x^2 - x - 7}{x - 2}$

25. $(f \cdot g)(x) = f(x) \cdot g(x) = (2x - 3)(x^2 + 1) = 2x^3 + 2x - 3x^2 - 3;$ $\left(\dfrac{f}{g}\right)(x) = \dfrac{f(x)}{g(x)} = \dfrac{2x - 3}{x^2 + 1}$

27. $(f \cdot g)(x) = f(x) \cdot g(x) = \left(\dfrac{2x^2 - x - 3}{x - 2}\right)(x^2 - x - 2) = \left(\dfrac{(2x - 3)(x + 1)}{x - 2}\right)\bigl((x - 2)(x + 1)\bigr)$

$= (2x - 3)(x + 1)^2 = 2x^3 + x^2 - 4x - 3$

$\left(\dfrac{f}{g}\right)(x) = \dfrac{f(x)}{g(x)} = \dfrac{\dfrac{2x^2 - 2x - 3}{x - 2}}{x^2 - x - 2} = \dfrac{2x^2 - x - 3}{x - 2} \cdot \dfrac{1}{x^2 - x - 2}$

$= \dfrac{(2x - 3)(x + 1)}{x - 2} \cdot \dfrac{1}{(x - 2)(x + 1)} = \dfrac{2x - 3}{(x - 2)^2}$ OR $\dfrac{2x - 3}{x^2 - 4x - 4}$

29. $(f \circ g)(x) = f[g(x)] = f[x^2 + 1] = 2(x^2 + 1) - 3 = 2x^2 + 2 - 3 = 2x^2 - 1$

$(g \circ f)(x) = g[f(x)] = g[2x - 3] = (2x - 3)^2 + 1 = 4x^2 - 12x + 9 + 1 = 4x^2 - 12x + 10$

31. $(f \circ g)(x) = f[g(x)] = f[x^2 - x - 2] = \dfrac{2(x^2 - x - 2)^2 - 3(x^2 - x - 2) - 2}{x - 2} =$

$\dfrac{2(x^4 - 2x^3 - 3x^2 + 4x + 4) - 3x^2 + 3x + 6 - 2}{x - 2} = \dfrac{2x^4 - 4x^3 - 6x^2 + 8x + 8 - 3x^2 + 3x + 4}{x - 2}$

$= \dfrac{2x^4 - 4x^3 - 9x^2 + 11x + 12}{x - 2}$

$(g \circ f)(x) = g[f(x)] = g\left(\dfrac{2x^2 - 3x - 2}{x - 2}\right) = \left(\dfrac{2x^2 - 3x - 2}{x - 2}\right)^2 - \left(\dfrac{2x^2 - 3x - 2}{x - 2}\right) - 2$

$\dfrac{4x^4 - 12x^3 + x^2 + 12x + 4}{(x - 2)^2} - \dfrac{(2x^2 - 3x - 2)(x - 2)}{(x - 2)^2} - \dfrac{2(x - 2)}{(x - 2)^2}$

$= \dfrac{4x^4 - 12x^3 + x^2 + 12x + 4}{(x - 2)^2} - \dfrac{2x^3 - 7x^2 + 4x + 4}{(x - 2)^2} - \dfrac{2x - 4}{(x - 2)^2}$

$= \dfrac{4x^4 - 14x^3 + 8x^2 + 6x + 4}{(x - 2)^2}$ OR $\dfrac{4x^4 - 14x^3 + 8x^2 + 6x + 4}{x^2 - 4x + 4}$

33. {(0,10), (2, 12), (4, 17), (10, 16.5), (14, 12), (16, 14), (20, 23.5)}

35. {(−19, 15), (−17, 9), (−15, 7.5), (−12, 12), (−10, 10), (−9, 24)}

37. {(0, 7), (2, 3.5), (4, 2.25), (6, 3.5), (8, 5.25), (10, 9.25), (13, 20.5), (16, 32.75)}

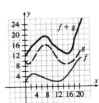
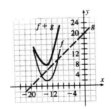
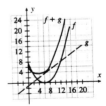

39. $f[u(x)] = 2\left(\sqrt[3]{x}\right)^2 + 1$ since $\sqrt[3]{x}$ is substituted for x.

41. $f[u(x)] = 3(x^3) + 5(x^3) + 1$ since x^3 is substituted for x.

43. $(f \circ g)(x) = f[g(x)] = f[2x-1] = (2x-1)^2 = 4x^2 - 4x + 1$
$(g \circ h)(x) = g[h(x)] = g[3x + 2] = 2(3x + 2) - 1 = 6x + 4 - 1 = 6x + 3$

45. $(f \circ g \circ h)(x) = f \circ [g(h(x))] = f[g(3x + 2)] =$ since $\left\{g(3x + 2) = 2(3x + 2) - 1 = 6x + 3\right\}$
$f[6x + 3] = (6x + 3)^2 = 36x^2 + 36x + 9$

47. $(f \circ g) \circ h = f[g(x)] \circ h = f[3x - 1] \circ h = (3x - 1)^2 \circ h = (9x^2 - 6x + 1) \circ h$
$9(x^2 + 1)^2 - 6(x^2 + 1) + 1 = 9(x^4 + 2x^2 + 1) - 6x^2 - 6 + 1 = 9x^4 + 18x^2 + 9 - 6x^2 - 5$
$= 9x^4 + 12x^2 + 4$

49. $(f \circ g)(x) = f[g(x)] = f[x^2 - 2] = \sqrt{x^2 - 2}$ with domain: $(\sqrt{2}, \infty)$
$(g \circ h)(x) = g[h(x)] = g[x + 2] = (x + 2)^2 - 2 = x^2 + 4x + 4 - 2 = x^2 + 4x + 2$

51. $f \circ (g \circ h) = f \circ g[h(x)] = f \circ [g(x + 2)] = f \circ [(x + 2)^2 - 2] = f \circ (x^2 + 4x + 4 - 2)$
$= f[x^2 + 4x + 2] = \sqrt{x^2 + 4x + 2}$; x is approximately $=$ to $-.5858$ and -3.4142; no solutions.

53. $(f \circ g) \circ h = f[g(x)] \circ h = f[x] \circ h = x$

55. If $u = 2x^2 - 1$ and if $g = x^4$, then $g[u(x)] = g[2x^2 - 1] = (2x^2 - 1)^4 = f(x)$.

57. If $u = 5x - 1$ and if $g = \sqrt{x}$, then $g[u(x)] = g[5x - 1] = \sqrt{5x - 1} = f(x)$.

59. If $u = x^2 - 1$ and if $g = x^3 + \sqrt{x} + 5$, then $g[u(x)] = g[x^2 - 1] = (x^2 - 1)^3 + \sqrt{x^2 - 1} + 5$

61. $V(h) = \frac{\pi h^3}{12}$, $h(t) = 2t$

    a. $V(h) = \frac{\pi(2t)^3}{12}$; $t=2$ then $V(h) = \frac{\pi[2(2)]^3}{12} = \frac{\pi(4)^3}{12} = \frac{64\pi}{12} = \frac{16\pi}{3}$

    b. $V \circ h = V[h(t)] = V[2t] = \frac{\pi[(2t)^3]}{12} = \frac{8\pi t^3}{12} = \frac{2\pi t^3}{3}$

    c. The permissible values of t or the domain of h: $(0, 1,4203]$.

63. $f(x) = x^2$ then $f\left(\frac{2}{x}\right) = \left(\frac{2}{x}\right)^2 = \frac{4}{x^2} \neq \frac{1}{x^2} = \frac{1}{f(x)}$.

65. a. $(f \circ f)(x) = f(f(x)) = f\left(1 + \frac{1}{x}\right) = 1 + \dfrac{1}{1 + \frac{1}{x}} = 1 + \left(\dfrac{1}{1 + \frac{1}{x}}\right)\left(\dfrac{x}{x}\right) = 1 + \dfrac{x}{x+1}$

$= \dfrac{x+1}{x+1} + \dfrac{x}{x+1} = \dfrac{2x+1}{x+1}$

b. $(f \circ f \circ f)(x) = f \circ f[f(x)] = f \circ \left(\dfrac{2x+1}{x+1}\right)\ \{\text{from part a}\} = 1 + \dfrac{1}{\frac{2x+1}{x+1}}$

$= 1 + \dfrac{x+1}{2x+1} = \dfrac{2x+1+x+1}{2x+1} = \dfrac{3x+2}{2x+1}$

c. $(f \circ f \circ f \circ f)(x) = f \circ (f \circ f \circ f)(x) = f \circ \left(\dfrac{3x+2}{2x+1}\right) = 1 + \dfrac{1}{\frac{3x+2}{2x+1}} = 1 + \dfrac{2x+1}{3x+2}$

$= \dfrac{3x+2+2x+1}{3x+2} = \dfrac{5x+3}{3x+2}$

67. Problem 66 shows that you will get the square root of the previous iteration. So for $f(x) = 2\sqrt{x}$ you will get the following:

$(f \circ f)(x) = f[f(x)] = f[2\sqrt{x}] = 2\sqrt{2\sqrt{x}}$

$(f \circ f \circ f) = f \circ (f \circ f) = f\left(2\sqrt{2\sqrt{x}}\right) = 2\sqrt{2\sqrt{2\sqrt{x}}}$

$(f \circ f \circ f \circ f)(x) = f \circ (f \circ f \circ f) = f\left(2\sqrt{2\sqrt{2\sqrt{x}}}\right) = 2\sqrt{2\sqrt{2\sqrt{2\sqrt{x}}}}$

69. $(f \circ f)(x) = f[k\sqrt{x}] = k\sqrt{k\sqrt{x}}$

$(f \circ f \circ f)(x) = f \circ (f \circ f) = \left(k\sqrt{k\sqrt{x}}\right) = k\sqrt{k\sqrt{k\sqrt{x}}}$

$(f \circ f \circ f \circ f)(x) = f \circ (f \circ f \circ f) = f \circ \left(k\sqrt{k\sqrt{k\sqrt{x}}}\right) = k\sqrt{k\sqrt{k\sqrt{k\sqrt{x}}}}$

PROBLEM SET 2.5

1. $(f \circ g)(x) = f[g(x)] = f\left(\frac{1}{5}x\right) = 5\left(\frac{1}{5}x\right) = x;\quad g \circ f)(x) = f[f(x)] = g[5x] = \frac{1}{5}(5x) = x;\ $ inverses

3. $(f \circ g)(x) = f[g(x)] = f\left(\frac{x-3}{5}\right) = 5\left(\frac{x-3}{5}\right) + 3 = x - 3 + 3 = x;$

$(g \circ f)(x) = g[f(x)] = g(5x+3) = \dfrac{5x+3-3}{5} = \dfrac{5x}{5} = x;\ $ inverses

5. $(f \circ g)(x) = f[g(x)] = f\left(\frac{5}{4}x+3\right) = \frac{4}{5}\left(\frac{5}{4}x+3\right) + 4 = x + \frac{12}{5} + 4 = x + \frac{32}{5};$

$(g \circ f)(x) = g[f(x)] = g\left(\frac{4}{5}x+4\right) = \frac{5}{4}\left(\frac{4}{5}x+4\right) + 3 = x + 5 + 3 = x + 8;\ $ not inverses

7. $(f \circ g)(x) = f[g(x)] = f[\sqrt{x}] = (\sqrt{x})^2 = x;\quad (g \circ f)(x) = g[f(x)] = g(x^2) = \sqrt{x^2} = x;\ $ inverses

9. $f^{-1} = \{(5, 4), (3, 6), (1, 7), (4, 2)\}$ 11. $f^{-1}:\ x = y + 3;\ y = x - 3$

13. $g^{-1}:\ x = 5y;\ y = \frac{1}{5}x$

15. $h(x) = x^2 - 5$ is the equation for a parabola. Since there is no restriction on the domain, there is no inverse.

17. f^{-1}: x 19. $f(x) = 6$; there is no inverse.

21. f^{-1}: $x = \dfrac{1}{y-2}$; $xy - 2x = 1$; $xy = 2x + 1$; $y = \dfrac{2x+1}{x}$, $x \neq 0$

23. $f(x) = y = \dfrac{2x-6}{3x+3}$; $x \neq -1$ f^{-1}: $x = \dfrac{2y-6}{3y+3}$; $x(3y+3) = 2y - 6$; $3xy + 3x = 2y - 6$;

$3xy - 2y = -3x - 6$; $y(3x-2) = -(3x+6)$; $y = \dfrac{-(3x+6)}{3x-2}$; $x \neq \dfrac{2}{3}$

25. a. $f(3) = -3$;   b. $f(4) = -2$;   c. $f(7) = 3$;   d. $f(11) = 5$;   e. $f(14) = 7$

27. a. $f(1) = -5$;   a. $f^{-1}(-1) = 5$;   c. $f^{-1}(-4) = 3$;   d. $f(2) = 4.5$; $f^{-1}(2) = 6.5$

29. a. $f^{-1}(4) = 8$;   b. $f(8) = 3.5$;   c. $f^{-1}(-3) = 3$;   d. $f(9) = 4$;   e. $f^{-1}(7) = 14$

31. graphing points: $f(0) = -6$,

$f(1) = -5$, $f(2) = -4$,

$f(3) = -3$, $f(4) = -1$,

$f(5) = -0.5$, $f(5.5) = 0$,

$f(6) = 1$, $f(7) = 7$,

$f(9) = 4$, $f(11) = 5$,

$f(12) = 6$, $f(14) = 8$ etc.

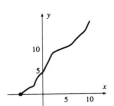

33. $f(x) = 2x^2 + 1$, $x \leq 0$; $g(x) = -\dfrac{1}{2}\sqrt{2x-2}$, $x \geq 1$

$= f[g(x)] = f\left(-\dfrac{1}{2}\sqrt{2x-2}\right) = 2\left(\dfrac{\sqrt{2x-2}}{2}\right)^2 + 1 = 2\left(\dfrac{2x-2}{4}\right) + 1 = 2\left(\dfrac{2(x-1)}{4}\right) + 1$

$= x - 1 + 1 = x$; $g[f(x)] = g[2x^2 + 1] = -\dfrac{1}{2}\sqrt{2(2x^2+1) - 2} = -\dfrac{1}{2}\sqrt{4x^2 + 2 - 2}$

$= -\dfrac{1}{2} \cdot 2x = -x$; so not inverses.

35. $f(x) = (x+1)^2$, $x \geq -1$; $g(x) = -1 + \sqrt{x}$, $x \geq 0$

$f[g(x)] = f[-1 + \sqrt{x}] = (-1 + \sqrt{x} + 1)^2 = (\sqrt{x})^2 = x$; $g[f(x)] = g[(x+1)^2]$

$= -1 + \sqrt{(x+1)^2} = -1 + x + 1 = x$; so inverses.

37. $f(x) = |x+1|$, $x \geq -1$; $g(x) = x - 1$, $x \geq 1$

$f[g(x)] = f[x-1] = |(x-1) + 1| = |x|$; $g[f(x)] = g[|x+1|] = |x+1| - 1$; not inverses.

40

39. $g(x) = x - 4$;

$\quad g^{-1}(x) = x + 4$

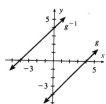

41. $f(x) = 7x + 3$;

$\quad f^{-1}(x) = \dfrac{x - 3}{7}$

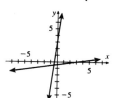

43. $g(x) = x^2, \ x \leq 0$;

$\quad g^{-1}(x) = \sqrt{x}\ , \ x \geq 0$

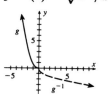

45. $g(x) = \mid x - 1 \mid, \ x \geq 1$;

$\quad g^{-1}(x) = \mid x + 1 \mid, \ x \geq 0$

47. $g(x) = \dfrac{1}{3}x + 1$;

$\quad g^{-1}(x) = 3x - 3$

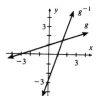

49. $k(x) = -\dfrac{1}{4}x + \dfrac{3}{4}$;

$\quad k^{-1}(x) = -4x + 3$

51. $m(x) = 3x + 6$;

53. $f(x) = x^2, \ (-\infty, 0]$; $\ f^{-1}$: $\ x = y^2, \ y = \sqrt{x}, \ [0, \infty)$

55. $f(x) = x^2 + 1, \ [0, \infty)$; $\ f^{-1}$: $\ x = y^2 + 1, \ x - 1 = y^2$,

$\quad y = \sqrt{x - 1}\ , \ [1, \infty)$

57. $f(x) = 2x^2, \ [-10, -1]$; $\ f^{-1}$: $\ x = 2y^2, \ \dfrac{x}{2} = y^2, \ y = \pm \sqrt{\dfrac{x}{2}}$,

$\quad y = -\dfrac{1}{2}\sqrt{2x}\ , \ [2, 200]$

59. $f(x) = \dfrac{1}{3x + 1}\ , \ (-\dfrac{1}{3}, \infty)$; $\ f^{-1}$: $\ x = \dfrac{1}{3y + 1}$,

$\quad x\,(3y + 1) = 1, \ 3xy + x = 1, \ 3xy = 1 - x$,

$\quad y = \dfrac{1 - x}{3x}, \ (0, \infty)$

61. $f(x) = x, \ [0, \infty)$; $\ f^{-1}$: $\ x = y, \ [0, \infty)$ so own inverse.

CHAPTER 2 SUMMARY PROBLEMS

1. function, one-to-one   3. not a function

5. a. $f(4) = 3(4) - 1 = 12 - 1 = 11$ b. $g(4) = 5 - 4^2 = 5 - 16 = -11$

7. a. $f(w) = 3w - 1$ b. $g(w + h) = 1 - (w + h)^2 = 1 - (w^2 + 2hw + h^2) = 1 - w^2 - 2hw - h^2$

9. $f(x) = 5 - x^2$; $\dfrac{f(w + h) - f(x)}{h} = \dfrac{\left[5 - (x + h)^2\right] - \left[5 - x^2\right]}{h} =$
$\dfrac{5 - (x^2 + 2hx + h^2) - 5 + x^2}{h} = \dfrac{5 - x^2 - 2hx - h^2 - 5 + x^2}{h} = \dfrac{-2hx - h^2}{h} = -2x - h$

11. $f(x) = 5$; $\dfrac{f(x + h) - f(x)}{h} = \dfrac{5 - 5}{h} = 0$

13. $x^2 - y + 5 = 0$, $y = x^2 + 5$, $V(0, 5)$; Domain: $(-\infty, \infty)$, Range: $[5, \infty)$

15. $y = \sqrt{\dfrac{x + 5}{1 - x}}$, $x \neq 1$; Domain: $[-5, 1)$, Range: $(0, \infty)$

17. $y = (x - 2)(3x + 5)(x + 2)$; x-intercepts: $(2, 0)$, $(-2, 0)$, $(-\frac{5}{3})$; y-intercept: $(0, -20)$

19. $2x^2 + xy - 4y^2 = 1$; $x = 0$, $0 - 4y^2 = 1$, $4y^2 = -1$ so no y-intercepts; $y = 0$, $2x^2 = 1$,
$x^2 = \frac{1}{2}$, $x = \pm\dfrac{\sqrt{2}}{2}$, x-intercepts: $\left(\pm\dfrac{\sqrt{2}}{2}, 0\right)$

21. $f(x) = \dfrac{x^2 + x}{x + 1}$, $x \neq -1$; $g(x) = x$; $f(x) \neq g(x)$ so not equal.

23. $f(x) = \sqrt{x^2}$; $g(x) = x$; $x = -2$, then $f(x) = (-2)^2 = 4$ and $\sqrt{4} = 2$ and $g(x) = -2$, not equal.

25. $f(x) = 3x^2 - 2x - 5$; $f(-x) = 3(-x)^2 - 2(-x) - 5 = 3x^2 + 2x - 5$, not even
$-f(x) = -(3x^2 - 2x - 5) = -3x^2 + 2x + 5$, not odd therefore neither.

27. $f(x) = \dfrac{1}{(x + 4)^2} = \dfrac{1}{x^2 + 8x + 16}$; $f(-x) = \dfrac{1}{(-x + 4)^2} = \dfrac{1}{x^2 - 8x + 16}$, $-f(x) = \dfrac{-1}{x^2 + 8x + 16}$,
$f(x) \neq f(-x)$ and $f(-x) \neq -f(x)$ so neither.

29. $y - 6 = f(x + \pi)$; $(h, k) = (-\pi, 6)$ so it shifts to the right π units and up 6 units.

31. $y = 5(x - 4)^2$; $(h, k) = (4, 0)$ so shifts left 4 units with no vertical shift.

33. $f(x) = 9x^2$, $(h, k) = \left(-\sqrt{2}, 3\right)$ so equation is: $y - 3 = 9\left(x + \sqrt{2}\right)^2$.

35. $f(x) = -2x^2$, $(h, k) = (\pi, 0)$ so equation is: $y = -2(x - \pi)^2$.

37. $y - 1 = 3(x + 3)^2$; $y' = 3(x')^2$ 39. $y = 5(x + 3)^2 + 1$, $y - 1 = 5(x + 3)^2$; $y' = 5(x')^2$

41. $y - 2 = f(x - 3)$;
   shifts 3 units right
   shifts 2 units up

43. $y = f(x - \pi)$
   shifts π units right
   no vertical shift

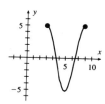

45. $y = -f(x)$
   reflects over the x-axis

47. $y = \frac{1}{2} f(x)$, shrinks or
   compresses the graph.

49. $(f + g)(x) = f(x) + g(x) = x^2 + 5x - 2$
51. $(f \cdot g)(x) = x^2(5x - 2) = 5x^3 - 2x^2$
53. $(f + g)(x)$ ordinates added:
   $\{(-4, 5), (-3, 2.5), 9 - 2, 1), (-1, .5), (0, 1)$
   $(1, 2.5), (2, 5), (3, 8.5), (4, 13)\}$

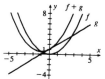

55. $\{(0, .5), (1, 1), (2, .6), (2.5, .2), (3, -.25),$
   $(4, 1.1), (4.5, -1), (5, -.9), (6, .4)$

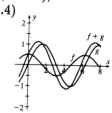

57. $(f \circ g)(x) = f[g(x)] = f[5 - x^2] = 3[5 - x^2] - 1 = 15 - 3x^2 - 1 = -3x^2 + 14$
59. $(f \circ h)(x) = f[h(x)] = f[6] = 3(6) - 1 = 18 - 1 = 17$
61. $(f \circ g)(x) = f[g(x)] = f[x - 2] = 3(x - 2) + 6 = 3x - 6 + 6 = 3x$ and $g[f(x)] = g[3x + 6]$
   $= [3x + 6] - 2 = 3x + 4$; not inverses since $f[g(x)] \neq g[f(x)]$.
63. $(f \circ g)(x) = f[g(x)] = f\left(\frac{x - 2}{3}\right) = 3\left(\frac{x - 2}{3}\right) - 2 = x - 2 - 2 = x - 4$;

   $(g \circ f)(x) = g[f(x)] = g[3x - 2] = \frac{(3x - 2) - 2}{3} = \frac{3x - 4}{3}$; not inverses since $f[g(x)] \neq g[f(x)]$.
65. $f(x) = 3x - 1$; f^{-1}: $x = 3y - 1$, $3y = x + 1$, $y = \frac{x + 1}{3} = f^{-1}(x)$.
67. $f(x) = \frac{1}{2}x + 5 = \frac{x + 10}{2}$; f^{-1}: $x = \frac{y + 10}{2}$, $2x = y + 10$, $y = 2x - 10 = f^{-1}(x)$.

CHAPTER 3

PROBLEM SET 3.1

1. $a = 3;$ $b = -9;$ $c = -6$

3. $a = -2;$ $b = -4;$ $c = -10;$ $d = 16;$ $e = 12;$ $f = -12;$ $g = 2x^2 - 8x + 6$

5. $a = -4;$ $b = 1;$ $c = 3;$ $d = 8;$ $e = -4;$ $f = -2;$ $g = -4;$

$h = x^2 + 3x - 4 - \dfrac{4}{x + 4}$

7. $a = 1;$ $b = 1;$ $c = 2;$ $d = 0;$ $e = -5;$ $f = -20;$ $g = 2;$ $h = 0;$

$i = x^3 + 3x^2 + 3x - 2$

9. $a = -2;$ $b = 1;$ $c = 0;$ $d = 0;$ $e = 0;$ $f = -8;$ $g = 0;$ $h = 16;$

$i = 16;$ $j = -32;$ $k = -32;$ $l = -64;$ $m = x^4 - 2x^3 + 4x^2 - 8x + 16 - \dfrac{64}{x + 2}$

11. $x = 3$

$$
\begin{array}{r|rrrr}
3 & 3 & -2 & 4 & -75 \\
  &   & 9 & 21 & 75 \\
\hline
  & 3 & 7 & 25 & 0
\end{array}
$$

$3x^2 + 7x + 25$

13. $x = -1$

$$
\begin{array}{r|rrrrr}
-1 & 1 & -6 & 1 & 0 & -8 \\
   &   & -1 & 7 & -8 & 8 \\
\hline
   & 1 & -7 & 8 & -8 & 0
\end{array}
$$

$x^3 - 7x^2 + 8x - 8$

15. $x = -3$

$$
\begin{array}{r|rrrrr}
-3 & 2 & 0 & -15 & 8 & -3 \\
   &   & -6 & 18 & -9 & 3 \\
\hline
   & 2 & -6 & 3 & -1 & 0
\end{array}
$$

$2x^3 - 6x^2 + 3x - 1$

17. $x = 1$

$$
\begin{array}{r|rrrrrr}
1 & 4 & -3 & -5 & 0 & 0 & 4 \\
  &   & 4 & 1 & -4 & -4 & -4 \\
\hline
  & 4 & 1 & -4 & -4 & -4 & 0
\end{array}
$$

$4x^4 + x^3 - 4x^2 - x - 4$

19. $x = 2$

$$
\begin{array}{r|rrrr}
2 & 3 & -2 & 4 & -24 \\
  &   & 6 & 8 & 24 \\
\hline
  & 3 & 4 & 12 & 0
\end{array}
$$

$3x^2 + 4x + 12$

21. $x = -4$

$$
\begin{array}{r|rrrr}
-4 & 1 & 5 & -2 & -24 \\
   &   & -4 & -4 & 24 \\
\hline
   & 1 & 1 & -6 & 0
\end{array}
$$

$x^2 + x - 6$

23. $x = 5$

$$
\begin{array}{r|rrrr}
5 & 1 & -4 & -17 & 60 \\
  &   & 5 & 5 & -60 \\
\hline
  & 1 & 1 & -12 & 0
\end{array}
$$

$x^2 + x - 12$

25. $x = 1$

$$
\begin{array}{r|rrrrr}
1 & 4 & 4 & -15 & 0 & 7 \\
  &   & 4 & 8 & -7 & -7 \\
\hline
  & 4 & 8 & -7 & -7 & 0
\end{array}
$$

$4x^3 + 8x^2 - 7x - 7$

27. $x = -1$

$$\begin{array}{r|rrrrr} -1 & 1 & -1 & 1 & -1 & -4 \\ & & -1 & 2 & -3 & 4 \\ \hline & 1 & -2 & 3 & -4 & 0 \end{array}$$

$x^3 - 2x^2 + 3x - 4$

29. $x = -4$

$$\begin{array}{r|rrrrr} -4 & 3 & 10 & -8 & -5 & -20 \\ & & -12 & 8 & 0 & 20 \\ \hline & 3 & -2 & 0 & -5 & 0 \end{array}$$

$3x^3 - 2x^2 - 5$

31.

$$\begin{array}{r}
2x^3 - 6x^2 + 2x - 2 \\
2x-1 \overline{\smash{\big)}\ 4x^4 - 14x^3 + 10x^2 - 6x + 2} \\
\mp 4x^4 \pm 2x^3 \\
\hline
-12x^3 + 10x^2 \\
\mp 12x^3 \mp 6x^2 \\
\hline
4x^2 - 6x \\
\mp 4x^2 \pm 2x \\
\hline
-4x + 2 \\
\mp 4x \pm 2 \\
\hline
0
\end{array}$$

$Q(x) = 2x^3 - 6x^2 + 2x - 2$
$R(x) = 0$

33.

$$\begin{array}{r}
x^3 - x^2 + 1 \\
x^2+x \overline{\smash{\big)}\ x^5 + 0x^4 - x^3 + x^2 + \quad 1} \\
\mp x^5 \mp x^4 \\
\hline
x^4 - x^3 \\
\pm x^4 \pm x^3 \\
\hline
0 + x^2 + \quad 1 \\
\mp x^2 \mp x \\
\hline
-x + 1
\end{array}$$

$Q(x) = x^3 - x^2 + 1$
$R(x) = 1 - x$

35.

$$\begin{array}{r}
3x^2 - 7x + 5 \\
2x^2+x+1 \overline{\smash{\big)}\ 6x^4 - 11x^3 + 6x^2 - 2x + 5} \\
\pm 6x^4 \mp 3x^3 \mp 3x^2 \\
\hline
-14x^3 + 3x^2 - 2x \\
\pm 14x^3 \pm 7x^2 \pm 7x \\
\hline
10x^2 + 5x + 5 \\
10x^2 \mp 5x \mp 5 \\
\hline
0
\end{array}$$

$Q(x) = 3x^2 - 7x + 5$
$R(x) = 0$

37. $x = 3$

$$\begin{array}{r|rrrr} 3 & 3 & -7 & -5 & 2 \\ & & 9 & 6 & 3 \\ \hline & 3 & 2 & 1 & 5 \end{array}$$

$Q(x) = 3x^2 + 2x + 1$

39. $x = 4$

$$\begin{array}{r|rrrrr} 4 & 1 & -3 & -4 & 2 & -5 \\ & & 4 & 4 & 0 & 8 \\ \hline & 1 & 1 & 0 & 2 & 3 \end{array}$$

$Q(x) = x^3 + x^2 + 2$

41. $x = -3$

$$\begin{array}{r|rrrrr} -3 & 5 & 10 & -20 & -12 & -2 \\ & & -15 & 15 & 15 & -9 \\ \hline & 5 & -5 & -5 & 3 & -11 \end{array}$$

$Q(x) = 5x^3 - 5x^2 - 5x + 3$

43. $x = -2$

$$\begin{array}{r|rrrrrr} -2 & 1 & -3 & 0 & 2 & 0 & -5 \\ & & -2 & 10 & -20 & 36 & -72 \\ \hline & 1 & -5 & 10 & -18 & 36 & -77 \end{array}$$

$Q(x) = x^4 - 5x^3 + 10x^2 - 18x + 36$

45. $x = -1$

$$\begin{array}{r|rrrr} -1 & 4 & -1 & 2 & -1 \\ & & -4 & 5 & -7 \\ \hline & 4 & -5 & 7 & -8 \end{array}$$

$Q(x) = 4x^2 - 5x + 7$

47. $x = -2$

$$\begin{array}{r|rrrrrr} -2 & 5 & 0 & 0 & 0 & -2 & 1 \\ & & -10 & 20 & -40 & 80 & -156 \\ \hline & 5 & -10 & 20 & -40 & 78 & -155 \end{array}$$

$Q(x) = 5x^4 - 10x^3 + 20x^2 - 40x + 78$

49. $x = 3$

$$\begin{array}{r|rrrrrr} 3 & 1 & -3 & 3 & -16 & 21 & -6 \\ & & 3 & 0 & 9 & -21 & 0 \\ \hline & 1 & 0 & 3 & -7 & 0 & -6 \end{array}$$

$Q(x) = x^4 + 3x^2 - 7x$

51. $x = -1, 3$

$$\begin{array}{r|rrrrr} -1 & 1 & 0 & -12 & 4 & 15 \\ & & -1 & 1 & 11 & -15 \\ \hline & 1 & -1 & -11 & 15 & 0 \end{array}$$

$$\begin{array}{r|rrrr} 3 & 1 & -1 & -11 & 15 \\ & & 3 & 6 & -15 \\ \hline & 1 & 2 & -5 & 0 \end{array}$$

$Q(x) = x^2 + 2x - 5$

53. $x = -2, 3$

$$\begin{array}{r|rrrr} -2 & 2 & -3 & -11 & 6 \\ & & -4 & 14 & -6 \\ \hline & 2 & -7 & 3 & 0 \end{array}$$

$$\begin{array}{r|rrr} 3 & 2 & -7 & 3 \\ & & 6 & -3 \\ \hline & 2 & -1 & 0 \end{array}$$

$Q(x) = 2x - 1$

55. $x = -4, 3$

$$\begin{array}{r|rrrrr} -4 & 2 & 5 & -25 & -40 & 48 \\ & & -8 & 12 & 52 & -48 \\ \hline & 2 & -3 & -13 & 12 & 0 \end{array}$$

$$\begin{array}{r|rrrr} 3 & 2 & -3 & -13 & 12 \\ & & 6 & 9 & -12 \\ \hline & 2 & 3 & -4 & 0 \end{array}$$

$Q(x) = 2x^2 + 3x - 4$

57. $x^2 + 4x - 5 = (x + 5)(x - 1)$ so $x = -5, 1$

$$\begin{array}{r|rrrrr} -5 & 1 & 7 & 5 & -23 & 10 \\ & & -5 & -10 & 25 & -10 \\ \hline & 1 & 2 & -5 & 2 & 0 \end{array}$$

$$\begin{array}{r|rrrr} 1 & 1 & 2 & -5 & 2 \\ & & 1 & 3 & -2 \\ \hline & 1 & 3 & -2 & 0 \end{array}$$

$Q(x) = x^2 + 3x - 2$

59. $x = -2$

$$\begin{array}{r|rrrrr} -2 & 1 & K & 7 & -2 & 8 \\ & & -2 & -10 & 6 & -8 \\ \hline & 1 & 5 & -3 & 4 & 0 \end{array}$$

$K = 7$

46

PROBLEM SET 3.2

1.　$m = \dfrac{6-3}{5-2} = \dfrac{3}{3} = 1$

3.　$m = \dfrac{11-(-2)}{4-(-1)} = \dfrac{13}{5}$

5.　$m = \dfrac{3-(-4)}{-9-(-6)} = -\dfrac{7}{3}$

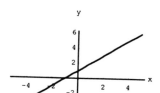

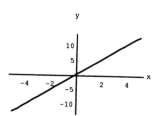

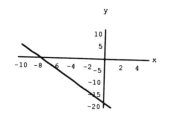

7.　$m = \dfrac{3}{0}$, undefined
　　so $x = 0$

9.　$m = \dfrac{2-2}{3-(-1)} = 0$

11.　$m = -\dfrac{4}{1}$; $b = -1$

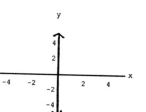

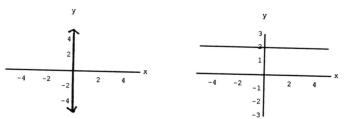

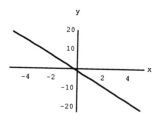

13.　$m = \dfrac{1}{5}$; $b = -\dfrac{6}{5}$

15.　$m = 300$; $b = 0$

17.　$m = 0$; $b = -2$

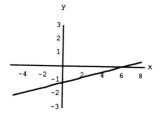

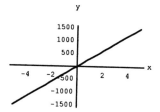

19. $x + 2 = 3y$

$y = \frac{1}{3}x + \frac{2}{3}$

$m = \frac{1}{3}$; $b = \frac{2}{3}$

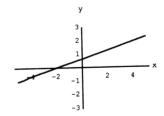

21. $5y = 2x - 1200$

$y = \frac{2}{5}x - 240$

$m = \frac{2}{5}$; $b = -240$

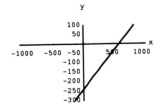

23. $2x - 2y + 6 = 0$; $5 \leq x \leq 9$

$2y = 2x + 6$

$y = x + 3$

$m = 1$; $b = 3$

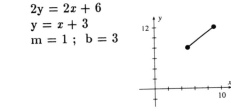

25. $y = 2|x|$; $b = 0$

x	0	1	-1	2	-2
y	0	2	2	4	4

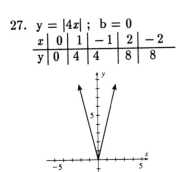

27. $y = |4x|$; $b = 0$

x	0	1	-1	2	-2
y	0	4	4	8	8

29. $m_{AN} = \dfrac{1 - (-1)}{4 - 1} = \dfrac{2}{3}$

   $m_{AG} = \dfrac{7 - (-1)}{0 - 1} = -8$

   $m_{NG} = \dfrac{7 - 1}{0 - 4} = -\dfrac{3}{2}$

   $m_{AN} \cdot m_{NG} = \dfrac{2}{3} \cdot -\dfrac{3}{2}$

   $= -1;$ a right
   triangle

31. $m_{RE} = \dfrac{7 - 0}{6 - 3} = \dfrac{7}{3}$

   $m_{EC} = \dfrac{9 - 7}{2 - 6} = -\dfrac{1}{2}$

   $m_{CT} = \dfrac{3 - 9}{-3 - 2} = \dfrac{6}{5}$

   not a parallelogram

33. $m_{PA} = \dfrac{5 - 10}{-4 - 1} = 1$

   $m_{AR} = \dfrac{-2 - 5}{-3 - (-4)} = -7$

   $m_{RL} = \dfrac{3 - (-2)}{2 - (-3)} = 1$

   $m_{LP} = \dfrac{3 - 10}{2 - 1} = -7$

   $m_{PA} = m_{RL}$, so parallel
   $m_{AR} = m_{LP}$, so parallel
   a parallelogram

35. $m_{AL} = \dfrac{3 - 5}{2 - (-4)} = -\dfrac{1}{3}$; $m_{PR} = \dfrac{-2 - 10}{-3 - 1} = 3$; $m_{AL} \cdot m_{PR} = -\dfrac{1}{3}(3) = -1$;
   so the lines are perpendicular

37. $y = 5x + 6$; $5x - y + 6 = 0$ 39. $y = 0$

41. $y = 3x + b$ and $P(2, 3)$; $3 = 3(2) + b$ so $b = -3$ and $3x - y - 3 = 0$

43. $y - 3 = \dfrac{1}{2}(x - 3)$; $y - 3 = \dfrac{1}{2}x - \dfrac{3}{2}$; $y = \dfrac{1}{2}x + \dfrac{3}{2}$; $x - 2y + 3 = 0$

45. $m = \dfrac{3 - (-1)}{4 - (-4)} = \dfrac{1}{2}$; $y - 3 = \dfrac{1}{2}(x - 4)$; $y - 3 = \dfrac{1}{2}x - 2$; $y = \dfrac{1}{2}x + 1$; $x - 2y + 2 = 0$

47. $m \dfrac{-2 - 6}{1 - 5} = 2$; $y - 6 = 2(x - 5)$; $y - 6 = 2x - 10$; $y = 2x - 4$; $2x - y - 4 = 0$

49. $3y = -2x + 6$; $y = -\dfrac{2}{3}x + 2$; new equation has $m = -\dfrac{2}{3}$ and $P(2, 4)$ so
   $y - 4 = -\dfrac{2}{3}(x - 2)$; $y - 4 = -\dfrac{2}{3}x + \dfrac{4}{3}$; $y = -\dfrac{2}{3}x + \dfrac{16}{3}$; $2x + 3y - 16 = 0$

51. $2y = x + 4$; $y = \dfrac{1}{2}x + 2$; $m = \dfrac{1}{2}$ so the new equation has $m = -2$ and $P(-1, -2)$;
   $y + 2 = -2(x + 1)$; $y + 2 = -2x - 2$; $y = -2x - 4$; $2x + y + 4 = 0$

53. a. $A(x_0, f(x_0))$; $B(x_0 + \Delta x, f(x_0 + \Delta x))$;   b. $m = \dfrac{f(x_0 + \Delta x) - f(x_0)}{\Delta x}$

55. a. $A(x_0, H(x_0))$; $B(x_0 + \Delta x, H(x_0 + \Delta x))$ b. $m = \dfrac{H(x_0 + \Delta x) - H(x_0)}{\Delta x}$

57. $x =$ price and $y =$ number of boxes; $A(1, 10)$; $B(2, 20)$
   $m = \dfrac{20 - 10}{2 - 1} = 10$; $y - 10 = 10(x - 1)$; $y - 10 = 10x - 10$; $y = 10x$; $10x - y = 0$

59. $x =$ years and $y =$ population; $1980 =$ base year so $x = 0$; $(0, 14.2$ million$)$;
   $(10, 16.8$ million$)$; $m = \dfrac{16.8 - 14.2}{10 - 0} = \dfrac{2.6}{10} = .26$mil.; $y - 14.2 = .26(x - 0)$;
   $y - 14.2 = .26x$; $y = .26x + 14.2$. If $x = 21$ then $y = .26(21) + 14.2 = 5.46 + 14.2 = 19.66$ mil.

61. $(50, 60)$ and $(260, 60)$ note that $y = 60$ so it will cost \$60 no matter how far you drive.

63. Given $y = mx + b$. Since the line passes through the point (h, k), these coordinates satisfy the equation so $k = mh + b$ or $b = k - mh$. Substitute this value of b into the given equation and you have $y = mx + (k - mh)$; $y - k = mx - mh$; $y - k = m(x - h)$

65. Given $y - y_1 = \left(\frac{y_2 - y_1}{x_2 - x_1}\right)(x - x_1)$. Since the line passes through $(0, b)$ and $(a, 0)$,

$a \neq 0$, $b \neq 0$, $y - 0 = \left(\frac{0 - b}{a - 0}\right)(x - a)$; $y = -\frac{b}{a}(x - a)$; $ay = -bx + ab$;

$bx + ay = ab$. Divide both sides by ab ($a \neq 0$, $b \neq 0$) and you have $\frac{x}{a} + \frac{y}{b} = 1$.

67. $y = mx + b$ and $y = -x + b$ if $m < 0$, $m =$ negative number than $-\frac{a}{b} = \frac{-a}{b}$ and the line is dropping from left to right which means that it is decreasing or $f(x_1) > f(x_2)$ whenever $x_1 < x_2$.

PROBLEM SET 3.3

1. $y = -x^2$, $V(0, 0)$

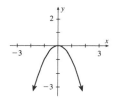

3. $y = 5x^2$, $V(0, 0)$

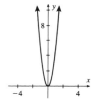

5. $y = -\frac{1}{10}x^2$, $V(0, 0)$

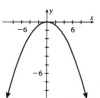

7. $y = (x - 3)^2$
   $V(3, 0)$

9. $y = (x + 2)^2$,
   $V(-2, 0)$

11. $y = \frac{1}{4}(x - 1)^2$
    $V(1, 0)$

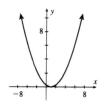

13. $y = \frac{1}{3}(x + 2)^2$
    $V(-2, 0)$

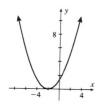

15. $y - 2 = 3(x + 2)^2$
    $V(-2, 2)$

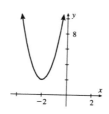

17. $y + 3 = \frac{2}{3}(x + 2)^2$
    $V(-2, -3)$

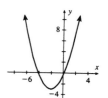

19. No, since $3x^3 + -3x^2 = 0$ and you actually have $f(x) = 5x + 2$ which is a linear equation.

21. No, since the coefficient on x^2 must be one to do the process of completing the square. This means that you have $y + 5 = 5(x^2 + 2x + \underline{1})$.

23. The vertex is $(5, 3)$. This parabola opens upward so the minimum value is $y = 3$.

25. $y = (x + 2)^2$
    $V(-2, 0)$

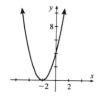

27. $y = x^2 + 2x - 3$
      $y + 4 = x^2 + 2x + 1$
      $y + 4 = (x + 1)^2$
      $V(-1, -4)$

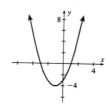

29. $y = 2x^2 - 4x + 4$
      $y - 2 = 2(x^2 - 2x + 1)$
      $y - 2 = 2(x - 1)^2$
      $V(1, 2)$

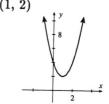

51

31. $y = 3x^2 - 12x + 10$

$y + 2 = 3(x^2 - 4x + 4)$

$y + 2 = 3(x - 2)^2$

V(2, -2)

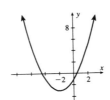

33. $y = \frac{1}{2} x^2 + 2x - 1$

$y + 3 = \frac{1}{2}(x^2 + 4x + 4)$

$y + 3 = \frac{1}{2}(x + 2)^2$

V(-2, -3)

35. $y = \frac{1}{2} x^2 - x + \frac{5}{2}$

$y - \frac{5}{2} = \frac{1}{2}(x^2 - 2x + 1)$

$y - 2 = \frac{1}{2}(x - 1)^2$

V(1, 2)

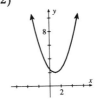

37. $x^2 - 6x - 2y - 1 = 0$

$2y = x^2 - 6x - 1$

$y = \frac{1}{2}x^2 - 3x - \frac{1}{2}$

$y + \frac{1}{2} + \frac{9}{2} = \frac{1}{2}(x^2 - 6x + 9)$

$y + 5 = \frac{1}{2}(x - 3)^2$

V(3, -5)

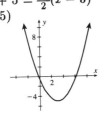

39. $x^2 - 6x - 3y - 3 = 0$

$3y + 3 = x^2 - 6x$

$3y + 12 = x^2 - 6x + 9$

$3(y + 4) = (x - 3)^2$

V(3, -4)

41. $y = -4x^2 - 8x - 1$; $y + 1 = -4(x^2 + 2x)$; $y + 1 - 4 = -4(x^2 - 2x + 1)$;

$y - 3 = -4(x - 1)^2$;   V(1, 3) so maximum at 3.

43. $10x^2 - 160x + y + 655 = 0$; $y = -10x^2 + 160x - 655$; $y + 655 = -10(x^2 - 16x)$;

$y + 655 - 640 = -10(x^2 + 16x + 64)$; $y + 15 = -10(x - 8)^2$;

V(8, -15) so maximum at -15.

45. $9x^2 + 6x + 81y - 53 = 0$; $\quad 81y = -9x - 6x + 53$; $\quad y = -\frac{9}{81}x^2 - \frac{6}{81}x + \frac{53}{81}$;

$y - \frac{53}{81} = -\frac{1}{9}x^2 - \frac{2}{27}x$; $\quad y - \frac{53}{81} - \frac{1}{81} = -\frac{1}{9}\left(x^2 - \frac{2}{3}x + \frac{1}{9}\right)$;

$y - \frac{2}{3} = -\frac{1}{9}\left(x - \frac{1}{3}\right)^2$; $\quad V\left(\frac{1}{3}, \frac{2}{3}\right)$ so maximum at $\frac{2}{3}$.

47. a. $P(x) = -10(x - 375)^2 + 1{,}156{,}250$

$\quad y - 1{,}156{,}250 = -10(x - 375)^2$ and $V(375, 1156250)$ so the number of boats is 375.

b. $P(0) = -10(0 - 375)^2 + 1{,}156{,}350$

$\quad = -249{,}900$ loss

c. $V(375, 1156250)$ so maximum profit is \$1,156,250.

49. $P(x) = -2(x - 25)^2 + 650$; $\quad y - 650 = -2(x - 25)^2$; $\quad V(25, 650)$; $\quad$ profit \$650.

51. $y - 18 = -\frac{2}{81}x^2$

a. $y - 18 = -\frac{2}{81}(9)^2$; $\quad y - 18 = -2$; $\quad y = 16$ so the height is 16 feet.

b. $y - 18 = -\frac{2}{81}(18)^2$; $\quad y - 18 = -8$; $\quad y = 10$ so height is 10 feet.

53. $L + W = 50$ meters so $L = 50 - W$; $\quad A = L \cdot W = $ maximum

$A = L \cdot W$; $\quad A = (50 - W)W$; $\quad A = 50W - W^2$; $\quad A = -W^2 + 50W$;

$A - 625 = -(W^2 - 50W + 625)$; $\quad A - 625 = -(W - 25)^2$; $\quad V(25, 625)$;

$A = 625$ feet and $L = W = 25$ feet.

55. $d(x) = -0.0005x^2 + 2.39x + 600$; $\quad y - 600 = -(0.0005x^2 - 2.39x)$;

$y - 600 - 2856 = -0.0005(x^2 - 4780x + 5712100)$; $\quad y - 3456 = -0.0005(x - 2390)^2$;

$V(2390, 3456)$; $\quad$ maximum height is 3,456 ft.

57. $2x^2 - x - y + 3 = 0$

$y = 2x^2 - x + 3$;

$y - 3 + \frac{1}{8} = 2\left(x^2 - \frac{1}{2}x + \frac{1}{16}\right)$;

$y - \frac{23}{8} = 2\left(x - \frac{1}{4}\right)^2$; $\quad V\left(\frac{1}{4}, \frac{23}{8}\right)$

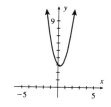

59. $4x^2 - 20x - 16y + 33 = 0$

$16y - 33 = 4x^2 - 20x$;

$y - \frac{33}{16} = \frac{1}{4}x^2 - \frac{5}{4}x$;

$y - \frac{33}{16} + \frac{25}{16} = \frac{1}{4}\left(x^2 - 5x + \frac{25}{4}\right)$;

$y - \frac{1}{2} = \frac{1}{4}\left(x - \frac{5}{2}\right)^2$; $\quad V\left(\frac{5}{2}, \frac{1}{2}\right)$

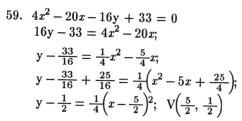

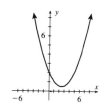

61. $25x^2 - 30x - 5y + 2 = 0$

$\quad\quad 5y - 2 = 25x^2 - 30x$

$\quad\quad y - \frac{2}{5} = 5x^2 - 6x$

$\quad y - \frac{2}{5} + \frac{9}{5} = 5\left(x^2 - \frac{6}{5}x + \frac{9}{25}\right)$

$\quad\quad y + \frac{7}{5} = 5\left(x - \frac{3}{5}\right)^2$

$\quad V\left(\frac{3}{5}, -\frac{7}{5}\right)$

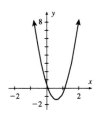

PROBLEM SET 3.4

1. quadratic   3. cubic   5. quartic   7. cubic   9. cubic   11. quartic   13. I   15. G   17. B

19. H   21. L   23. M   25. Q   27. T   29. E   31. U

33. $y = -f(x)$,
    refects over x-axis

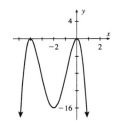

35. $y = -h(x)$,
    reflects over x-axis

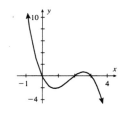

37. $y = g(x) - 4$,
    shifts down 4 units

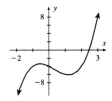

39. $y = f(x - 2)$,
    shifts right 2 units

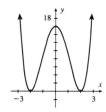

41. $y = h(x + 1)$,
    shifts left 1 unit

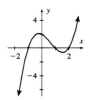

43. $y = g(x + 2) - 4$,
    shifts left 2 unit, down 4 units

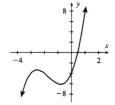

45. $y = -g(x)$,
    reflects over x-axis
    and stretches

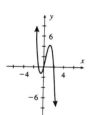

47. $y |f(x)|$,
    same as $f(x)$

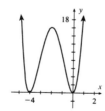

49. $y = |h(x)|$,
    graph has all values of $y \geq 0$

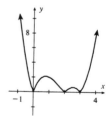

51. $y = g(|x|)$, graph has
    all values of $x \geq 0$.

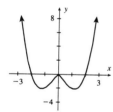

53. The graph is not
    complete.

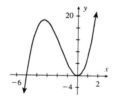

55. The graph is not
    complete.

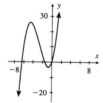

57. The cubic graphs have
    smooth curve, no
    straight lines or points.

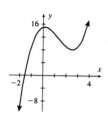

59. $n = 3$, $a < 0$ so
    rises on left and
    falls on on right.

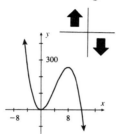

61. $n = 4$, $a > 0$,
    rises on left and
    right.

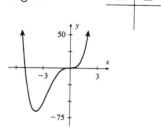

55

63. n = 4, a > 0
      rises on left and right.

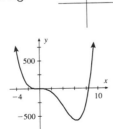

PROBLEM SET 3.5

1. a. $\begin{array}{r|rrrr} 1 & 5 & -7 & 3 & -4 \\ & & 5 & -2 & 1 \\ \hline & 5 & -2 & 1 & -3 \end{array}$

    $f(1) = -3$

    b. $\begin{array}{r|rrrr} -1 & 5 & -7 & 3 & -4 \\ & & -5 & 12 & -15 \\ \hline & 5 & -12 & 15 & -19 \end{array}$

    $f(-1) = -19$

    c. $\begin{array}{r|rrrr} 0 & 5 & -7 & 3 & -4 \\ & & 0 & 0 & 0 \\ \hline & 5 & -7 & 3 & -4 \end{array}$

    $f(0) = -4$

    d. $\begin{array}{r|rrrr} 6 & 5 & -7 & 3 & -4 \\ & & 30 & 138 & 846 \\ \hline & 5 & 23 & 141 & 842 \end{array}$

    $f(6) = 842$

    e. $\begin{array}{r|rrrr} -4 & 5 & -7 & 3 & -4 \\ & & -20 & 108 & -444 \\ \hline & 5 & -27 & 111 & -448 \end{array}$

    $f(-4) = -448$

3. a. $\begin{array}{r|rrrrr} 1 & 1 & -10 & 20 & -23 & -812 \\ & & 1 & -9 & 11 & -12 \\ \hline & 1 & -9 & 11 & -12 & -824 \end{array}$

    $P(1) = -824$

    c. $P(0) = -812$

3. b. $\begin{array}{r|rrrrr} -1 & 1 & -10 & 20 & -23 & -812 \\ & & -1 & 11 & -31 & 54 \\ \hline & 1 & -11 & 31 & -54 & -758 \end{array}$

    $P(-1) = -758$

    d. $\begin{array}{r|rrrrr} 3 & 1 & -10 & 20 & -23 & -812 \\ & & 3 & -21 & -3 & -78 \\ \hline & 1 & -7 & -1 & -26 & -890 \end{array}$

    $P(3) = -890$

    e. $\begin{array}{r|rrrrr} 7 & 1 & -10 & 20 & -23 & -812 \\ & & 7 & -21 & -7 & -210 \\ \hline & 1 & -3 & -1 & -30 & -1022 \end{array}$

    $P(7) = -1022$

5. a. $g(0) = -10$

    b. $\begin{array}{r|rrrrr} -1 & 4 & -3 & 0 & 5 & -10 \\ & & -4 & 7 & -7 & 2 \\ \hline & 4 & -7 & 7 & -2 & -8 \end{array}$

    $P(-1) = -8$

    c. $\begin{array}{r|rrrrr} -2 & 4 & -3 & 0 & 5 & -10 \\ & & -8 & 22 & -44 & 78 \\ \hline & 4 & -11 & 22 & -39 & 68 \end{array}$

    $P(-2) = 68$

d.
$$5 \,\rvert\, \begin{array}{ccccc} 4 & -3 & 0 & 5 & -10 \\ & 20 & 85 & 425 & 2150 \\ \hline 4 & 17 & 85 & 430 & 2140 \end{array}$$
$P(5) = 2140$

e.
$$-3 \,\rvert\, \begin{array}{ccccc} 4 & -3 & 0 & 5 & -10 \\ & -12 & 45 & -135 & 390 \\ \hline 4 & -15 & 45 & -130 & 380 \end{array}$$
$P(-3) = 380$

7. a. $f(0) = -3$

b.
$$1 \,\rvert\, \begin{array}{ccccc} 8 & -6 & 5 & 4 & -3 \\ & 8 & 2 & 7 & 11 \\ \hline 8 & 2 & 7 & 11 & 8 \end{array}$$
$P(1) = 8$

c.
$$1/2 \,\rvert\, \begin{array}{ccccc} 8 & -6 & 5 & 4 & -3 \\ & 4 & -1 & 2 & 3 \\ \hline 8 & -2 & 4 & 6 & 0 \end{array}$$
$P(\tfrac{1}{2}) = 0$

d.
$$-1/2 \,\rvert\, \begin{array}{ccccc} 8 & -6 & 5 & 4 & -3 \\ & -4 & 5 & -5 & 1/2 \\ \hline 8 & -10 & 10 & -1 & -5/2 \end{array}$$
$P(-\tfrac{1}{2}) = -\tfrac{5}{2}$

e.
$$-3 \,\rvert\, \begin{array}{ccccc} 8 & -6 & 5 & 4 & -3 \\ & -24 & 90 & -285 & 843 \\ \hline 8 & -30 & 95 & -281 & 840 \end{array}$$
$P(-3) = 840$

9. a.
$$-1 \,\rvert\, \begin{array}{ccccc} 4 & -8 & -43 & 29 & 60 \\ & -4 & 12 & 31 & -60 \\ \hline 4 & -12 & -31 & 60 & 0 \end{array}$$
$P(-1) = 0$

b.
$$1 \,\rvert\, \begin{array}{ccccc} 4 & -8 & -43 & 29 & 60 \\ & 4 & -4 & -47 & -18 \\ \hline 4 & -4 & -47 & -18 & 42 \end{array}$$
$P(1) = 42$

c.
$$4 \,\rvert\, \begin{array}{ccccc} 4 & -8 & -43 & 29 & 60 \\ & 16 & 32 & -44 & -60 \\ \hline 4 & 8 & -11 & -15 & 0 \end{array}$$
$P(4) = 0$

d.
$$3/2 \,\rvert\, \begin{array}{ccccc} 4 & -8 & -43 & 29 & 60 \\ & 6 & -3 & -69 & -60 \\ \hline 4 & -2 & -46 & -40 & 0 \end{array}$$
$P(\tfrac{3}{2}) = 0$

e.
$$-5/2 \,\rvert\, \begin{array}{ccccc} 4 & -8 & -43 & 29 & 60 \\ & -10 & 45 & -5 & -60 \\ \hline 4 & -18 & 2 & 24 & 0 \end{array}$$
$P(-\tfrac{5}{2}) = 0$

11. a. $g(1) = (1-1)(1-4)(2+1)$
$$= 0(-3)(3) = 0$$
b. $g(4) = (4-1)(4-4)(8+1)$
$$= 3(0)(9) = 0$$

11. c. $g(-\tfrac{1}{2}) = (-\tfrac{1}{2}-1)(-\tfrac{1}{2}-4)(-1+1) = -\tfrac{3}{2}(-\tfrac{9}{2})(0) = 0$

d. $g(-2) = (-2-1)(-2-4)(-4+1) = -3(-6)(-3) = -54$

e. $g(-1) = (-1-1)(-1-4)(-2+1) = -2(-5)(-1) = -10$

13. $f(x) = x^3 - 3x^2 + 10$
    $\{(0, 10), (1, 8), (-1, 6),$
    $(2, 6), (-2, -10),$
    $(3, 10), (5, 60)\}$

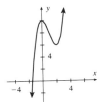

15. $f(x) = 2x^3 - 3x^2 - 12x + 3$
    $\{(0, 3), (1, -10), (-1, 10),$
    $(2, -17), (-2, -1),$
    $(3, -6), (-3, -42)\}$

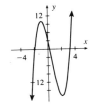

17. $f(x) = x^3 - 2x^2 + x - 5$
    $\{(0, -5), (1, -5),$
    $(-1, -9), (2, -3), (3, 7),$
    $(\frac{1}{2}, -\frac{39}{8}), (-\frac{1}{2}, -\frac{49}{8})\}$

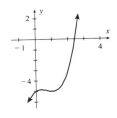

19. One part of the graph is missing on the left side. The local maximum is at $(-2.85, 74)$.

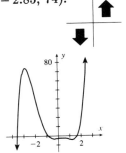

21. There is only one zero. There is an abrupt change of direction at the local maximum, $(2.4, 26.31)$.

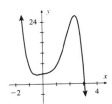

23. This graph shows the zeros, the range needs to be Y[$-150, 450$] by increments of 50.

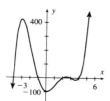

25. $f(x) = x^3 - 6x^2 + 9x - 9$; $\{(0, -9), (1, -5),$
$\{(-1, -25), (2, -7), (3, -9), 4, -5), (3, 11),$
$(6, 45), (4.425, 0)\}$

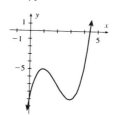

27. $f(x) = x^4 - 8x^3 - 43x^2 + 29x + 60$
By synthetic division -1 and 4 are zeroes, this gives you the equation: $4x^2 + 4x - 15 = 0$. By any method you can find the other two zeroes: $-\frac{5}{2}$ and $\frac{3}{2}$. $\{(0, 60), (-1, 0),$
$(4, 0), (-\frac{5}{2}, 0), (\frac{3}{2}, 0), (1, 42), (2, -54),$
$(-2, -42), (3, -132), (-3, 126), (-2.41, 0)\}$

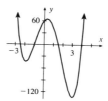

29. $f(x) = x^4 - 7x^2 - 2x + 2$
$\{(0, 2), (1, -6), (-1, -2), (2, -14)$
$(-2, -6), (3, 14), (-3, 26)\}$

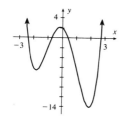

31. $f(x) = x^6 - 4x^4 - 4x^2 + 4$
$\{(0, 4), (1, -3), (-1, -3), (2, -12),$
$(2, -12), (3, 373), (-3, 373)\}$

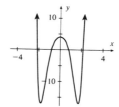

33. $y = 3x^4 - x^3 - 14x^2 + 4x + 8$; $\{(0, 8), (2, 0), (-\frac{2}{3}, 0), (-1, -6), (2, 0), (-\frac{3}{2}, 10.9), (\frac{3}{2}, -5.7)\}$

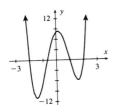

35. $y = x^4 - x^3 - 3x^2 + 2x + 4$; $\{(0,4), (1, 3), (-1, 1), (2, 4), (-2, 12, (3, 37)\}$

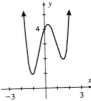

37. $y = (x - 1)(x + 1)(x + 3)$
$\{(0, -3), (1, 0), (-1, 0), (-3, 0) (2, 15), (-2, 3), (-4, -15)\}$

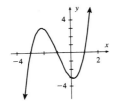

39. $y = (x - 1)(x + 3)(2x - 5)$
$\{(0, -15), (-1, 0), (-3, 0), (\frac{5}{2}, 0), (1, -24), (2, -15), (-2, 9)\}$

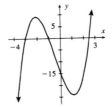

41. $y = (x + 1)(x + 2)(3x - 1)$; $\{(0, -22), (-1, 0), (-2, 0), (\frac{1}{3}, 0), (1, 12), (2, 60), (-3, -20)\}$

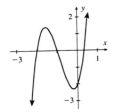

43. $y = 3x^2(x - 3)(x + 1)$
$\{(0, 0), (3, 0), (-1, 0), (1, -12), (\frac{1}{2}, -\frac{21}{16}), (2, -36), (-2, 60)\}$

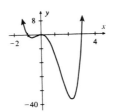

45. $y = x^2(x^2 - 1)$
  {(0, 0), (−1, 0), (1, 0), $(\frac{1}{2}, -\frac{3}{16})$
  $(-\frac{1}{2}, -\frac{3}{16})$, (2, 12), (−2, 12)}

47. $f(x) = 3x^4 - 7x^3 + 5x^2 + x - 10$
  {(0, −10), (1, −8), (−1, 4), (2, 4), $(\frac{1}{3}, -\frac{28}{3})$
  $(-\frac{1}{3}, -9.5)$, $(\frac{2}{3}, -8.6)$, $(-\frac{2}{3}, -5.8)$}

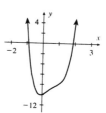

49. $f(x) = x^5 - 3x^4 + 2x^3 - 7x + 15$
  {(0, 15), (1, 8), (−1, 16), (2, 1),
  (−2, −67), (3, 48), (−1.44, 0)}

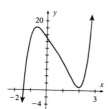

51. a. $y = x^5$
  $x = 0$ with mult-
  iplicity of 5.

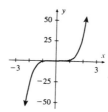

  b. $y = (x - 1)^5 + 10$
  shifts right one and
  up ten.

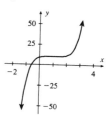

53. a. $y = x^7$
  $x = 0$ with mult-
  iplicity of 7.

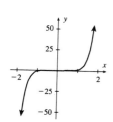

  b. $y = -0.01x^7$
  $x = 0$ with multi-
  of 7 and (2, −1.3),
  (−2, 0.158)

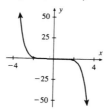

55. $y = |x^3 + 3x^2 - x - 3|$
  {(−3, 0), (−1, 0), (1, 0)}
  and all values of $y \geq 0$.

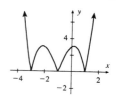

57. $y = |-x|^3 + 2|x|^2 + |x| - 2$
$\{(-2, 0), (0, 0), (2, 0)\}$

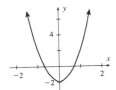

59. $y = -|x|^3 + 2|x|^2 + |x| - 2$
$\{(-2, 0), (-1, 0), (1, 0), (2, 0), (0, -2)\}$

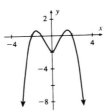

PROBLEM SET 3.6

1. $2, -3$ (mult. 2)   3. 0 (mult. 3), $\frac{3}{2}$ (mult. 2)   5. $-2, 1$ (mult. 2), -1 (mult. 2)

7. -5 (mult. 2), 3 (mult. 2)   9. 0 (mult. 2), 4 (mult. 2)

11. 0(mult. 2), 3 (mult. 2), -3 (mult. 2)

13.

P	N	I
4	0	0
2	0	2
0	0	4

15.

P	N	I
2	1	2
0	1	4

17.

P	N	I
1	2	0
1	0	2

19.

P	N	I
1	2	0
1	0	2

21.

P	N	I
2	2	0
0	2	2
2	0	2
0	0	4

23.

P	N	I
2	1	0
0	1	2

25. $\frac{15}{1}$ so $1, 3, 5, 15$; $\pm (1, 3, 5, 15)$

27. $\frac{12}{2}$ so $\frac{1, 2, 3, 4, 6, 12}{1, 2}$;

$\pm (1, 2, 3, 4, 6, 12, \frac{1}{2}, \frac{3}{2})$

29. $\frac{12}{1}$; $\pm (1, 2, 3, 4, 6, 12)$ 31. $\frac{15}{2}$ so $\frac{1, 3, 5, 15}{1, 2}$; $\pm (1, 3, 5, 15, \frac{1}{2}, \frac{3}{2}, \frac{5}{2}, \frac{15}{2})$

33. $\frac{18}{1}$; $\pm (1, 2, 3, 6, 9, 18)$ 35. $\frac{1}{5}$ so $\frac{1}{1, 5}$; $\pm (1, \frac{1}{5})$

37. $x^3 - x^2 - 4x + 4 = 0$; $\pm (1, 2, 4)$;

$$\begin{array}{r|rrrr} \underline{1} & 1 & -1 & -4 & 4 \\ & & 1 & 0 & -4 \\ \hline & 1 & 0 & -4 & 0 \end{array}$$

$x^2 - 4 = (x - 2)(x + 2)$, $x = \pm 2$; $\{-2, -1, 1\}$

39. $x^3 - 2x^2 - 9x + 18 = 0;$ $\pm$ (1, 2, 3, 6, 9, 18) ;

$x^2 - 9 = 0$; $x^2 = 9$; $x = \pm 3$

$\{-3, 3, 2\}$

$$\underline{2}\lfloor 1 \quad -2 \quad -9 \quad 18$$
$$\phantom{\underline{2}\lfloor 1} \quad 2 \quad 0 \quad -18$$
$$\overline{\phantom{\underline{2}\lfloor} 1 \quad 0 \quad -9 \quad 0}$$

41. $x^3 + 3x^2 - 4x - 12 = 0;$ $\pm$ (1, 2, 3, 4, 6, 12);

$x^2 - 4 = (x + 2)(x - 2)$; $x = \pm 2$

$\{-3, -2, 2\}$

$$\underline{-3}\lfloor 1 \quad 3 \quad -4 \quad -12$$
$$\phantom{\underline{-3}\lfloor 1} \quad -3 \quad 0 \quad 12$$
$$\overline{\phantom{\underline{-3}\lfloor} 1 \quad 0 \quad -4 \quad 0}$$

43. $2x^3 - 3x^2 - 32x - 15 = 0;$ $\pm$ (1, 3, 5, 15, $\frac{1}{2}$, $\frac{3}{2}$, $\frac{5}{2}$, $\frac{15}{2}$);

$$\underline{-3}\lfloor 2 \quad -3 \quad -32 \quad -15$$
$$\phantom{\underline{-3}\lfloor 2} \quad -6 \quad 27 \quad 15$$
$$\overline{\phantom{\underline{-3}\lfloor} 2 \quad -9 \quad -5 \quad 0}$$

$2x^2 - 9x - 5 = (2x + 1)(x - 5)$; $x = -\frac{1}{2}, 5$

$\{-3, -\frac{1}{2}, 5\}$

45. $x^4 + 3x^3 - 19x^2 - 3x + 18 = 0;$ $\pm$ (1, 2, 3, 6, 9, 18)

$$\underline{1}\lfloor 1 \quad 3 \quad -19 \quad -3 \quad 18$$
$$\phantom{\underline{1}\lfloor 1} \quad 1 \quad 4 \quad -15 \quad -18$$
$$\overline{\phantom{\underline{1}\lfloor} 1 \quad 4 \quad -15 \quad -18 \quad 0}$$

$$\underline{-1}\lfloor 1 \quad 4 \quad -15 \quad -18$$
$$\phantom{\underline{-1}\lfloor 1} \quad -1 \quad -3 \quad 18$$
$$\overline{\phantom{\underline{-1}\lfloor} 1 \quad 3 \quad -18 \quad 0}$$

$x^2 + 3x - 18 = (x + 6)(x - 3)$; $x = -6, 3$; $\{-6, -1, 1, 3\}$

47. $x^3 + 15x^2 + 71x + 105 = 0;$ $\pm$ (1, 3, 5, 7, 15, 21, 35, 105);

$$\underline{-3}\lfloor 1 \quad 15 \quad 71 \quad 105$$
$$\phantom{\underline{-3}\lfloor 1} \quad -3 \quad -36 \quad -105$$
$$\overline{\phantom{\underline{-3}\lfloor} 1 \quad 12 \quad 35 \quad 0}$$

$x^2 + 12x + 35 = (x + 5)(x + 7)$;

$x = -7, -5$

$\{-7, -5, -3\}$

49. $8x^3 - 12x^2 - 66x + 35 = 0;$ $\pm$ (1, 5, 7, 35, $\frac{1}{2}$, $\frac{5}{2}$, $\frac{7}{2}$, $\frac{35}{2}$, $\frac{1}{4}$, $\frac{5}{4}$, $\frac{7}{4}$, $\frac{35}{4}$, $\frac{1}{8}$, $\frac{5}{8}$, $\frac{7}{8}$, $\frac{35}{8}$);

$$\underline{-5/2}\lfloor 8 \quad -12 \quad -66 \quad 35$$
$$\phantom{\underline{-5/2}\lfloor 8} \quad -20 \quad 80 \quad -35$$
$$\overline{\phantom{\underline{-5/2}\lfloor} 8 \quad -32 \quad 14 \quad 0}$$

$8x^2 - 32x + 14 = 2(4x^2 - 16x + 7)$

$= (2x - 1)(2x - 7)$; $x = \frac{1}{2}, \frac{7}{2}$; $\{-\frac{5}{2}, \frac{1}{2}, \frac{7}{2}\}$

51. $x^5 + 8x^4 + 10x^3 - 60x^2 - 171x - 108 = 0;$

$\pm$ (1, 2, 3, 4, 6, 9, 12, 18, 27, 36, 54, 108); $x = -4, -3, -1, 3;$

$\{-4, -3, -1, 3\}$, -3 has multiplicity of 2

53. $x^7 + 2x^6 - 4x^5 - 2x^4 + 3x^3 = 0;$ $x^3(x^4 + 2x^3 - 4x^2 - 2x + 3) = 0$

By synthetic division: $x = 1, 1$; $x^2 + 4x + 3 = (x + 3)(x + 1)$, $x = -3, -1$

$\{-3, -1, 0, 1\}$, 1 (mult. 2) and 0 (mult. 3)

55. $x^6 - 12x^4 + 48x^2 - 64 = 0;$ $\pm$ (1, 2, 4, 8, 16, 32, 64); By synthetic

division: $x = -2, 2$; $\{-2, 2\}$, -2 (mult. 3) and 2 (mult. 3)

57. Yes, let x be the number; then $x - 1 = x^3$ or $x^3 - x + 1 = 0$. Check the possible rational roots to find there are none. By Descartes' Rule of signs, there are 2 or 0 positive zeroes and 1 negative root. Test these values synthetically to find by the Location Theorem that there is at least one real root 0 and -2, so there exists a real number that exceeds its cube by 1.

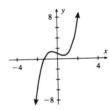

59. a = length of the sides so $(a - 2)(a^2) = 384$ cm^2; $a^3 - 2a^2 = 384$; $a^3 - 2a^2 + 384 = 0$. Use synthetic division and divide by 8 and you have $a^2 + 6a + 48 = 0$; $a = -3 \pm \sqrt{39}$ irrational so $a = 8$ cm. The dimensions are 8 cm. on each side.

61. $x^3 + px^2 = n$; $p = 5$ and $n = 21$; $x^3 + 5x^2 = 21$ and $x^3 + 5x^3 - 21 = 0$; If $5x^2 > x^3$ then $5x^2 = 21$ and $x^2 \approx 4$. If $x = 2$, then $8 + 20 = 28 > 21$. Not acceptable so try $1 < x < 2$. By trial and error and squeezing you find that $x \approx 1.76224...$ or $x = 1.8$

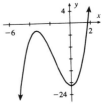

63. $x^4 + 2x^3 - 13x^2 + 2x + 1 = 0$; Use sythetic division and round to the nearest tenth. $x \approx -4.8, -0.2, 0.4, 2.6$; $\{-4.8, -0.2, 0.4, 2.6\}$

65. a. By synthetic division: Upper Bound at $x = 3$ and Lower Bound at $x = -2$ (one answer).
    b. You get values of y which get larger very fast. This happens because the graph is going away from the x-axis. The bounds tell you that the graph is not going to turn again and and cross the x-axis.
    c. Yes.
    d. The y-values get closer to zero as the graph approaches the x-axis.
    e. Yes.

PROBLEM SET 3.7

1.
$$
\begin{array}{r|rrrrr}
i & 1 & -6 & 15 & -2 & -10 \\
 & & i & -1-6i & 6+14i & 4+4i \\
\hline
 & 1 & -6+i & 14-6i & 4+14i & -24+4i
\end{array}
$$
$f(i) = -24 + 4i$

3.
$$
\begin{array}{r|rrrrr}
\sqrt{2} & 1 & -6 & 15 & -2 & -10 \\
 & & \sqrt{2} & 2-6\sqrt{2} & -12+17\sqrt{2} & 34-14\sqrt{2} \\
\hline
 & 1 & \sqrt{2}-6 & 17-6\sqrt{2} & -14+17\sqrt{2} & 24-14\sqrt{2}
\end{array}
$$
$f(\sqrt{2}) = 24 - 14\sqrt{2}$

5.
$$
\begin{array}{r|rrrrr}
2+i & 1 & -6 & 15 & -2 & -10 \\
 & & 2+i & -9-2i & 14+2i & 22+16i \\
\hline
 & 1 & -4+i & 6-2i & 12+2i & 12+16i
\end{array}
$$
$f(2 + i) = 12 + 16i$

7.
$$
\begin{array}{r|rrrrr}
i & 1 & -10 & 36 & -58 & 35 \\
 & & i & -1-10i & 10+35i & -35-48i \\
\hline
 & 1 & -10+i & 35-10i & -48+35i & -48i
\end{array}
$$
$f(i) = -48i$

64

9. $2-i$ | 1 -10 36 -58 35

           $2-i$ $-17+6i$ $44-7i$ -35

         1     $-8-i$ $19+6i$ $-14-7i$ 0 $f(2-i)=0$

11. $3+\sqrt{2}$ | 1 -10 36 -58 35

          $3+\sqrt{2}$ $-19-4\sqrt{2}$ $43+5\sqrt{2}$ -35

         1     $-7+\sqrt{2}$ $17-4\sqrt{2}$ $-15+5\sqrt{2}$ 0 $f(3+\sqrt{2})=0$

13. 2 | 1 -6 18 -30 25

          2 -8 20 -20

       1   -4 10 -10 5 $f(2)=5$

15. $2+i$ | 1 -6 18 -30 25

          $2+i$ $-9-2i$ $20+5i$ -25

         1     $-4+i$ $9-2i$ $-10+5i$ 0 $f(2+i)=0$

17. $1+2i$ | 1 -6 18 -30 25

          $1+2i$ $-9-8i$ $25+10i$ -25

         1     $-5+2i$ $9-8i$ $-5+10i$ 0 $f(1+2i)=0$

19. 2 | 2 -1 -13 5   15

          4   6 -14 -18

       2   3 -7 -9 -3 $f(2)=-3$

21. $\sqrt{5}$ | 2 -1 -13 5   15

          $2\sqrt{5}$ $10-\sqrt{5}$ $-5-3\sqrt{5}$ -15

         2     $-1+2\sqrt{5}$ $-3-\sqrt{5}$ $-3\sqrt{5}$ 0 $f(\sqrt{5})=0$

23. i | 2 -1 -13 5   15

          $2i$ $-2-i$ $1-15i$ $15+6i$

         2     $-1+2i$ $-15-i$ $6-15i$ $30+6i$ $f(i)=30+6i$

25. $1+\sqrt{2}$ | 1 -2 0   1

          $1+\sqrt{2}$ 1 $1+\sqrt{2}$

         1     $-1+\sqrt{2}$ 1 $2+\sqrt{2}$ No, $1+\sqrt{2}$ is not a root.

27. $1-2i$ | 1 -1 3   5

          $1-2i$ $-4-2i$ -5

         1     $-2i$ $-1-2i$ 0   Yes, $1+2i$ is also a root.

29. $\underline{1 + 2i}\,|$ $\quad 1 \qquad\quad -7 \qquad\quad 14 \qquad\quad 2 \qquad\quad -20$

$\qquad\qquad\qquad 1 + 2i \quad -10 - 10i \quad 24 - 2i \quad 30 + 50i$

$\qquad\qquad\overline{\quad 1 \quad -6 + 2i \quad\; 4 - 10i \quad\; 26 - 2i \quad 10 + 50i}$ $\qquad$ Not a root

31. $P(x) = x^3 - 8$
$\quad (x - 2)(x^2 + 2x + 4)$
$\quad \{2,\ -1 \pm i\sqrt{3}\}$

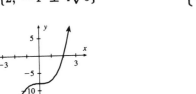

33. $P(x) = x^3 - 125$
$\quad (x - 5)(x^2 + 5x + 25)$
$\quad \left\{5,\ \dfrac{-5 \pm 5i\sqrt{3}}{2}\right\}$

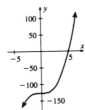

35. $P(x) = x^4 - 81$
$\quad (x^2 + 9)(x - 3)(x + 3)$
$\quad \{-3,\ 3,\ -3i,\ 3i\}$

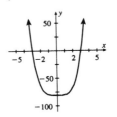

37. $P(x) = x^4 + 9x^2 + 20 =$
$\quad (x^2 + 5)(x^2 + 4)$
$\quad \{-i\sqrt{5},\ i\sqrt{5},\ -2i,\ 2i\}$

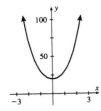

39. $P(x) = x^4 + 13x^2 + 36$
$\quad (x^2 + 9)(x^2 + 4)$
$\quad \{-3i,\ 3i,\ -2i,\ 2i\}$

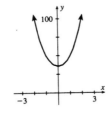

41. $P(x) = (x^2 - 4x - 1)$
$\quad (x^2 - 8x + 17)$
$\quad \{4 \pm i,\ 3 \pm i\}$

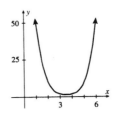

43. $P(x) = (x^2 - 4x - 1)$
$(x^2 - 3x + 5)$
$\left\{2 \pm \sqrt{5}, \dfrac{3 \pm i\sqrt{11}}{2}\right\}$

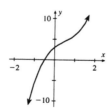

45. $P(x) = 2x^3 - 3x^2 + 4x + 3$
$\left\{-\dfrac{1}{2}, 1 \pm i\sqrt{2}\right\}$

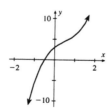

47. $P(x) = x^4 - 2x^3 + 4x^2 + 2x - 5$
$\{-1, 1, 1 \pm 2i\}$

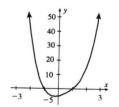

49. $P(x) = 27x^4 - 180x^3 + 213x^2 + 62x - 10$
$\left\{-\dfrac{1}{3}, 5, \dfrac{3 \pm \sqrt{7}}{3}\right\}$

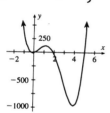

51. $P(x) = x^5 + 5x^2$

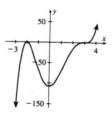

53. $P(x) = (x + 4)(x + 3)(x + 2)^2(x - 5)$

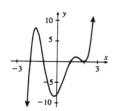

55. $P(x) = (x - 2)^2(x + 3)^2(x - 1)^2$

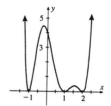

57. $-3i$ is a given root then $3i$ is also a root.

$$-3i \,|\; \begin{array}{ccccc} 1 & 0 & 13 & 0 & 36 \\ & -3i & -9 & -12i & -36 \\ \hline 1 & -3i & 4 & -12i & 0 \end{array}$$

$$3i \,|\; \begin{array}{cccc} 1 & -3i & 4 & -12i \\ & 3i & 0 & 12i \\ \hline 1 & 0 & 4 & 0 \end{array}$$

$x^2 + 4 = 0,$
$x = \pm\, 2i$
$\{\pm 2i,\ \pm 3i\}$

59. Using $2 + i$ as the divisor gives $x^3 + (-2+i)x^2 - 2x + (4 - 2i).$

$$2 - i \,|\; \begin{array}{cccc} 1 & -2+i & -2 & 4-2i \\ & 2-i & 0 & -4+2i \\ \hline 1 & 0 & -2 & 0 \end{array}$$

$x^2 - 2 = 0$ so $x^2 = 2$ and $x = \pm\sqrt{2}$
$\{\pm\sqrt{2},\ 2-i,\ 2+i\}$

61. Using $-\sqrt{5}$ as the divisor gives the following, then divide by $\sqrt{5}$:

$$\sqrt{5} \,|\; \begin{array}{cccc} 2 & -1-2\sqrt{5} & -3+\sqrt{5} & 3\sqrt{5} \\ & 2\sqrt{5} & -\sqrt{5} & -3\sqrt{5} \\ \hline 2 & -1 & -3 & 0 \end{array}$$

$2x^2 - x - 3 = 0,$
$(2x-3)(x+1) = 0, \quad \{\pm\sqrt{5},\ -1, -\tfrac{3}{2}\}$

63.
$$i\sqrt{2} \,|\; \begin{array}{cccccc} 2 & 9 & 0 & -3 & -8 & -42 \\ & 2i\sqrt{2} & -4+9i\sqrt{2} & -18-4i\sqrt{2} & 8-21i\sqrt{2} & 42 \\ \hline 2 & 9+2i\sqrt{2} & -4+9i\sqrt{2} & -21-4i\sqrt{2} & -21i\sqrt{2} & 0 \end{array}$$

$$-i\sqrt{2} \,|\; \begin{array}{ccccc} 2 & 9+2i\sqrt{2} & -4+9i\sqrt{2} & -21-4i\sqrt{2} & -21i\sqrt{2} \\ & -2i\sqrt{2} & -9i\sqrt{2} & 4i\sqrt{2} & 21i\sqrt{2} \\ \hline 2 & 9 & -4 & -21 & 0 \end{array}$$

$2x^3 + 9x^2 - 4x - 21 = 0, \quad \pm\, (1, 3, 7, 21, \tfrac{1}{2}, \tfrac{3}{2}, \tfrac{7}{2}, \tfrac{21}{2})$

$$3/2 \,|\; \begin{array}{cccc} 2 & 9 & -4 & -21 \\ & 3 & 18 & 21 \\ \hline 2 & 12 & 14 & 0 \end{array}$$

$2x^2 + 12x + 14 = 2(x^2 + 6x + 7)$

$\{\tfrac{3}{2}, -3,\ -i\sqrt{2}, i\sqrt{2},\ \pm\sqrt{2}\}$

65. Use the same method as in problem # 63. $-\sqrt{2}$ has multiplicity of 2 as does $\sqrt{2}$.

The other two roots are $\dfrac{-1 \pm i\sqrt{3}}{2}$.

67. The answers will vary but all the graphs will have vertices at $(0, 0)$. The higher the power the more zeroes at the vertex. The graphs will all open downward and as the powers get higher the graphs will become narrower around the y-axis.

CHAPTER THREE SUMMARY

1. $x = -2$

$$\begin{array}{r|rrrrr} -2 & 1 & 0 & 2 & -1 & 26 \\ & & -2 & 4 & -12 & 26 \\ \hline & 1 & -2 & 10 & -13 & 0 \end{array}$$

3. $x = 2$

$$\begin{array}{r|rrrr} 2 & 3 & 2 & -12 & -8 \\ & & 6 & 16 & 8 \\ \hline & 3 & 8 & 4 & 0 \end{array}$$

5. $y = 5x - 3$

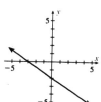

7. $2x + 3y + 6 = 0$
$y = -\frac{2}{3}x - 2,$
$\{(-3, 0), (0, -2)\}$

9. $m = \frac{-3-2}{6-2} = -\frac{5}{4}$

11. $7x - 5y + 3 = 0$
$5y = 7x + 3$
$y = \frac{7}{5}x + \frac{3}{5}; \quad m = \frac{7}{5}$

13. $4x - 3y + 9 = 0$
$-3 < x \le 4$

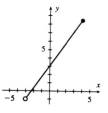

15. $y = \frac{3}{5}x - 3$
$-5 \le x \le 5$

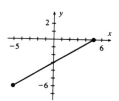

17. A(12, -7), B(4,3), C(-3, -2)

$$m_{AB} = \frac{-7-3}{12-4} = \frac{-10}{8} = -\frac{5}{4}$$

$$m_{AC} = \frac{-7+2}{12+3} = \frac{-5}{15} = -\frac{1}{3}$$

$$m_{BC} = \frac{-2-3}{-3-4} = \frac{-5}{-7} = \frac{5}{7}$$

Not a right triangle.

19. $6x - 4y + 3 = 0$

$4y = 6x + 3; y = \frac{3}{2}x + \frac{3}{4};$

$m = \frac{3}{2}$ so new slope is $-\frac{2}{3}$

21. $ax + by + c = 0;$ (x, y) is a point on the line and a, b, c are constants with a $\ne$ 0 and b $\ne$ 0.

23. $y = mx + b;$ (x, y) is a point on the line, m is the slope of the line, and b is the y-intercept.

25. $y = \frac{2}{3}x - 9;$ $3y = 2x - 27;$ $0 = 2x - 3y - 27.$

27. $m = \dfrac{-6 - (-1)}{5 - (-3)} = \dfrac{-6 + 1}{5 + 3} = \dfrac{-5}{8}$; $y = -\dfrac{5}{8}x + b$; $-6 = -\dfrac{5}{8}(5) + b$;

$-6 + \dfrac{25}{8} = b$; $b = -\dfrac{23}{8}$, therefore, $y = -\dfrac{5}{8}x - \dfrac{23}{8}$ and $8y = -5x - 23$, $5x + 8y + 23 = 0$.

29. $y - 1 = \dfrac{3}{4}(x + 3)^2$, $V(-3, 1)$

31. $y + 1 = -2(x + 2)^2$, $V(-2, -1)$

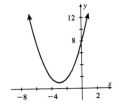

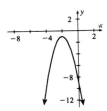

33. $x^2 + 2x - y - 1 = 0$,
   $y + 1 = x^2 + 2x$,

   $y + 1 + \underline{1} = x^2 + 2x\ \underline{1}$,

   $y + 2 = (x + 1)^2$, $V(-1, -2)$

35. $x^2 + 4x - 2y - 2 = 0$, $2y + 2 = x^2 + 4x$,
   $y + 1 = \dfrac{1}{2}x^2 + 2x$, $y + 1 = \dfrac{1}{2}(x^2 + 4x)$,

   $y + 1 + \underline{\ 2\ } = \dfrac{1}{2}(x^2 + 4x + \underline{\ 4\ })$,

   $y + 3 = \dfrac{1}{2}(x + 2)^2$, $V(-2, -3)$

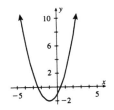

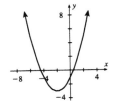

37. $2x^2 + 24x + y = 178$; $y - 178 = -2(x^2 + 12x)$, $y - 178 - 72 = -2(x^2 + 12x + \underline{\ 36\ })$,
   $y - 250 = -2(x + 6)^2$, $V(-6, 250)$; therefore you have a maximum value at 250 when $x = -6$.

39. $2x^2 + 20x + y + 190 = 0$; $y + 190 = -2x^2 - 20x$, $y + 190 - 50 = -2(x^2 + 10x + 25)$,
   $y + 140 = -2(x + 5)^2$, $V(-5, -140)$; therefore you have a maximum at -140 when $x\ -5$.

70

41. zeroes at $(0, 0)$ and $(6, 0)$, for a complete graph use the range: $X[-5, 10]$ by 5's and $Y[-35, 10]$ by 5's.

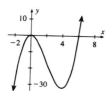

43. Since n = 4 and a > 0, the values of of y are increasing on the left and right right sides of the graph.

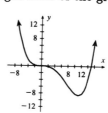

45. $y = -f(x)$, reflection over the x-axis

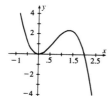

47. $y = f(2x)$, stretches the graph.

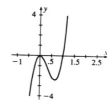

49.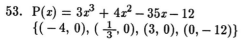

$$\begin{array}{r|rrrr} 2 & 3 & 4 & -35 & -12 \\ & & 6 & 20 & -30 \\ \hline & 3 & 10 & -15 & -42 \end{array}$$
$P(2) = -42$

53. $P(x) = 3x^3 + 4x^2 - 35x - 12$
$\{(-4, 0), (\frac{1}{3}, 0), (3, 0), (0, -12)\}$

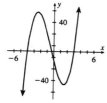

51. $$\begin{array}{r|rrrr} 3 & 3 & 4 & -35 & -12 \\ & & 9 & 39 & 12 \\ \hline & 3 & 13 & 4 & 0 \end{array}$$
$P(3) = 0$

55. $f(x) = 3x^4 - 8x^3 - 48x^2 + 492$
$\{(3.6, 0), (4.4, 0), (0, 492)\}$

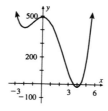

57. $y = (x-1)^3(x+2)^2$; 1(mult. 3) and -2(mult. 2).

59. $y = (x^2-1)^2(x+1)^3$; $(x-1)^2$ has 1(mult. 2)and-1(mult. 2) and $(x+1)^3$ has -1 (mult. 3) therefore $(x^2-1)^2(x+1)^3$ has 1(mult. 2) and -1(mult. 5).

61. $3x^3 + 4x^2 - 35x - 12 = 0$; $\frac{12}{3}$, $\pm(1, 2, 3, 4, 6, 12, \frac{1}{3}, \frac{2}{3}, \frac{4}{3})$.

63. $6x^4 - 13x^3 + 3x^2 + 3x - 2 = 0$; $\frac{2}{6}$, $\pm(1, 2, \frac{1}{2}, \frac{1}{3}, \frac{2}{3}, \frac{1}{6})$.

65.

P	N	I
1	0	2
1	0	2

67.

P	N	I
2	1	0
0	1	2

69. $3x^3 + 4x^2 - 35x - 12 = 0$; $\pm(1, 2, 3, 4, 6, 12, \frac{1}{3}, \frac{2}{3}, \frac{4}{3})$; $x = -4, -\frac{1}{3}, 3$; $\{-4, -\frac{1}{3}, 3\}$

71. $6x^4 - 13x^3 + 3x^2 + 9x - 5 = 0$; $\pm(1, 5, \frac{1}{2}, \frac{5}{2}, \frac{1}{3}, \frac{5}{3}, \frac{1}{6}, \frac{5}{6})$; $x = -\frac{5}{6}, 1$ (mult. 3).

73.
```
i | 1      1       -2        -2       -3     -3
  |        i    -1+i     -1-3i    3-3i      3
  ----------------------------------------------
    1    1+i    -3+i    -3-3i     -3i      0
```
$f(i) = 0$

75.
```
√3 | 1      1         -2         -2        -3     3
   |     1+√3     3+√3      3+√3    3+√3     3
   ------------------------------------------------
     1    2+√3     1+√3     1+√3       √3    0
```
$f(\sqrt{3}) = 0$

77. $x^4 - 3x^3 - 9x^2 + 25x - 6 = 0$; $\pm(1, 2, 3, 6)$
```
2 | 1    -3      -9      25     -6
  |        2      -2     -22      6
  ---------------------------------
    1    -1     -11       3      0
```
```
-3 | 1    -1    -11      3
   |       -3     12     -3
   ----------------------
     1    -4      1      0
```
$x^2 - 4x + 1 = 0$;

$x = \dfrac{4 \pm \sqrt{(-4)^2 - 4(1)(1)}}{2(1)} = \dfrac{4 \pm \sqrt{12}}{2} = \dfrac{4 \pm 2\sqrt{3}}{2} = 2 \pm \sqrt{3}$;

$\{-3, 2, 2 \pm \sqrt{3}\}$

79. $x^4 + x^3 - 14x^2 - 14x = x(x^3 + x^2 - 14x - 14) = 0$; $x = 0$; $\pm(1, 2, 7, 14)$
```
-1 | 1     1    -14    -14
   |      -1      0     14
   ----------------------
     1     0    -14      0
```
$x^2 - 14 = 0$, $x^2 = 14$, $x \pm \sqrt{14}$

$\{-1, 0, \pm\sqrt{14}\}$

81. $\underline{1+i}\ \lfloor\ 1$ -2 5 -6 6

$$\begin{array}{r} 1+i \quad -2 \quad 3+3i \quad -6 \\ \hline 1 \quad -1+i \quad 3 \quad -3+3i \quad 0 \end{array}$$

$\underline{1-i}\ \lfloor\ 1\quad -1+i\quad 3\quad -3+3i$

$$\begin{array}{r} 1-i \quad 0 \quad 3-3i \\ \hline 1 \quad 0 \quad 3 \quad 0 \end{array}$$

$x^2 + 3 = 0$ so $x^2 = -3$;
and $x = \pm\ \sqrt{3}i$

$\{1+i,\ 1-i,\ -\sqrt{3},\ \sqrt{3}\}$

83. $\underline{1+\sqrt{3}i}\ \lfloor\ 1\qquad -2\quad -1\qquad 10\quad -20$

$$\begin{array}{r} 1+\sqrt{3} \quad -4 \quad -5-5\sqrt{3}i \quad 20 \\ \hline 1 \quad -1+\sqrt{3}i \quad -5 \quad 5-5\sqrt{3}i \quad 0 \end{array}$$

$\underline{1-\sqrt{3}i}\ \lfloor\ 1\quad -1+\sqrt{3}i\quad -5\qquad 5-5\sqrt{3}i$

$$\begin{array}{r} 1-\sqrt{3}i \quad 0 \quad -5+5\sqrt{3}i \\ \hline 1 \quad 0 \quad -5 \quad 0 \end{array}$$

$x^2 - 5 = 0$ so $x^2 = 5$
and $x = \pm\ \sqrt{5}$
$\{1 \pm \sqrt{3}i,\ \pm\ \sqrt{5}\}$

CHAPTER 4

PROBLEM SET 4.1

1. $y = \frac{3}{x}$; $x \neq 0$
   Vertical: $x = 0$
   Horizontal: $y = 0$

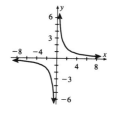

3. $y = \frac{-2}{x}$; $x \neq 0$
   Vertical: $x = 0$
   Horizontal: $y = 0$

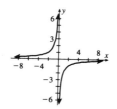

5. $y = \frac{1}{x} + 1$
   Vertical: $x = 0$
   Horizontal: $y = 0$

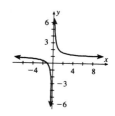

7. $y = -\dfrac{1}{x} + 2$
   Vert: $x = 0$
   Horiz: $y = 2$

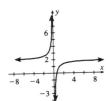

9. $y = -\dfrac{1}{x} - 3$
   Vert: $x = 0$
   Horiz: $y = -3$

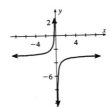

11. $y = \dfrac{1}{x-3}$
    Vert: $x = 3$
    Horiz: $y = 0$

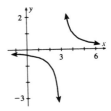

13. $y = \dfrac{-1}{x-3}$
    Vert: $x = -3$
    Horiz: $y = 0$

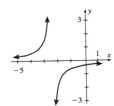

15. $y = \dfrac{-2}{x-2}$
    Vert: $x = 1$
    Horiz: $y = 0$

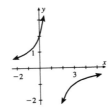

17. $y = \dfrac{-1}{x+2} - 1$
    Vert: $x = -2$
    Horiz: $y = -1$

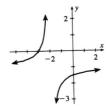

19. $y = \dfrac{(x+1)(x-2)(x+2)}{(x+1)(x-2)}$;
    $y = x + 2$, straight line with holes at $x = -1, 2$.

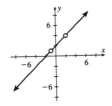

21. $y = \dfrac{(x+2)(x-1)(3x+2)}{(x+2)(x-1)}$;
    $y = 3x + 2$, straight line with holes at $x = -2, 1$.

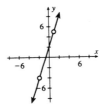

23. $y = \dfrac{(x+1)(x+2)(x-1)}{(x+2)}$;

$y = x^2 - 1$, parabola with a hole at $x = -2$.

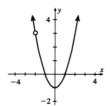

25. $y = \dfrac{(x-1)(x+2)(x-3)(x+4)}{(x+2)(x-1)}$;

$y = x^2 + x - 12$, parabola with holes at $x = -2, 1$.

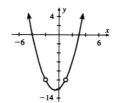

27. $y = \dfrac{(x-4)(x+3)}{x+3}$; $y = x - 4$,

straight line, hole at $x = -3$.

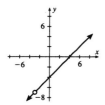

29. $y = \dfrac{(x+2)(x-3)}{x+2}$; $y = x - 3$, straight

line with hole at $x = -2$.

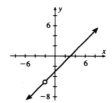

31. $y = \dfrac{4}{x^2}$; Vert: $x = 0$, Horiz: $y = 0$

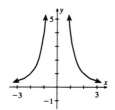

33. $y = \dfrac{-3}{x^2}$; Vert: $x = 0$, Horiz: $y = 0$.

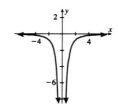

35. $y = \dfrac{-1}{(x-1)^2}$; $x \neq 1$,
Vert: $x = 1$, Horiz: $y = 3$

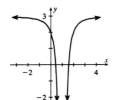

37. $y = \dfrac{2x^2 + 2}{x^2}$; $x \neq 0$, $y \neq 2$
Vert: $x = 0$, Horiz: $y = 2$

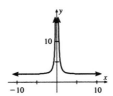

39. $y = \dfrac{x^2}{x-4}$; $x \neq 4$
Vertical: $x = 4$
Horizontal: none
Slant: $y = x + 4$

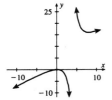

41. $y = \dfrac{-x^2}{x-1}$; $x \neq 1$
Vertical: $x = 1$
Horizontal: none
Slant: $y = -x - 1$

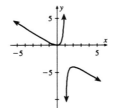

43. $y = \dfrac{x^2}{x-2}$; $x \neq 2$
Vertical: $x = 2$
Horiz: none
Slant: $y = x + 2$

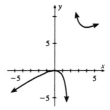

45. $y = \dfrac{(2x+1)(3x-4)}{2x+1}$;

$y = 3x - 4$, straight line with hole at $x = -\frac{1}{2}$.

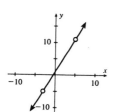

47. $y = \dfrac{(x+2)(x^2+10x+20)}{x+2}$;

$y = x^2 + 10x + 20$, $x \neq -2$, parabola.

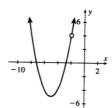

49. $y = \dfrac{(x+3)(x^2+6x-3)}{x+3}$

$y = x^2 + 6x, -3$, $x \neq -3$, parabola.

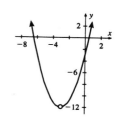

51. $y = \dfrac{(x+3)(2x+1)(x-5)}{(x+3)(x-5)}$;

$y = 2x + 1$, staight line with hole at $x = -3, 5$.

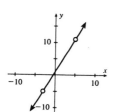

53. $y = \dfrac{2x+1}{3x-2}$; rational function with a hole at $x = \frac{2}{3}$.

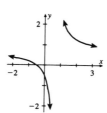

55. $y = \dfrac{2x^2+3x-1}{x-1}$;

rational function with asymptote at $x = 1$.

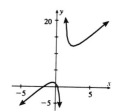

PROBLEM SET 4.2

1. x gets very large so no limit

3. $\dfrac{x}{2x+1} \cdot \dfrac{1/x}{1/x} = \dfrac{1}{2+1/x} = \dfrac{1}{2}$ so $\dfrac{x}{2x+1} \to \dfrac{1}{2}$

5. $\dfrac{x^2+3x-10}{x^2-2} \cdot \dfrac{1/x^2}{1/x^2} = \dfrac{x^2/x^2 + 3x/x^2 - 10/x^2}{x^2/x^2 - 2/x^2} = \dfrac{1}{1}$; $\dfrac{x^2+3x-10}{x^2-2} \to 1$

7. $\dfrac{x^2+3x-10}{2x^2+5} \cdot \dfrac{1/x^2}{1/x^2} = \dfrac{x^2/x^2 + 3x/x^2 - 10/x^2}{2x^2/x^2 + 5/x^2} = \dfrac{1}{2}$; $\dfrac{x^2+3x-10}{2x+5} \to \dfrac{1}{2}$

9. $\dfrac{2x^2-5x-12}{x^2-4} \cdot \dfrac{1/x^2}{1/x^2} = \dfrac{2x^2/x^2 - 5x/x^2 - 12/x^21}{x^2/x^2 - 4/x^2} = \dfrac{2}{1}$ so the lim $\to 2$

77

11. $\dfrac{x^2 + 2x + 4}{x^2 - 8} \cdot \dfrac{1/x^2}{1/x^2} = \dfrac{x^2/x^2 + 2x/x^2 + 4/x^2}{x^2/x^2 - 8/x^2} = \dfrac{1}{1}$ so the $\lim \to 1$

13. $\dfrac{6 - x}{2x - 15} = \dfrac{-x + 6}{2x - 15} \cdot \dfrac{1/x}{1/x} = \dfrac{-x/x + 6/x}{2x/x - 15/x} = \dfrac{-1}{2}$ so the $\lim \to -\dfrac{1}{2}$

15. $\dfrac{5x + 10{,}000}{x - 1} \cdot \dfrac{1/x}{1/x} = \dfrac{5x/x + 10{,}000/x}{x/x - 1/x} = \dfrac{5}{1}$ so the $\lim \to 5$

17. $\dfrac{4x^4 - 3x^3 + 2x + 1}{3x^4 - 9} \cdot \dfrac{1/x^4}{1/x^4} = \dfrac{4 - 3/x + 2/x^3 + 1/x^4}{3 - 9/x^4} = \dfrac{4}{3}$; so the $\lim \to \dfrac{4}{3}$

19. $\dfrac{3x^3 - 2x^2 + 1}{5x^3 + 3x - 100} \cdot \dfrac{1/x^3}{1/x^3} = \dfrac{3 - 2/x + 1/x^3}{5 + 3/x^2 - 100/x^3} = \dfrac{3}{5}$; so the $\lim \to \dfrac{3}{5}$

21. $y = \dfrac{1}{x}$; Vertical: $x = 0$; Horizontal: $y = 0$

23. $y = -\dfrac{1}{x} + 1$, $y = \dfrac{x - 1}{x}$; Vertical: $x = 0$; $\dfrac{x - 1}{x} \cdot \dfrac{1/x}{1/x} = \dfrac{1 - 1/x}{1} = 1$, Horizontal: $y = 1$

25. $\dfrac{2x^2 + 2}{x^2} \cdot \dfrac{1/x^2}{1/x^2} = \dfrac{2 + 2/x^2}{1} = 2$; Vertical: $x = 0$; Horizontal: $y = 2$

27. $\dfrac{-1}{x + 3} \cdot \dfrac{1/x}{1/x} = \dfrac{-1/x}{1 + 3/x} = \dfrac{0}{1} = 0$; Vertical: $x = -3$; Horizontal: $y = 0$

29. $\dfrac{x^2}{x - 4} \cdot \dfrac{1/x^2}{1/x^2} = \dfrac{1}{1/x - 4/x^2}$ is undefined ; Vertical: $x = 4$; Horizontal: does not exist;
by long division, Slant: $y = x + 4$

31. $\dfrac{-x^2}{x - 1} \cdot \dfrac{1/x^2}{1/x^2} = \dfrac{-1}{1/x - 1/x^2}$ is undefined ; Vertical: $x = 1$; Horizontal: does not exist;

by long division, Slant: $y = -x - 1$ or $y = -(x + 1)$

33. $y = \dfrac{x^3 - 2x^2 + x - 2}{(x - 2)(x^2 + 1)} = \dfrac{x^3 - 2x^2 + x - 2}{x^3 - 2x^2 + x - 2} = 1$; graph of a straight line so no asymptotes.

35. $y = \dfrac{(15x^2 + 13x - 6)(x - 1)}{3x^2 - 4x + 1} = \dfrac{(3x - 1)(5x + 6)(x - 1)}{(3x - 1)(x - 1)} = 5x + 6$; straight line so no asymptotes.

37. $y = \dfrac{3x^3 + 5x^2 - 26x + 8}{x^2 + 2x - 8} = \dfrac{(x + 4)(3x - 1)(x - 2)}{(x + 4)(x - 2)} = 3x - 1$; straight line so none.

39. $y = \dfrac{x^3 + 9x^2 + 15x - 9}{x + 3} = \dfrac{(x + 3)(x^2 + 6x - 3)}{x + 3} = x^2 + 6x - 3$; parabola so none.

41. $y = \dfrac{x + 3}{x - 2}$; $(0, -\frac{3}{2})$

$(-3, 0)$; Vert: $x = 2$,

Horiz: $y = 1$

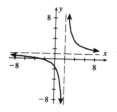

43. $y = \dfrac{3x + 5}{3x - 2}$; $(0, -\frac{5}{2})$,

$(-\frac{5}{3}, 0)$, Vert: $x = \frac{2}{3}$,

Horiz: $y = 1$

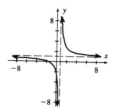

45. $y = \dfrac{x^2 + x - 6}{x^2 + 2x - 8} = \dfrac{x + 3}{x + 4}$

$(0, \frac{3}{4})$, $(-3, 0)$; Horiz: $y = 1$

Vert: $x = -4$

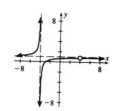

47. $y = \dfrac{6x^2 - 6x - 12}{3x^2 + 4x + 5}$;

$(0, -\frac{12}{5})$, $(-1, 0)$,

$(2, 0)$, Vert: none,
Horiz: $y = 2$

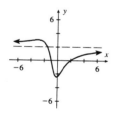

49. $y = \dfrac{x}{(x + 3)(x - 2)}$

$= (0, -6)$, $(0, 0)$;

Vert: $x = -3, 2$,
Horiz: $y = 0$

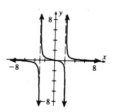

51. $y = \dfrac{x^2}{x(x^2 - x - 20)}$

$= \dfrac{x}{(x - 5)(x + 4)}$; hole at

$(0, 0)$; $(-4, 0)$, $(5, 0)$;

Vert: $x = -4, 5$,

Horiz: $y = 0$

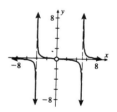

53. $y = \dfrac{(x+3)(x-2)}{x+3}$;
straight line, hole at
$x = -3$

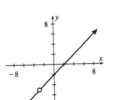

55. $y = \dfrac{x^3}{x^2+4}$; $(0, 0)$;
no Vert., no Horiz.,
Slant: $y = x$

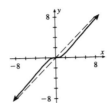

57. $y = \dfrac{x^3}{x^2+9}$; $(0, 0)$;
no Vert., no Horiz.,
Slant: $y = x$

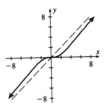

59. $y = \dfrac{x^3 + x^2 + 2x + 2}{x^2 + 9}$;
$(0, \frac{2}{9})$, $(-1, 0)$; no Vert.,
no Horiz., Slant: $y = x$

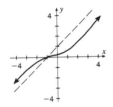

61. $y = \dfrac{x^3}{x^2 - 1}$; $(0, 0)$;
Vert: $x \neq 1$, no Horiz.,
Slant: $y = x$

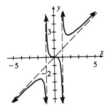

63. All three problems have the following: the point, $(0, 0)$; Vertical asymptotes at $x = \neq 1$. Number 60 has Horizontal asymptote at $y = 1$ and a parabolic shape between the asymptotes. Number 61 has a Slant asymptote at $y = x$ and a cubic shape between the asymptotes. Also 0 is a double root. Number 62 has a parabolic shape between the asymptotes and zero is multiple root of three. If you have a graphing calculator try continuing the pattern with x^5, x^6, etc. to see if there are other patterns.

80

PROBLEM SET 4.3

1. $(x)^{1/4}$ on $[0, 25]$

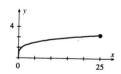

3. $y = x^{3/5}$ on $[0, 32]$

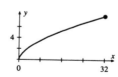

5. $y = (x)^{1/4}$ on $[0, 1]$

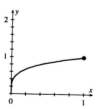

7. $y = x^{3/5}$ on $[0, 1]$

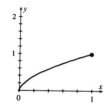

9. $y = -2(x)^{1/4}$ on $[0, 1]$

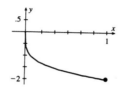

11. $y = -\frac{1}{2}(x)^{3/5}$ on $[0, 1]$

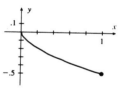

13. $y = (x)^{2/3}$

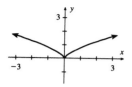

15. $y = -2(x)^{1/4}$

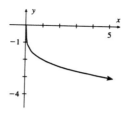

17. $y = (x)^{3/5} - 2$

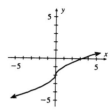

81

19. $y = (x-2)^{3/5}$

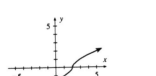

21. $y = (x+1)^{2/5} + 2$

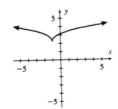

23. $y = (x+3)^{1/3} - 1$

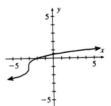

25. $y = x + \sqrt{x}$
$(0, 0), (1, 2)$
$(2, 3.4), (4, 6)$

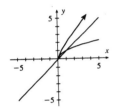

27. $y = x^3 + \sqrt{x}$
$(0, 0), (1, 2),$
$(2, 8.4), (3, 28.7)$

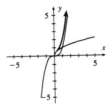

29. $y = x^{-1} + \sqrt{x}$
$(1, 2), (2, 1.9),$
$(3, \ 2.03), (4, 2.25)$

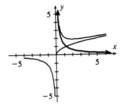

31. $y = \sqrt{x^2 - 1}$
$(\pm 1, 0), (2, 1.7),$
$(3, 2\sqrt{2}), (4, \sqrt{15})$

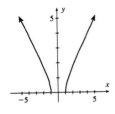

33. $y = \sqrt{x^2 + 1}$
$(0, 1), (1, \sqrt{2}),$
$(-1, \sqrt{2}), (\pm 2, \sqrt{5})$

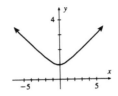

35. $y = \sqrt{(x-2)(x-3)}$
$(2, 0), (3, 0), (1, \sqrt{2}),$
$(-1, 2\sqrt{3}),$

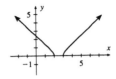

37. $y = \sqrt{x^2 - 6x + 8}$
    $x \geq 4$ and $x \leq 2$

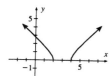

39. $y = \sqrt{10(x^2 - 1)}$
    $x \geq 1$ and $x \leq -1$

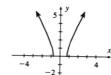

41. $y = \sqrt{4(x^2 + 1)}$
    all x-values

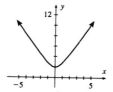

43. $y = \sqrt{(x-1)(x-3)(x-5)}$
    $1 \leq x \leq 3$ and $x \geq 5$

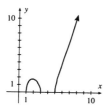

45. $y = \sqrt{x(x-2)(x-4)}$
    $0 \leq x \leq 2$ and $x \geq 4$

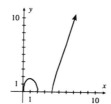

47. $y = \dfrac{x}{\sqrt{x^2 - 9}}$; $x \leq -3$
    and $x \geq 3$

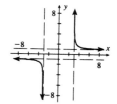

49. $y = \dfrac{x}{\sqrt{x^2 + 4}}$, all x's

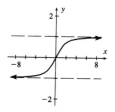

51. $y = -\dfrac{2}{3}\sqrt{x^2 - 9}$
    $x \geq 3$ and $x \leq -3$

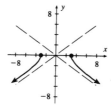

53. $y = \dfrac{3}{5}\sqrt{x^2 - 4}$
    $x \geq 2$ and $x \leq -2$

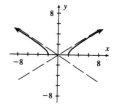

55. $y = \dfrac{x}{\sqrt{1-x^2}}$

$-1 < x < 1$

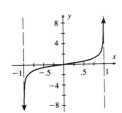

57. $y = \sqrt{\dfrac{x^2-1}{x-2}}$

$-1 \le x \le 1$ and $x > 2$

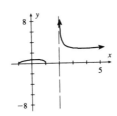

59. $y = \sqrt{\dfrac{3}{x^2-1}}$

$x > 1$ and $x < -1$

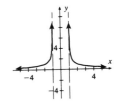

63. p > q, q is even; then concave up (or it opens up).

61. $y = \sqrt{\dfrac{x^2+1}{x}}$, $x > 0$

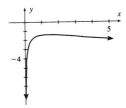

65. p > q, q is odd and p is even; then concave up.

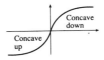

67. p < q, q is odd and p is odd; going left to right it is concave up and at zero it changes to concave down.

PROBLEM SET 4.4

1. $\left(\dfrac{x^2}{12} - \dfrac{x}{3} = \dfrac{3}{4}\right)$ 12; $x^2 - 4x = 9$; $x^2 - 4x - 9 = 0$; use the quadratic formula $x = 2 \pm \sqrt{13}$.

3. $\left(\dfrac{3}{4y} + \dfrac{7}{16} = \dfrac{4}{3y}\right)$ 48y; $3(12) + 7(3y) = 4(16)$; $36 + 21y = 64$; $21y = 28$, $y = \dfrac{4}{3}$.

5. $\left(\dfrac{4}{y} - \dfrac{3(2-y)}{y-1} = 3\right)$ [y(y − 1)]; $4(y-1) - 3y(2-y) = 3y(y-1)$; $4y - 4 - 6y + 3y = 3y^2 - 3y$;

$-4 - 2y = -3y$; y = 4.

7. $\left(\dfrac{z-1}{z} = \dfrac{z-1}{6}\right)$ 6z; $6(z-1) = z(z-1)$; $6z - 6 = z^2 - z$; $0 = z^2 - 7z - z$; z = 1, 6.

9. $\left(\dfrac{1}{x+2} - \dfrac{1}{2-x} = \dfrac{3x+8}{x^2-4}\right)$ $x^2 - 4$; $1(x-2) + 1(x+2) = 3x + 8$; $x - 2 + x + 2 = 3x + 8$;

$2x = 3x + 8$; $x = -8$.

84

11. $\left(\dfrac{x+2}{3x-1} - \dfrac{1}{x} = \dfrac{x+1}{3x^2-x}\right) x(3x-1);$ $x(x+2)-1(3x-1) = x+1;$ $x^2+2x-3x+1 = x+1;$
$x^2-2x = 0;$ $x(x-2) = 0;$ $x = 0$ is extraneous so $x = 2.$

13. $\left(\dfrac{x+1}{x-1} - \dfrac{x-1}{x+1} = \dfrac{5}{6}\right) 6(x+1)(x-1);$ $6(x+1)^2 - 6(x-1)^2 = 5(x+1)(x-1);$
$6(x^2+2x+1) - 6(x^2-2x+1) = 5(x^2-1);$ $6x^2+12x+2-6x^2+12x-6 = 5x^2-5;$
$24x-4 = 5x^2-5;$ $0 = 5^2-24x-1;$ $x = -\dfrac{1}{5}, 5.$

15. $\left(\dfrac{z-1}{z-3} + \dfrac{z+2}{z+3} = \dfrac{3}{4}\right) 4(z-3)(z+3);$ $4(z+3)(z-1) + 4(z-3)(z+2) = 3(z-3)(z+3);$
$4(z^2+2z-3) + 4(z^2-z-6) = 3(z^2-9);$ $4z^2+8z-12+4z^2-4z-24 = 3z^2-27;$
$8z^2+4z-36 = 3z^2-27;$ $5z^2+4z-9 = 0;$ $z = -\dfrac{9}{5}, 1.$

17. $(2\sqrt{x})^2 = (x+1)^2;$ $4x = x^2+2x+1;$ $0 = x^2-2x+1;$ $0 = (x-1)^2;$ $x = 1.$

19. $(\sqrt{z+2})^2 = 3^2;$ $z+2 = 9;$ $z = 7.$

21. $\left\{(4x+4)^{1/3}\right\}^3 = 2^3;$ $4x+4 = 8;$ $4x = 4;$ $x = 1.$

23. $x - \sqrt{x} - 2 = 0;$ $x-2 = \sqrt{x};$ $(x-2)^2 = (\sqrt{x})^2;$ $x^2-4x+4 = x;$ $x^2-5x+4 = 0;$
1 is extranous so $x = 4.$

25. $x = 6 - 3\sqrt{x-2};$ $x-6 = -3\sqrt{x-2};$ $(x-6)^2 = (-3\sqrt{x-2})^2;$ $x^2-12x+36 = 9(x-2);$
$x^2-12z+36 = 9x-18;$ $x^2-21x+54;$ 18 is extraneous so $x = 3.$

27. $(\sqrt{x-3})^2 = (\sqrt{4x-5})^2;$ $x-3 = 4x-5;$ $3x = 2;$ $x = \dfrac{2}{3}$ which does not work so $\{ \ \}.$

29. $\left(\sqrt{y^2+4y-5}\right)^2 = \left(\sqrt{2-2y}\right)^2;$ $y^2+4y-5 = 2-2y;$ $y^2+6y-7 = 0;$ $y = -7, 1.$

31. $\left(\dfrac{3}{2x+1} - \dfrac{2x+1}{2x-1} = 1 - \dfrac{8x^2}{(2x+1)(2x-1)}\right)(2x+1)(2x-1);$
$3(2x-1) - (2x+1)(2x+1) = 1(2x+1)(2x-1) - 8x^2;$
$6x-3 - (4x^2+4x+1) = 4x^2-1-8x^2;$ $6x-3-4x^2-4x-1 = -4x^2-1;$ $2x-3 = 0,$ $x = \dfrac{3}{2}.$

33. $\left(\dfrac{x-2}{x+3} - \dfrac{1}{x-2} = \dfrac{x-4}{(x+3)(x-2)}\right)(x+3)(x-2);$ $(x-2)(x-2) - 1(x+3) = x-4;$
$x^2-4x+4-x-3 = x-4;$ $x^2-5x+1 = x-4;$ $x^2-6x+5 = 0;$ $x = 1, 5.$

35. $\left(\dfrac{2x+1}{x+2} - \dfrac{x+2}{x+1} = -1\right)(x+2)(x+1);$ $(2x+1)(x+1) - (x+2)(x+2) = -(x+2)(x+1);$
$2x^2+3x+1-(x^2+4x+4) = -(x^2+3x+2);$ $2x^2+3x+1-x^2-4x-4 = -x^2-3x-2;$
$x^2-x-3 = -x^2-3x-2;$ $2x^2+2x-1 = 0;$ using the quadratic formula: $x = \dfrac{-1\pm\sqrt{3}}{2}.$

37. $\left(\dfrac{x-1}{x-2} + \dfrac{x+4}{2x+1} = \dfrac{1}{2x^2 - 3x - 2}\right)(2x+1)(x-2);$

$(x-1)(2x+1) + (x+4)(x-2) = 1; \ 2x^2 - x - 1 + x^2 + 2x - 8 = 1; \ 3x^2 + x - 9 = 1;$

$3x^2 + x - 10 = 0; \ x = -2, \dfrac{5}{3}.$

39. $\left(\dfrac{3}{x+2} + \dfrac{x-1}{x+5} = \dfrac{5(x+4)}{6(x+4)}\right)(6(x+2)(x+5);$

$3(6)(x+5) + 6(x-1)(x+2) = 5(x+2)(x+5); \ 18(x+5) + 6(x^2 + x - 2) = 5(x^2 + 7x + 10);$

$18x + 90 + 6x^2 + 6x - 12 = 5x^2 + 35x + 50; \ 6x^2 + 24x + 78 = 5x^2 + 35x + 50;$

$x^2 - 11x + 28 = 0; \ x = 4, 7.$

41. $(2 - \sqrt{3x+1})^2 = (\sqrt{x-1})^2; \ 4 - 4\sqrt{3x+1} + 3x + 1 = x - 1; \ 5 + 3x - 4\sqrt{3x+1} = x - 1;$

$6 + 2x = 4\sqrt{3x+1}; \ 3 + x = 2\sqrt{3x+1}; \ (3+x)^2 = (2\sqrt{3x+1})^2; \ 9 + 6x + x^2 = 4(3x+1);$

$x^2 + 6x + 9 = 12x + 4; \ x^2 - 6x + 5 = 0; \ 5 \text{ is extraneous so } x = 1.$

43. $(\sqrt{4x+1})^2 = (2 + \sqrt{2x+1})^2; \ 4x + 1 = 4 + 4\sqrt{2x+1} + 2x + 1; \ 2x - 4 = 4\sqrt{2x+1};$

$(x-2)^2 = (2\sqrt{2x+1})^2; \ x^2 - 4x + 4 = 4(2x+1); \ x^2 - 4x + 4 = 8x + 4; \ x^2 - 12x = 0;$

$0 \text{ is extraneous so } x = 12.$

45. $(1 + \sqrt{x+2})^2 = (\sqrt{x})^2; \ 1 + 2\sqrt{x+2} + x + 2 = x; \ 2\sqrt{x+2} = -3 \text{ is not true so } \{ \ \}.$

47. $\left\{(u+1)^{1/2}\right\}^2 = \left\{(u+1)^{1/4}\right\}^2; \ u + 1 = (u+1)^{1/2} \ \text{OR} \ u + 1 = \sqrt{u+1};$

$(u+1)^2 = (\sqrt{u+1})^2; \ u^2 + 2u + 1 = u + 1; \ u^2 + u = 0, \ u = -1, 0.$

49. $(\sqrt{2x+3} + 3)^2 = (3\sqrt{x+1})^2; \ 2x + 3 + 9 + 6\sqrt{2x+3} = 9(x+1);$

$2x + 12 + 6\sqrt{2x+3} = 9x + 9; \ 6\sqrt{2x+3} = 7x - 3; \ (6\sqrt{2x+3})^2 = (7x-3)^2;$

$36(2x+3) = 49x^2 - 42x + 9; \ 72x + 108 = 49x^2 - 42x + 9; \ 0 = 49x^2 - 114x - 99;$

solve by the quadratic formula, $x = 3.$

51. $x^6 + 7x^3 - 8 = 0; \ (x^3 + 8)(x^3 - 1) = 0; \ (x+2)(x^2 - 2x + 4)(x-1)(x^2 + x + 1) = 0;$

$(x^2 - 2x + 4) \text{ and } (x^2 + x + 1) \text{ give complex roots so } x = -2, 1.$

53. $u = x^2; \ 4u^2 - 35u - 9 = 0; \ (4u+1)(u-9) = 0; \ u = -\dfrac{1}{4}, 9, \ -\dfrac{1}{4} \text{ is extraneous};$

$u = 9, \ x^{-2} = 9; \ \text{so } x = \pm\dfrac{1}{3}.$

55. $(x^2 + 4x)^2 + 7(x^2 + 4x) + 12 = 0, \ u = x^2 + 4x; \ u^2 + 7u + 12 = 0, \ (u+4)(u+3) = 0 \text{ so}$

$u = -4, -3; \ x^2 + 4x = -4, \ x^2 + 4x + 4 = 0, \ x = -2 \text{ with multiplicity of two; and}$

$x^2 + 4x = -3, \ x^2 + 4x + 3 = 0, \ x = -3, -1; \ x = -3, -2, -1.$

57. $(\sqrt{x-1} + 2)^2 = \{3(x-1)^{1/4}\}^2$; $x-1+4+4\sqrt{x-1} = 9\sqrt{x-1}$; $x+3 = 5\sqrt{x-1}$; $(x+3)^2 = (5\sqrt{x-1})^2$; $x^2+6x+9 = 25(x-1)$; $x^2+6x+9 = 25x-25$; $x^2-19x+34 = 0$; $x = 2, 17$.

59. $\sqrt{x} + \dfrac{6}{\sqrt{x}} - 5 = 0$; $\left(\sqrt{x} + \dfrac{6}{\sqrt{x}} - 5 = 0\right)\sqrt{x}$; $x+6-5\sqrt{x} = 0$; $x-5\sqrt{x}+6 = 0$; $u = \sqrt{x}$, $u^2-5u+6 = 0$, $u = 2, 3$; $\sqrt{x} = 2$, $x = 4$; $\sqrt{x} = 3$, $x = 9$, $x = 4, 9$.

61. $\dfrac{1}{\sqrt{3}} = \dfrac{\sqrt{2w+4}}{\sqrt{w}} - \dfrac{\sqrt{3w+4}}{\sqrt{3w}}$, multiply by the LCM $\sqrt{3w}$ and you will have the following: $\sqrt{w} = \sqrt{6w+12} - \sqrt{3w+4}$. Square both sides to obtain: $w = 6w+12+3w+4-2\sqrt{(6w+12)(3w+4)}$; $2\sqrt{18w^2+60w+48} = 8w+16$; $\sqrt{18w^2+60w+48} = 4w+8$. Square both sides again and you get: $18w^2+60w+48 = 16w^2+64w+64$; $2w^2-4w-16 = 0$ and $w^2-2w-8 = 0$ so $w = -2, 4$ but -2 is not an acceptable answer so $w = 4$.

63. $\sqrt{2x-1} = \sqrt{7x^2+2} - \sqrt{x+3}$, square both sides. $2x-1 = 7x+2+x+3-2\sqrt{(7x+2)(x+3)}$; $2x-1 = 8x+5-2\sqrt{7x^2+23x+6}$; $2\sqrt{7x^2+23x+6} = 6x+6$; $\sqrt{7x^2+23x+6} = 3x+3$, square both sides and obtain: $7x^2+23x+6 = 9x^2+18x+9$; $0 = 2x^2-5x+3$, so $x = 1, \dfrac{3}{2}$.

65. $x^2-2x-8 = 3\sqrt{x^2-2x+2}$, let $u = \sqrt{x^2-2x+2}$ and let $u^2 = x^2-2x+2$. $(x^2-2x+2)-10 = 3\sqrt{x^2-2x+2}$ becomes $u^2-10 = 3u$ and $u^2-3u-10 = 0$; Solve and $u = -2, 5$ but -2 is not an acceptable answer so $u = 5$. Substitute into the original equation: $x^2-2x-8 = 3u$ and you have $x^2-2x-8 = 3(5)$, $x^2-2x-23 = 0$; solving this equation gives: $x = 1 \pm 2\sqrt{6}$.

67. $\sqrt{\dfrac{x}{1-x}} + \sqrt{\dfrac{1-x}{x}} = \dfrac{10}{3}$, multiply by the LCD which is $3\sqrt{x}\sqrt{1-x}$ an you obtain: $3\sqrt{x^2} + 3\sqrt{(1-x)^2} = 10\sqrt{x}\sqrt{1-x}$; $3x+3(1-x) = 10\sqrt{x}\sqrt{1-x}$; $3x+3-3x = 10\sqrt{x(1-x)}$; $3 = 10\sqrt{x-x^2}$ and square both sides for $9 = 100(x-x^2)$; $9 = -100x^2+100x$, $100x^2-100x+9 = 0$, now solve for $x = \dfrac{1}{10}, \dfrac{9}{10}$.

PROBLEM SET 4.5

1. $y = \frac{2}{x}$ and $x = \frac{2}{y}$ so $D = (-\infty, 0) \cup (0, \infty)$

    and $R = (-\infty, 0) \cup (0, \infty)$; y-intercept: no value,

    x-intercept: no value ; $(-x, -y)$: $(-x)(-y) = 2$,

    $xy = 2$ so origin symmetry; no x or y symmetry

    since you will get $-xy = 2$; Vertical asymptote:

    $x = 0$; Horizontal asymptote: $y = 0$

31.

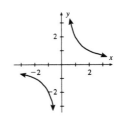

3. $y = \frac{x+1}{x}$ and $x = \frac{1}{y-1}$ so

    $D = (-\infty, 0) \cup (0, \infty)$ and $R = (-\infty, 1) \cup (1. \infty)$;

    y-intercept: y is undefined so no value;

    x-intercept: $(-1, 0)$;

    no symmetry ;

    Vertical: $x = 0$; Horizontal: $y = 1$

33.

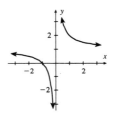

5. $y = \sqrt{4 - x}$, $D = (-\infty, 4]$ and $R = [0, \infty)$;

    y-intercept: $y = 2$, $(0, 2)$;

    x-intercept: $x = 4$, $(4, 0)$;

    no symmetry ;

    no asymptotes

35.

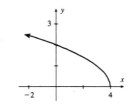

88

7. $y = -\sqrt{x-6}$, $D = [6, \infty)$ and $R = (-\infty, 0]$;

y-intercept: no value ;

x-intercept: $x = 6$, $(6, 0)$;

no symmetry ;

no asymptotes

37.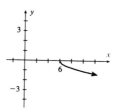

9. $y = \dfrac{1}{x^2 - 4}$, $D = (-\infty, -2) \cup (-2, 2) \cup (2, \infty)$

and $R = (-\infty, 0) \cup (0, \infty)$;

y-intercept: $y = -\frac{1}{4}$, $(0, -\frac{1}{4})$;

x-intercept: no value ;

$(-x, y)$: $y = \dfrac{1}{(-x)^2 - 4} = \dfrac{1}{x^2 - 4}$ so y-axis

Vertical: $x = \pm 2$; Horizontal: $y = 0$

39.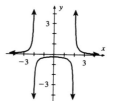

11. $y = \dfrac{2x^2 + 9x + 10}{x + 2}$, $D = (-\infty, -2) \cup (-2, \infty)$

and $R = (-\infty, 1) \cup (1, \infty)$

y-intercept: $y = 5$, $(0, 5)$;

x-intercept: $x = -\frac{5}{2}$, $(-\frac{5}{2}, 0)$;

no symmetry ; straight line with

at $x = -\dfrac{5}{2}$ or -2.5

41.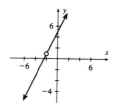

13. $y = \dfrac{2x^3 - 3x^2 - 2}{2x + 1} = \dfrac{x(2x + 1)(x - 2)}{2x + 1} = x^2 - 2x,$

   $D = (-\infty, -\frac{1}{2}) \cup (-\frac{1}{2}, \infty)$ and $R = (-1, \infty)$

   y-intercept: $y = 0$, $(0, 0)$

   x-intercept: $x = 0, 2$; $(0, 0)$, $(2, 0)$

   no symmetry ;

   Vertical: none ; Horizontal: none

43.

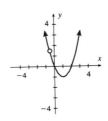

15. $y = \sqrt{x} - x$, $D = [0, \infty)$, $R = (-\infty, \frac{1}{4}]$;

   y-intercept: $y = 0$, $(0, 0)$;

   x-intercept: $x = 0, 1$, $(0, 0)$, $(1, 0)$;

   no symmetry ;

   Vertical: none ;

   Horizontal: none

45.

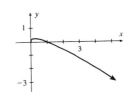

17. $y = \sqrt{x^2 + 2x - 3}$, $x^2 + 2x - 3 \ge 0$

   $\begin{array}{c|c|c|c} x + 3 & -\, - & -\, - & +\ + \\ x - 1 & -\, - & +\ + & +\ + \\ \hline & -3 & 1 & \end{array}$

   $D = (-\infty, -3] \cup [1, \infty)$

   $R = [0, \infty)$

   y-intercept: none

   x-intercept: $x = -3, 1$, $(-3, 0)$, $(1, 0)$

   no symmetry ; Vertical: none ; Horizontal: none

47.

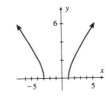

90

19. $y = \dfrac{-x}{\sqrt{4 - x^2}}$; $D = (-2, 2)$; $R = (-\infty, \infty)$

   y-intercept: $y = 0$, $(0, 0)$;

   x-intercept: $x = 0$, $(0, 0)$;

   no symmetry ;

   Vertical: $x = \pm 2$; Horizontal: none

49.

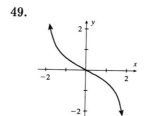

21. $|x| + |y| = 5$

   $D = [-5, 5]$; $R = [-5, 5]$;

   y-intercept: $y = \pm 5$, $(0, -5)$, $(0, 5)$;

   x-intercept: $x = \pm 5$, $(-5, 0)$, $(5, 0)$;

   symmetry: origin, x-axis, and y-axis ;

   Vertical: none ; Horizontal: none

51.

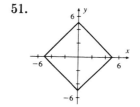

23. $9x^2 + 16y^2 = 144$, not a function

   $D = [-4, 4]$; $R = [-3, 3]$;

   y-intercept: $y = \pm 3$; $(0, -3)$, $(0, 3)$

   x-intercept: $x = \pm 4$; $(-4, 0)$, $(4, 0)$

   symmetry: origin, x-axis, and y-axis ;

   Vertical: none ; Horizontal: none

53.

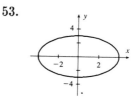

25. $y = \dfrac{x^2 + 1}{x}$

   $D = (-\infty, 0) \cup (0, \infty)$;

   $R = (-\infty, -2] \cup [2, \infty)$;

   y-intercept: none ;

   x-intercept: none ;

   symmetry: origin ;

   Vertical: $x = 0$; Horizontal: none ; Slant: $y = x$

55.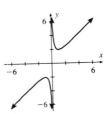

27. $y^2 = \dfrac{x^2 - 1}{x - 4}$, not a function

   $D = [-1, 1] \cup (4, \infty)$; $R = \left(-\infty, \sqrt{8 + 2\sqrt{15}}\,\right) \cup$
   $\left(-\sqrt{8 - 2\sqrt{15}},\ \sqrt{8 - 2\sqrt{15}}\,\right) \cup \left(\sqrt{8 + 2\sqrt{15}}, \infty\right)$;

   y-intercept: $y = \pm \tfrac{1}{2}$, $(0, -\tfrac{1}{2})$, $(0, \tfrac{1}{2})$,

   $\qquad\qquad y^2 = \tfrac{-1}{-4} = \tfrac{1}{4}$ so $y = \pm \tfrac{1}{2}$;

   x-intercept: $x = \pm 1$, $(-1, 0)$, $(1, 0)$,

   $\qquad\qquad x^2 - 1 = 0$, so $x = \pm 1$;

   symmetry: x-axis since $(-y)^2 = y^2$;

   Vertical: $x = 4$; Horizontal: none

57.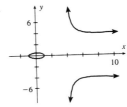

29. $y^2 = \dfrac{x^2 - 9}{x^2 - 2x - 15}$, not a function

   $D = (-\infty, -3) \cup (-3, 3] \cup (5, \infty)$;

   $R = (-\infty, -1) \cup (-1, 1) \cup (1, \infty)$;

   y-intercept: $y = \pm\tfrac{9}{5}$, $(0, -\tfrac{9}{5})$, $(0, \tfrac{9}{5})$;

   x-intercept: $x = 3$, $(3, 0)$;

   symmetry: none ;

   Vertical: $x = -3, 5$; Horizontal: $y = \pm 1$

59.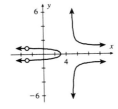

61. $x^{2/3} + y^{2/3} = 4$, not a function

    $D = [-8, 8]$; $R = [-8, 8]$;

    y-intercept: $y = \pm 8$, $(0, -8)$, $(0, 8)$;

    x-intercept: $x = \pm 8$, $(-8, 0)$, $(8, 0)$;

    symmetry: origin, x-axis, and y-axis ;

    Vertical: none ; Horizontal: none

61.

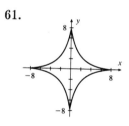

63. $x^2 y^2 - x^2 - 9y^2 + 4 = 0$, not a function

    $x^2 y^2 - 9y^2 = x^2 - 4$

    $y^2(x^2 - 9) = x^2 - 4$

    $y^2 = \frac{x^2 - 4}{x^2 - 9}$, $x \neq \pm 3$

    $D = (-\infty, -3) \cup [-2, 2] \cup (3, \infty)$;

    $R = (-\infty, -1) \cup [-\frac{2}{3}, \frac{2}{3}] \cup (1, \infty)$;

    y-intercept: $y = \pm \frac{2}{3}$, $(0, -\frac{2}{3})$, $(0, \frac{2}{3})$;

    x-intercept: $x = \pm 2$, $(-2, 0)$, $(2, 0)$;

    symmetry: origin, x-axis, and y-axis ;

    Vertical: $x = \pm 3$; Horizontal: $y = \pm 1$

63.

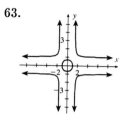

65. $x^4 + y^2 - 4x^2 = 0$, not a function

    $y^2 = -x^4 + 4x^2$ so $y^2 = -x^2(x^2 - 4)$

    $D = [-2, 2]$, $x^2 - 4$ is negative on the

        interval $[-2, 2]$ so $-x^2(x^2 - 4)$ is

        positive on the interval $[-2, 2]$.

    $R = [-2, 2]$, $x^2 = \frac{4 \pm \sqrt{16 - 4y^2}}{2}$,

      $16 - 4y^2 \geq 0$ so $-2 \leq y \leq 2$;

    y-intercept: $y = 0$, $(0, 0)$; x-intercept: $x = \pm 2$,

    $(-2, 0)$, $(2, 0)$; Vertical: none; Horizontal: none.

65.

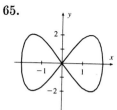

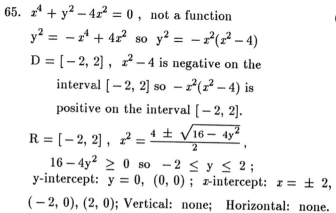

PROBLEM SET 4.6

1. $x^2 + 7x + 12 = (x + 4)(x + 3)$ 3. $2x^2 - 5x - 3 = (2x + 1)(x - 3)$

5. $x^2 - 14x + 48 = (x - 6)(x - 8)$ 7. $x^3 - x = x(x^2 - 1) = x(x - 1)(x + 1)$

9. $x^3 - 4x = x(x^2 - 4) = x(x - 2)(x + 2)$ 11. $(x - 5); \ (x - 5)^2$

13. $x; \ (x - 1); \ (x - 1)^2$ 15. $(x - 2); \ (x - 2)^2; \ (x - 2)^3$

17. $x^2 - x^2 - 3 = (x - 3)(x + 1); \ \ (x - 3); \ (x + 1)$

19. $x^3 - x^2 - 2x = x(x^2 - x - 2) = x(x - 2)(x + 1); \ \ x; \ (x - 2); \ (x + 1)$

21. $x^3 - 5x^2 + 4x = x(x^2 - 5x + 4) = x(x - 4)(x - 1); \ \ x; \ (x - 4); \ (x - 1)$

23. $1 - x^4 = (1 - x^2)(1 + x^2) = (1 - x)(1 + x)(1 + x^2); \ \ (1 - x); \ (1 + x); \ (1 + x^2)$

25. $\dfrac{x^2 + 2x + 5}{x^3} = \dfrac{A}{x} + \dfrac{B}{x^2} + \dfrac{C}{x^3};$

    $\dfrac{x^2 + 2x + 5}{x^3} = \dfrac{Ax^2 + Bx + C}{x^3};$

    $x^2 + 2x + 5 = Ax^2 + Bx + C;$

    $Ax^2 = 1x^2, \ A = 1; \ \ Bx = 2x,$

    $B = 2; \ \ C = 5 \ ; \ \dfrac{1}{x} + \dfrac{2}{x^2} + \dfrac{5}{x^3}$

27. $\dfrac{2x^2 - 5x + 4}{x^3} = \dfrac{A}{x} + \dfrac{B}{x^2} + \dfrac{C}{x^3};$

    $\dfrac{2x^2 - 5x + 4}{x^3} = \dfrac{Ax^2 + Bx + C}{x^3}$

    $2x^2 - 5x + 4 = Ax^2 + Bx + C$

    $Ax^2 = 2x^2, \ A = 2; \ \ Bx = -5x,$

    $B = -5; \ C = 4; \ \dfrac{2}{x} - \dfrac{5}{x^2} + \dfrac{4}{x^3}$

29. $\dfrac{1}{(x + 4)(x + 5)} = \dfrac{A}{(x + 4)} + \dfrac{B}{(x + 5)};$

    $1 = A(x + 5) + B(x + 4);$

    $x = -5: \ 1 = B(-5 + 4), \ B = -1$

    $x = -4: \ 1 = A(-4 + 5), \ A = 1$

    $\dfrac{1}{x + 4} - \dfrac{1}{x + 5}$

31. $\dfrac{7x - 10}{(x - 2)(x - 1)} = \dfrac{A}{x - 2} + \dfrac{B}{x - 1};$

    $7x - 10 = A(x - 1) + B(x - 2);$

    $x = 1: \ 7 - 10 = B(1 - 2), \ B = 3;$

    $x = 2: \ 14 - 10 = A(2 - 1), A = 4;$

    $\dfrac{4}{x - 2} + \dfrac{3}{x - 1}$

33. $\dfrac{7x + 2}{(x + 2)(x - 4)} = \dfrac{A}{x + 2} + \dfrac{B}{x - 4};$

    $7x + 2 = A(x - 4) + B(x + 2);$

    $x = -2: \ -14 + 2 = A(-2 - 4),$

    $A = 2; \ x = 4: \ 28 + 2 = B(4 + 2),$

    $B = 5; \ \dfrac{2}{x + 2} + \dfrac{5}{x - 4}$

35. $\dfrac{2x - 14}{(x + 3)(x - 2)} = \dfrac{A}{x + 3} + \dfrac{B}{x - 2};$

    $2x - 14 = A(x - 2) + B(x + 3);$

    $x = -3: \ -6 - 14 = -5A, \ A = 4;$

    $x = 2: \ 4 - 14 = 5B, \ B = -2;$

    $\dfrac{4}{x + 3} - \dfrac{2}{x - 2}$

37. $\dfrac{4x-4}{(x+2)(x-2)} = \dfrac{A}{x+2} + \dfrac{B}{x-2}$

$4x-4 = A(x-2) + B(x+2)$;

$x = 2$: $8 - 4 = B(2+2)$, $B = 1$;

$x = -2$: $-8 - 4 = -4A$, $A = 3$;

$\dfrac{3}{x+2} + \dfrac{1}{x-2}$

39. $\dfrac{x-7}{(x-5)(x-4)} = \dfrac{A}{x-5} + \dfrac{B}{x-4}$;

$x - 7 = A(x-4) + B(x-5)$;

$x = 5$: $5 - 7 = 1A$, $A = -2$;

$x = 4$: $4 - 7 = -B$, $B = 3$;

$\dfrac{3}{x-4} - \dfrac{2}{x-5}$

41. $\dfrac{4x^2 - 7x - 3}{x(x-1)(x+1)} = \dfrac{A}{x} + \dfrac{B}{x-1} + \dfrac{C}{x+1}$; $4x^2 - 7x - 3 = A(x^2 - 1) + Bx(x-1) + Cx(x+1)$;

$x = 0$: $-3 = -A$, $A = 3$; $x = -1$: $4 + 7 - 3 = 2B$, $B = 4$;

$x = 1$: $4 - 7 - 3 = 2C$, $C = -3$; $\dfrac{3}{x} + \dfrac{4}{x-1} - \dfrac{3}{x+1}$

43. $\dfrac{2x-1}{(x-2)^2} = \dfrac{A}{x-2} + \dfrac{B}{(x-2)^2}$; $2x - 1 = A(x-2) + B$; $2x - 1 = Ax - 2A + B$;

$2x = Ax$, $A = 2$; $-1 = -2A + B$, $-1 = -4 + B$, $B = 3$; $\dfrac{2}{x-2} + \dfrac{3}{(x-2)^2}$

45. $\dfrac{x^2 + 5x + 1}{x(x+1)^2} = \dfrac{A}{x} + \dfrac{B}{(x+1)} + \dfrac{C}{(x+1)^2}$; $x^2 + 5x + 1 = A(x+1)^2 + Bx(x+1)$

Cx; $x = 0$: $1 = 1A$, $A = 1$; $x = -1$: $1 - 5 + 1 = -C$, $C = 3$;

$x^2 + 5x + 1 = A(x+1)^2 + Bx(x+1) + Cx$,

$= x^2 + 5x + 1 = Ax^2 + 2Ax + A + Bx^2 + Bx + Cx$

$x^2 + 5x + 1 = (A + B)x^2 + (2A + B + C)x + A$;

$1x^2 = (A + B)x^2$, $1 = A + B$, $1 = 1 + B$, $B = 0$;

$\dfrac{1}{x} + \dfrac{0}{x+1} + \dfrac{3}{(x+1)^2}$

$\dfrac{1}{x} + \dfrac{3}{(x+1)^2}$

47. $\dfrac{2x^2 + 8x + 3}{(x+1)^3} = \dfrac{A}{(x+1)} + \dfrac{B}{(x+1)^2} + \dfrac{C}{(x+1)^3}$; $2x^2 + 8x + 3 = A(x+1)^2 + $

$B(x+1) + C$; $x = -1$: $2 - 8 + 3 = C$, $C = -3$

$2x^2 + 8x + 3 = A(x^2 + 2x + 1) + Bx + B + C$

$2x^2 + 8x + 3 = Ax^2 + 2Ax + A + Bx + B + C$

$2x^2 + 8x + 3 = Ax^2 + (2A + B)x + (A + B + C)$

$C = -3$; $A = 2$; $8 = 2A + B$, $8 = 4 + B$, $B = 4$

$\dfrac{2}{x+1} + \dfrac{4}{(x+1)^2} - \dfrac{3}{(x+1)^3}$

49. $\dfrac{x}{(x+5)(x-1)} = \dfrac{A}{x+5} + \dfrac{B}{x-1}$; $x = A(x-1) + B(x+5)$; $x = 1$: $6B = 1$, $B = \frac{1}{6}$;

$x = -5$: $-5 = -6A$, $A = \frac{5}{6}$; $\dfrac{\frac{5}{6}}{x+5} + \dfrac{\frac{1}{6}}{x-1} = \dfrac{5}{6(x+5)} + \dfrac{1}{6(x-1)}$

51. $\dfrac{7x-1}{(x-2)(x+1)} = \dfrac{A}{x-2} + \dfrac{B}{x+1}$; $7x - 1 = A(x+1) + B(x-2)$;

$x = -1$: $-7 - 1 = -3B$, $B = \dfrac{8}{3}$; $x = 2$: $14 - 1 = 3A$, $A = \dfrac{13}{3}$;

$A = \dfrac{13}{3}$, $B = \dfrac{8}{3}$ so $\dfrac{\frac{13}{3}}{x-2} + \dfrac{\frac{8}{3}}{x+1} = \dfrac{13}{3(x-2)} + \dfrac{8}{3(x+1)}$

53. $\dfrac{-17x-6}{x(x^2+x-6)} = \dfrac{17x-6}{x(x+3)(x-2)} = \dfrac{A}{x} + \dfrac{B}{x+3} + \dfrac{C}{x-2}$; $17x - 6 = A(x+3)(x-2)$

$+ Bx(x-2) + Cx(x+3)$; $x = 0$: $-6 = -6A$, $A = 1$; $x = -3$: $51 - 6 = 15B$,

$45 = 15B$, $B = 3$; $x = 2$: $-34 - 6 = 10C$, $-40 = 10C$, $C = 4$;

$A = 1$, $B = 3$, $C = -4$ so $\dfrac{1}{x} + \dfrac{3}{x+3} - \dfrac{4}{x-2}$

55. $\dfrac{5x^2-6x+7}{(x-1)(x^2+1)} = \dfrac{A}{x-1} + \dfrac{Bx+C}{x^2+1}$; $5x^2 - 6x + 7 = A(x^2+1) + (Bx+C)(x-1)$;

$x = 1$: $5 - 6 + 7 = 2A$, $6 = 2A$, $A = 3$;

$5x^2 - 6x + 7 = A(x^2+1) + (Bx+C)(x-1)$,

$5x^2 - 6x + 7 = Ax^2 + A + Bx^2 - Bx + Cx - C$

$5x^2 = (A+B)x^2$, $5 = A + B$, $A = 3$ so $5 = 3 + B$ and $B = 2$

$-6x = (-B+C)x$, $B = 2$ so $-6 = -2 + C$ and $C = -4$

$A = 3$, $B = 2$, $C = -4$ so $\dfrac{3}{x-1} + \dfrac{2x-4}{x^2+1}$

57. $\dfrac{x^3}{(x-1)^2} = \dfrac{x^3}{x^2-2x+1}$; by long division:

$$
\begin{array}{r}
x + 2 \\
x^2 - 2x + 1 \,\overline{)\, x^3 + 0x^2 + 0x + 0} \\
\mp x^3 \pm 2x^2 \mp 1x \\
\hline
2x^2 - 1x + 0 \\
\mp 2x^2 \pm 4x \mp 2 \\
\hline
3x - 2
\end{array}
$$

$x + 2 + \dfrac{3x-2}{x^2-2x+1}$

$\dfrac{3x-2}{(x-1)^2} = \dfrac{A}{x-1} + \dfrac{B}{(x-1)^2}$

$3x - 2 = A(x-1) + B$

$3x - 2 = Ax + (-A+B)$

$3x = Ax$ so $A = 3$

$-2 = -A + B$, $-2 = -3 + B$,

$B = 1$

$x + 2 + \dfrac{3}{x-1} + \dfrac{1}{(x-1)^2}$

59. $\dfrac{2x^3 - 3x^2 + 6x - 1}{1 - x^4} = \dfrac{2x^3 - 3x^2 + 6x - 1}{(1-x)(1+x)(1+x^2)} = \dfrac{A}{1-x} + \dfrac{B}{1+x} + \dfrac{Cx+D}{1+x^2}$

$2x^3 - 3x^2 + 6x - 1 = A(1+x)(1+x^2) + B(1-x)(1+x^2) + (Cx+D)(1-x^2)$

$2x^3 - 3x^2 + 6x - 1 = A(1 + x^2 + x + x^3) + B(1 + x^2 - x - x^3) + Cx - Cx^3 +$

$D - Dx^2 = A + Ax^2 + Ax + Ax^3 + B + Bx^2 - Bx - Bx^3 + Cx - Cx^3 + D - Dx^2$

$2x^3 - 3x^2 + 6x - 1 = (A - B - C)x^3 + (A + B - D)x^2 + (A - B + C)x + (A + B + D)$

Equation 1: $2 = A - B - C$ Add equations 1 and 3 and also 2 and 4.

Equation 2: $-3 = A + B - D$

Equation 3: $6 = A - B + C$

Equation 4: $-1 = A + B + D$

$$1A - 1B - C = 2 \qquad 1A + 1B - D = -3$$
$$\underline{1A - 1B + C = 6} \qquad \underline{1A + 1B + D = -1}$$
$$2A - 2B \quad = 8 \qquad\quad 2A + 2B \quad = -4$$
$$A - B = 4 \qquad\qquad A + B = -2$$

$$1A - 1B = 4$$
$$\underline{1A + 1B = -2}$$
$$2A = 2$$
$$A = 1$$

$A = 1$: $A - B = 4$; $1 - B = 4$, $B = -3$

$A = 1, B = -3$: $A + B + D = -1$; $1 - 3 + D = -1$, $D = 1$

$A = 1, B = -3$: $A - B + C = 6$; $1 + 3 + C = 6$, $C = 2$

$$\dfrac{1}{1-x} - \dfrac{3}{1+x} + \dfrac{2x+1}{1+x^2}$$

CHAPTER FOUR SUMMARY

1. $f(x) = \dfrac{1}{x-2} + 2;$

   $x \ne 2$

   Vertical: $x = 2$

   Horizontal: $y = 2$

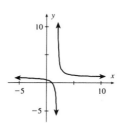

3. $f(x) = \dfrac{2x^2 - 3x - 1}{x^2 - x - 2};$

   $x \ne -1, 2$

   Vertical: $x = -1, 2$

   Horizontal: $y = 2$

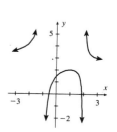

5. $\dfrac{3x^5 - 2x^3 + 1}{4x^5 - 1} \cdot \dfrac{1/x^5}{1/x^5} = \dfrac{3 - 2/x^2 + 1/x^5}{4 - 1/x^5} = \dfrac{3}{4};\ \dfrac{3x^5 - 2x^3 + 1}{4x^5 - 1} \to \dfrac{3}{4}$

7. $\dfrac{5x^3 - 2x^2 + 1}{4x^5 - 1} \cdot \dfrac{1/x^5}{1/x^5} = \dfrac{5/x^2 - 2x^3 + 1/x^5}{4 - 1/x^5} = \dfrac{0}{4};\ \dfrac{5x^3 - 2x^2 + 1}{4x^5 - 1} \to 0$

9. $y = \dfrac{6x^2 - 11x}{2x - 1},\ x \ne \dfrac{1}{2};$ Vertical: $x = \dfrac{1}{2};$ Slant: $y = 3x - 4$

97

11. $y = \dfrac{3x}{\sqrt{9-x^2}}$, $x \neq \pm 3$; Vertical: $x = \pm 3$; Horizontal: none

13. $y = \sqrt{x-4x+3}$;

(0, $\sqrt{3}$), (1, 0), (3, 0),

$x \geq 3$ and $x \leq 1$

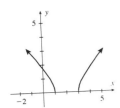

15. $y = \dfrac{x}{\sqrt{x^2+5}}$,

$x^2 + 5$ always positive
Domain: $(-\infty, \infty)$

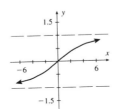

17. $\dfrac{x+1}{x+3} = \dfrac{2x-1}{2x+1}$; $(x+1)(2x+1) = (x+3)(2x-1)$; $2x^2 + 3x + 1 = 2x^2 + 5x - 3$;

$-2x = -4$; $x = 2$ and $x \neq -3, -\dfrac{1}{2}$.

19. $\dfrac{5}{x} - \dfrac{x-5}{x-3} = \dfrac{x-6}{2(x-6)}$; $\left(\dfrac{5}{x} - \dfrac{x-5}{x-3} = \dfrac{1}{2}\right) 2x(x-3)$;

$5(2)(x-3) - 2x(x-5) = x(x-3)$; $10x - 30 - 2x^2 + 10x = x^2 - 3x$; $0 = 3x^2 - 23x + 30$;

6 is extraneous so $x = \dfrac{5}{3}$, $x \neq 0, 3, 6$.

21. $x - \sqrt{3-x} = 3$; $(x-3)^2 = (\sqrt{3-x})^2$; $x^2 - 6x + 9 = 3 - x$; $x^2 - 5x + 6 = 0$;

2 is an extraneous root so $x = 3$.

23. $(\sqrt{x+5})^2 = (\sqrt{x}+1)^2$; $x + 5 = x + 1 + 2\sqrt{x}$; $4 = 2\sqrt{x}$; $16 = 4x$, $x = 4$.

25. $x^2 - x^2 y^2 - y^2 + 25 = 0$;

$x^2(1 - y^2) = y^2 - 25$;

$x^2 = \dfrac{y^2 - 25}{1 - y^2}$; $x = \pm\sqrt{\dfrac{y^2 - 25}{1 - y^2}}$,

$y \neq \pm 1$;

$x^2 + 25 = x^2 y^2 + y^2$

$x^2 + 25 = y^2(x^2 + 1)$

$y^2 = \dfrac{x^2 + 25}{x^2 + 1}$; $y = \pm\sqrt{\dfrac{x^2 + 25}{x^2 + 1}}$;

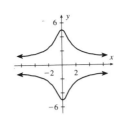

D: $(-\infty, \infty)$; R: $[-5, -1) \cup (1, 5]$.

y-intercept: y= ± 5, x-intercept: none; symmetry: origin, x-axis, and y-axis.

27. $x^{2/3} + y^{2/3} = 1$

  $D = [-1, 1];$ R: $[-1, 1]$

  y-intercepts: $y \pm 1,$ $(0, \pm 1);$

  $y^{2/3} = 1,$ $y^2 = 1,$ so $y = \pm 1;$

  x-intercept: $x = \pm 1,$ $(\pm 1, 0);$

  symmetry: origin, x-axis, and y-axis.

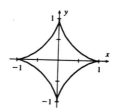

29. $\dfrac{5x^2 - 19x + 17}{(x-1)(x-2)^2} = \dfrac{A}{(x-1)} + \dfrac{B}{(x-2)} + \dfrac{C}{(x-2)^2}$

  $5x^2 - 19x + 17 = A(x-2)^2 + B(x-1)(x-2) + C(x-1)$

  $5x^2 - 19x + 17 = Ax^2 - 4Ax + 4A + Bx^2 - 3Bx + + 2B + Cx - C$

  $5x^2 - 19x + 17 = (A+B)x^2 + (-4A - 3B + C)x + (4A + 2B - C)$

  $\begin{aligned} -19 &= -4A - 3B + C \\ 17 &= 4A + 2B - C \\ \hline -2 &= -B \end{aligned}$

  Substitute $\underline{B=2}$ into: $A + B = 5,$ $A + 2 = 5,$ $\underline{A = 3};$ substitute A and B into: $4A + 2B - C = 17,$ $12 + 4 - C = 17,$ $\underline{C = -1}.$

  $A = 3,$ $B = 2,$ $C = -1,$ then the answer is $\dfrac{3}{x-1} + \dfrac{2}{x-2} - \dfrac{1}{(x-2)^2}$

31. $\dfrac{2x^2 + 13x - 9}{x^2 + 2x - 15} = \dfrac{2x^2 + 13x - 9}{(x+5)(x-3)}$

  $\begin{array}{r} 2 \\ x^2 + 2x - 15 \overline{\smash{\big)}\ 2x^2 + 13x - 9} \\ \underline{\mp 2x^2 \mp 4x \pm 30} \\ 9x + 21 \end{array}$ so $2 + \dfrac{9x + 21}{x^2 + 2x - 15} = 2 + \dfrac{9x + 21}{(x+5)(x-3)}$

  $\dfrac{9x + 21}{(x+5)(x-3)} = \dfrac{A}{x+5} + \dfrac{B}{x-3}$

  $9x + 21 = A(x-3) + B(x+5)$

  $x = -5:$ $-45 + 21 = -8A,$ $-24 = -8A,$ $\underline{A = 3};$ $x = 3;$ $27 + 21 = 8B,$ $48 = 8B,$ $\underline{B = 6};$

  $\dfrac{2x^2 + 13x - 9}{x^2 + 2x - 15} = 2 + \dfrac{3}{x+5} + \dfrac{6}{x-2}$

CHAPTER 5

PROBLEM SET 5.1

1. $y = 3^x$

$\{(-2, 9), (-1, \frac{1}{3}),$
$(0, 1), (1,3), 2, 9)\}$

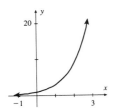

3. $y = 5^x$

$\{(-2, \frac{1}{25}), (-1, \frac{1}{5}),$
$(0, 1), (1, 5), (2, 25)\}$

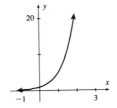

5. $y = 4^{-x}$

$\{(-2, 16), (-1, 4),$
$(0,1), (1, \frac{1}{4}), (2, \frac{1}{16})\}$

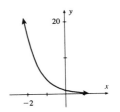

7. $y = 2^{(x-2)}$

$\{(2, 1), (3, 2),$
$(4, 4), (1, \frac{1}{2}),$
$(0, \frac{1}{4}), (-1, \frac{1}{8})\}$

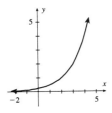

9. $y = 2^{(x+3)}$

$\{(-3, 1), (-2, 2),$
$(-1, 4), (-4, \frac{1}{2}),$
$(-5, \frac{1}{4}), (-6, \frac{1}{8})\}$

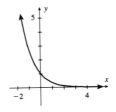

11. $f(x) = e^{(x+1)}$

$\{(-1, 1), (0, 2.72),$
$(1, 7.39), (2, 20.09),$
$(-2, 0, 0.37), (-3, 0.14)\}$

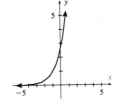

13. $y = e^{(x + 2)} + 2$
$\{(-2, 3), (-1, 4.72),$
$(0, 9.39), (1, 22.09),$
$(-3, 2.37), (-4, 2.14)\}$

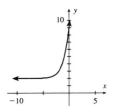

15. $y = 2^{(x + 3)}$
$\{(-3, 1), (-4, \frac{1}{2}),$
$(-5, \frac{1}{4}), (-2, 2),$
$(-1, 4), (0, 8)\}$

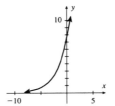

17. $y = 2^x + 3$
$\{(0, 4), (1, 5), (2, 7),$
$(-1, 3.5), (-2, 3.25),$
$(-3, 3.125)\}$

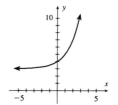

19. $y = 2^{(x + 4)} + 5$
$\{(-4, 6), (-3, 7),$
$(-2, 9), (-5, 5.5),$
$(-6, 5.25), (-7, 5.125)\}$

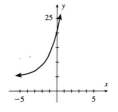

21. $y = 2^{(x - 3)} - 10$
$\{(3, -9), (4, -8),$
$(5, -6), (2, -9.5),$
$(1, -9.75), (0, -9.875)\}$

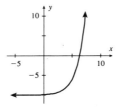

23. $y = 3^{|x|}$
$\{(0, 1), (1, 3),$
$(2, 9), -1, 3),$
$(-2, 9)\}$

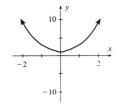

25. $y = 3^{x^2}$
$\{(0, 1), (1, 3)$
$(2, 81)\ (-1, 3),$
$(-2, 81)\}$

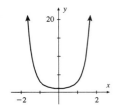

27. $y = 10^{x^2}$
$\{(0, 1), (1, 10),$
$(2, 10000), (-1, 10),$
$(-2, 10000)\}$

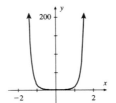

29. $y = 5^{-x^2}$
$\{(0, 1), (1, \frac{1}{5}),$
$(2, 0.0016), (-1, \frac{1}{5}),$
$(-2, 0.0016)\}$

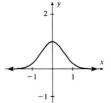

31. $e^3 = 20.0855$ 33. $e^{0.12} = 1.3910$ 35. $e^{0.15(5)} = 2.1170$ 37. $(1 + \frac{0.18}{12})^{24} = 1.4295$

39. $(1 + \frac{0.12}{365})^{365} = 1.1275$ 41. $(1 + \frac{1}{1,000})^{1,000} = 2.7169$

43. $2^{\sqrt{2}}$ is approximately equal to 2.7 so by using the squeezing process we get:

$$1 < \sqrt{2} < 2 \qquad\qquad 2^1 < 2^{\sqrt{2}} < 2^2$$
$$1.4 < \sqrt{2} < 1.5 \qquad\qquad 2^{1.4} < 2^{\sqrt{2}} < 2^{1.5}$$
$$1.41 < \sqrt{2} < 1.42 \qquad\qquad 2^{1.41} < 2^{\sqrt{2}} < 2^{1.42}$$
$$1.414 < \sqrt{2} < 1.415 \qquad\qquad 2^{1.414} < 2^{\sqrt{2}} < 2^{1.415}$$

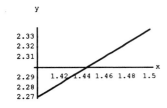

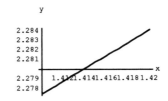

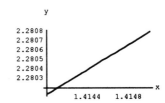

$1.4 < x < 1.5$
$\{(1, 2), (1.4, 2.64),$
$(1.5, 2.83), (2, 4)\}$

$1.41 < x < 1.42$
$\{(1.4, 2.6), (1.41, 2.65),$
$(1,42, 2.675), (1.5, 2.8)\}$

$1.414 < x < 1.415$
$\{(1.41, 2,65), (1.414, 2.66),$
$(1.415, 2.67), (1.42, 2.675)\}$

45. $10^{\sqrt{2}} \approx 26.0$; use the same method as in problem #43.

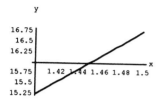 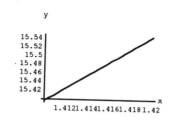 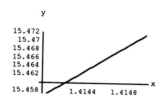

$1.4 < x < 1.5$
$\{(1, 10), (1.4, 25.1),$
$(1.5, 31.6)\}$

$1.41 < x < 1.42$
$\{(1.4, 25.1), (1.414, 25,7),$
$(1.42, 26.3), (1.5, 31.6)\}$

$1.414 < x < 1.415$
$\{(1.41, 25.7), (1.414, 25.9),$
$(1.415, 26), (1.42, 26.3)\}$

47. a. $y = b^x$; $b = 1$, then $y = 1$ which is a constant function, algebraic.

b. $y = b^x$; $b = 0$, then $y = 0$ which is a constant function, algebraic.

49. P = \$1000, r = .07, compounded annually, t = 25 years, $A = P(1 + r)^t$.
$A = 1000(1 + .07)^{25} = 1000(1.07)^{25} = 1000(5.42743264) = \$5427.43$.

51. P = \$1,000, r= .16, compounded continuously, t = 25, $A = Pe^{rt}$. $A = 1000\ e^{(.16)(25)}$
$= 1000e^4 = 1000(54.59815003) = \$54598.15$.

53. P = \$8500, r = .18, compounded monthly, t = 3, $A = P(1+ \frac{r}{n})^{nt}$. $A = 8500(\ 1 + \frac{.18}{12})^{(12)(3)}$
$= 8500(1 + 0.015)^{36} = 8500(1.015)^{36} = 8500(1.709139538) = \$14,527.69$.

55. P = \$10,000, r = .14, compounded daily or 360 days, t = 6 months.
$A = 10000(1 + \frac{.14}{360})^{180} = 10000(1.00038889)^{180} = 10000(1.072493609) = \$10,724.94$

57. P = \$110,000; r = .12 monthly so .01; n = 360; t = 30 years; $M = \dfrac{Pr}{1 - (1 + r)^{-n}}$.

$M = \dfrac{110,000(0.01)}{1 - (1 + 0.01)^{-360}} = \dfrac{1, 100}{1 - 0.278166892} = \dfrac{1,100}{0.9721833108} = \$1131.47$.

59. P = \$125,000; r = .145; monthly so .01020833333; n = 360; t= 30 years; same formula as #57.

$M = \dfrac{125,000(.0120833333)}{1 - (1 + .0120833333)^{-360}} = \dfrac{1510.41667}{.9867522632} = \$1530.69$.

61. a) $c(x) = \dfrac{e^x + e^{-x}}{2}$ b) $s(x) = \dfrac{e^x - e^{-x}}{2}$ 63. $A = 100(\frac{1}{2})^{t/4}$, $t \geq 0$

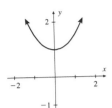   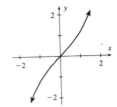   

PROBLEM SET 5.2

1. $\log_2 64 = 6$ 3. $\log_{10} 1000 = 3$ 5. $\log_5 125 = 3$ 7. $\log_n m = p$ 9. $\log_{1/3} 9 = -2$

11. $\log_9 \frac{1}{3} = -\frac{1}{2}$ 13. $10^4 = 10,000$ 15. $10^0 = 1$ 17. $e^2 = e^2$ 19. $e^5 = x$

21. $2^{-3} = \frac{1}{8}$ 23. $4^{1/2} = 2$ 25. $m^p = n$ 27. $x^3 = 8$ 29. $\log_b b^2 = 2 \log_b b = 2(1) = 2$

31. $\log_e e^4 = 4 \log_e e = 4$ 33. $\log_\pi (\frac{1}{\pi}) = \log_\pi \pi^{-1} = -1 \log_\pi \pi = -1$

35. $\log_3 9 = \log_3 3^2 = 2 \log_3 3 = 2$ 37. $\log 4.27 \approx 0.6304$ 39. $\log 8.43 \approx 0.9258$

41. $\log 71,600 \approx 4.8549$ 43. $\log 1.321 \approx 0.1209$ 45. $\ln 2.27 \approx 0.8198$

47. $\ln 2.431 \approx 0.8883$ 49. $\ln 13 \approx 2.5649$ 51. $\ln 7.3 \approx 1.9879$ 53. $\log_5 304 = 3.5522$

55. $\log_6 .10 = -1.2851$ 57. $\log_\pi 100.0 = 4.0229$ 59. $\log_\pi 25 = 2.8119$

61. There is no way that N can be negative since by definition the base has to be positive and a positive base with a positive or negative exponent always gives you a positive answer.

63. Then $\log_b$ N is the exponent.   65. The common is base 10   67. This is a true statement

69. $N = 1500 + 300 \ln a$, $a \geq 1$
   a. $a = 1,000$, $N = 1500 + 300 \ln 1000 \approx 1500 + 2072.326584 = 3,572$
   b. $a = 50,000$, $N = 1500 + 300 \ln 50,000 \approx 1500 + 3245.933485 = 4,746$
   c. $N = 5000$, $5000 = 1500 + 300 \ln a$, $\frac{3500}{300} = \ln a$, $\ln a \approx 11.\overline{6}$, $a = e^{11.\overline{6}}$,
      $a \approx 116618.904$, $a = \$116,619.$

71. $M = \dfrac{\log E - 11.8}{1.5}$; a. When $E = 15^{15}$; $M = \dfrac{\log 15^{15} - 11.8}{1.5} = \dfrac{15 \log 15 - 11.8}{1.5}$
    $\approx \dfrac{17.64136889}{1.5} \approx 3.894245925$; m ≈ 3.89 so 3.89 on the Richter scale
    b. $E = 10^{25}$; $M = \dfrac{\log 10^{25} - 11.8}{1.5} = \dfrac{25(1) - 11.8}{1.5} = \dfrac{13.2}{1.5} = 8.8$; so 8.8 on the Richter scale.

PROBLEM SET 5.3

1. a. $\log_5 25 = x$, $5^x = 25$, $5^x = 5^2$, $x = 2$ b. $\log_2 128 = x$, $2^x = 128$, $2^x = 2^7$, $x = 7$

3. a. $\log \frac{1}{10} = x$, $10^x = 10^{-1}$, $x = -1$ b. $\log 10{,}000 = x$, $10^x = 10^4$, $x = 4$

5. a. $\log_x 28 = 2$, $x^2 = 28$, $x = 2\sqrt{7}$ b. $\log_x 81 = 4$, $x^4 = 81$, $x^4 = 3^4$, $x = 3$

7. a. $\log_x e = 2$, $x^2 = e$, $x = \sqrt{e}$ b. $\log_x e = 1$, $x^1 = e$, $x = e$

9. a. $\ln x = 4$, $e^4 = x$ b. $\ln x = \ln 14$, $x = 14$

11. a. $\log_3 x^2 = \log_3 125$, $x^2 = 125$, $x = \pm 5\sqrt{5}$ b. $\ln x^2 = \ln 12$, $x^2 = 12$, $x \pm 2\sqrt{3}$

13. a. $\log_3 27\sqrt{3} = x$, $3^x = 27\sqrt{3}$, $27\sqrt{3} = 3^3 \cdot 3^{1/2} = 3^{7/2}$ so $3^x = 3^{7/2}$, $x = \frac{7}{2}$

     b. $\log_x 1 = 0$, $x^0 = 1$, so x can have any value

15. a. $\log_2 x = 5$, $2^5 = x$, $x = 32$ b. $\log_{10} x = 5$, $10^5 = x$, $x = 100{,}000$

17. This is a true statement.

19. Put this graph into your calculator and you will see that it is concave down.

21. $\log_b AB = \log_b A + \log_b B$ by definition.

23. $b > 1$, concave down   25. $0 < b < 1$, concave up

27. $y = \log_3 x$ or $3^y = x$
     asymptote at $x = 0$

29. $y = \log_\pi x$ or $\pi^y = x$
     asymptote at $x = 0$

31. $y = \log (x - 4)$ or
     $10^y = x - 4$
     asymptote: $x - 4$

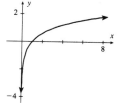

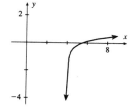

33. $y = \log x + 3$

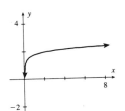

35. $y - 1 = \log (x + 2)$

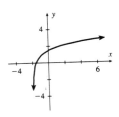

37. $y = \log_{1/2} x + 3$

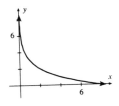

39. $y = \log |x|$

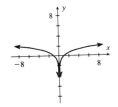

41. $y = -3 \log |x|$

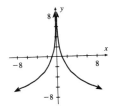

43. $y - 1 = \log x^2$

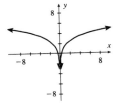

45. $\log 2 = \frac{1}{4} \log 16 - x$; $\log 2 = \log 16^{1/4} - x$; $\log 2 = \log 2 - x$; $x = 0$

47. $\ln 2 + \frac{1}{2} \ln 3 + 4 \ln 5 = \ln x$, $\ln 2 + \ln \sqrt{3} + \ln 5^4 = \ln x$, $\ln x = \ln 2\sqrt{3}(625)$
$= \ln x = \ln 1250\sqrt{3}$, $x = 1250\sqrt{3}$

49. $1 = 2 \log x - \log 1{,}000$, $1 = \log x^2 - \log 10^3$, $1 = \log x^2 - 3$,
$4 = \log x^2$, $10^4 = x^2$, $x = 100$

51. $\log_7 x - \frac{1}{2} \log_7 4 = \frac{1}{2} \log_7 (2x-3)$, $\log_7 x - \log_7 \sqrt{4} = \log_7 (2x-3)^{1/2}$,
$\log_7 x - \log_7 2 = \log_7 (2x-3)^{1/2}$, $\log_7 \frac{x}{2} = \log_7 \sqrt{2x-3}$, $x = 2\sqrt{2x-3}$,
$x^2 = 4(2x-3)$, $x^2 - 8x + 12 = 0$, $(x-2)(x-6) = 0$, $x = 2, 6$

53. $\frac{1}{2} \ln x = 3 \ln 5 - \ln x$, $\quad \ln \sqrt{x} + \ln x = \ln 5^3$, $\quad \ln x\sqrt{x} = \ln 125$, $\quad x\sqrt{x} = 125$,

$x^2(x) = 125^2$, $\quad x^3 = 15{,}625$, $\quad x = 25$

55. $2 \ln x - \frac{1}{2} \ln 9 = \ln 3(x-2)$, $\quad \ln x^2 - \ln 3 = \ln 3(x-2)$, $\quad \ln \frac{x^2}{3} = \ln 3(x-2)$,

$\frac{x^2}{3} = 3(x-2)$, $\quad x^2 = 9(x-2)$, $\quad x^2 - 9x + 18 = 0$, $\quad (x-3)(x-6) = 0$, $\quad x = 3, 6$

57. $3 \log 3 - \frac{1}{2} \log 3 = \log \sqrt{x}$, $\quad \log 27 - \log \sqrt{3} = \log \sqrt{x}$, $\quad \frac{27}{\sqrt{3}} = \sqrt{x}$,

$x = \frac{27^2}{3} = \frac{729}{3} = 243$

59. $5 \ln \frac{e}{\sqrt[5]{5}} = 3 - \ln x$, $\quad \ln \left(\frac{e}{\sqrt[5]{5}}\right)^5 + \ln x = 3$, $\quad \ln \frac{e^5}{5}(x) = 3$, $\quad e^3 = \frac{e^5 x}{5}$,

$\frac{5e^3}{e^5} = x$, $\quad x = \frac{5}{e^2}$ OR $x = 5e^{-2}$

61. $\log_x (x+6) = 2$, $\quad x^2 = x + 6$, $\quad x^2 - x - 6 = 0$, $\quad (x-3)(x+2) = 0$, $\quad x \neq -2$, $\quad x = 3$

63. $t = -62.5 \ln \left(1 - \frac{N}{80}\right)$

a. $n = 30$, $t = -62.5 \ln \left(1 - \frac{30}{80}\right)$, $t = -62.5 \ln (1 - 0.375)$,

$t = -62.5 \ln (0.625)$, $t \approx 29.37$, approximately 29 days

b. $n = 80$, $t = -62.5 \ln \left(1 - \frac{80}{80}\right)$, $t = -62.5 \ln (0)$ so $t = 0$, the answer is no.

c. $t = -62.5 \ln \left(1 - \frac{N}{80}\right)$, $\frac{-t}{62.5} = \ln \left(1 - \frac{N}{80}\right)$, $e^{-t/62.5} = 1 - \frac{N}{80}$,

$\frac{N}{80} = 1 - e^{-t/62.5}$, $N = 80 \left(1 - e^{-t/62.5}\right)$

65. $M = \frac{\log E - 11.8}{1.5}$

a. $1.5 M = \log E - 11.8$, $1.5 M + 11.8 = \log E$, $E = 10^{(1.5M + 11.8)}$

b. Richter scale measure is 7.5 then $E = 10^{(1.5(7.5) + 11.8)} = 10^{23.05} = 1.122 \times 10^{23}$ ergs.

67. Prove the Change of Base Theorem:

$\log_b x = y$

$b^y = x$

$\log_a b^y = \log_a x$

$y \log_a b = \log_a x$

$y = \frac{\log_a x}{\log_a b}$

69. $3x^2 - nx + 27 = 0$; $3(x^2 - \frac{n}{3}x + 9) = 0$; and $(x-p)(x-q) = x^2 - \frac{n}{3}x + 9$,

$x^2 - (p+q)x + \underline{pq} = x^2 - \frac{n}{3} + \underline{9}$ so $pq = 9$; therefore: $\log_3 p + \log_3 q = \log_3 pq = \log_3 9 = 2$.

PROBLEM SET 5.4

1. $2^x = 128$, $2^x = 2^7$, $x = 7$ 3. $8^x = 32$, $2^{3x} = 2^5$, $3x = 5$, $x = \frac{5}{3}$

5. $125^x = 25$, $5^{3x} = 5^2$, $x = \frac{2}{3}$ 7. $128^x = 8$, $2^{7x} = 2^3$, $7x = 3$, $x = \frac{3}{7}$

9. $4^x = \frac{1}{16}$, $4^x = 2^{-4}$, $2^{2x} = 2^{-4}$, $x = -2$ 11. $(\frac{1}{2})^x = 8$, $2^{-x} = 2^3$, $x = -3$

13. $(\frac{2}{3})^x = \frac{9}{4}$, $(\frac{2}{3})^x = (\frac{2}{3})^{-2}$, $x = -2$ 15. $8^{3x} = 2^1$, $2^{9x} = 2^1$, $x = \frac{1}{9}$

17. $2^{(3x+1)} = \frac{1}{2}$, $2^{(3x+1)} = 2^{-1}$, $3x+1 = -1$, $3x = -2$, $x = -\frac{2}{3}$

19. $27^{(2x+1)} = 3$, $3^{(6x+3)} = 3^1$, $6x+3 = 1$, $6x = -2$, $x = -\frac{1}{3}$

21. $125^{(2x+1)} = 25$, $5^{(6x+3)} = 5^2$, $6x+3 = 2$, $6x = -1$, $x = -\frac{1}{6}$

23. $10^x = 42$, $\log 10^x = \log 42$, $x \log 10 = \log 42$, $x = \log 42$, $x \approx 1.62324929$

25. $10^x = 0.00325$, $\log 10^x = \log 0.00325$, $x = \log 0.00325$, $x \approx -2.48811664$

27. $10^{(x+3)} = 214$, $(x+3) \log 10 = \log 214$, $x + 3 = \log 214$,
     $x = \log 214 - 3$, $x \approx -0.66958623$

29. $10^{(x-1)} = 0.613$, $(x-1) \log 10 = \log 0.613$, $x-1 = \log 0.613$,
     $x = \log 0.613 + 1$, $x \approx 0.78746047$

31. $10^{(5-3x)} = 0.041$, $(5-3x) \log 10 = \log 0.041$, $5-3x = \log 0.041$,
     $-3x = -5 + \log 0.041$, $x = -\frac{\log 0.041 - 5}{3}$, $x \approx 2.12907205$

33. $e^{2x} = 10$, $2x \ln e = \ln 10$, $2x = \ln 10$, $x = \frac{\ln 10}{2}$, $x \approx 1.15129 \approx 1.15129255$

35. $e^{4x} = \frac{1}{10}$, $4x \ln e = \ln 10^{-1}$, $4x = -1 \ln 10$, $x = -\frac{\ln 10}{4}$,
     $x \approx -0.575646 \approx -0.57564627$

37. $e^{(1-2x)} = 3$, $1 - 2x = \ln 3$, $2x = 1 - \ln 3$, $x = \frac{1 - \ln 3}{2}$, $x \approx -0.049306144$

39. $8^x = 300$, $x \log 8 = \log 300$, $x = \frac{\log 300}{\log 8} \approx 2.7429396$

41. $5^x = 10$, $x \log 5 = \log 10$, $x = \frac{1}{\log 5}$, $x \approx 1.430676 \approx 1.4306766$

43. $2^{-x} = 5$, $-x \log 2 = \log 5$, $x = -\frac{\log 5}{\log 2}$, $x \approx -2.321928$

45. $5^{-x} = 8$, $-x \log 5 = \log 8$, $x = -\frac{\log 8}{\log 5}$, $x \approx -1.2920297$

47. $P = 1000$, $r = 0.12$ compounded semiannually, $A = 2000$;
     $2000 = 1000 \left(1 + \frac{0.12}{2}\right)^t$, $2 = (1.06)^t$, $\log 2 = t \log 1.06$, $t = \frac{\log 2}{\log 1.06}$,
     $t \approx 11.8966$ periods with 2 periods/year so $\frac{11.8966}{2} \approx 5.95 \approx 6$ years

49. $P = 1000$, $r = 0.12$ compounded quarterly, $A = 2500$;
     $2500 = 1000 \left(1 + \frac{0.12}{4}\right)^t$, $2.5 = (1.03)^t$, $\log 2.5 = t \log 1.03$, $t = \frac{\log 2.5}{\log 1.03}$,
     $t \approx 30.9989$ periods with 4 periods/year so $\frac{30.9989}{4} \approx 7.7497 \approx 7.75$ years

51. $A = A_0 e^{-kt}$, $\frac{1}{2} = e^{-30k}$, $\ln 0.5 = \ln e^{-30k}$, $\ln 0.5 = -30k$, $k = \frac{-0.6931}{-30}$, $k \approx 0.023104906$

53. $k = 0.0641$, $\frac{1}{2} = e^{-0.0641t}$, $\ln 0.5 = \ln e^{-0.064t}$, $\ln 0.5 = -0.0641t$, $t = \frac{-0.69315}{-0.0641}$, $t \approx 10.81352856 \approx 10.813529$

55. $P = 25\%$; $P = (\frac{1}{2})^{t/5700}$, $0.25 = (\frac{1}{2})^{t/5700}$, $\log 0.25 = \log (\frac{1}{2})^{t/5700}$, $\log 0.25 = \frac{t}{5700} \log 0.5$, $t = \frac{5700 \log 0.25}{\log 0.5}$, $t \approx 11{,}400$ years old

57. $P = (\frac{1}{2})^{t/5700}$, $P = 0.12$; $0.12 = (\frac{1}{2})^{t/5700}$, $\ln 0.12 = \frac{t}{5700} \ln 0.5$, $t = \frac{5700 \ln 0.12}{\ln 0.5}$, $t \approx 17435.69 \approx 17{,}400$ years old

59. $P = 14.7 e^{-0.21a}$, $P = 13.23$; $13.23 = 14.7 e^{-0.21a}$, $\frac{13.23}{14.7} = e^{-0.21a}$, $\ln 0.9 = \ln e^{-0.21a}$, $\ln 0.9 = -0.21a$, $a = \frac{\ln 0.9}{-0.21}$, $a \approx 0.501716741 \approx 0.5$ miles

61. a. $T = A + (B-A)e^{-kt}$; $B = 120°$, $A = 75°$, $t = 30$ minutes, $k = 0.01$; $T = 75° + (120° - 75°)e^{-0.01(30)}$, $T = 75° + 45°(e^{-.3})$, $T = 108°$

   b. $t = 60$ minutes, $T = 75°$, $A = 72°$, $B = 375°$, find k; $75° = 72° + (375° - 72°)e^{-60k}$, $3° = 303°(e^{-60k})$, $\frac{3°}{303°} = e^{-60k}$ now take the natural log of both side and you have $-4.615120517 = -60k$, $k = -0.08$

   c. $T = A + (B-A)e^{-kt}$, $T - A = (B-A)e^{-kt}$, $\frac{T-A}{B-A} = e^{-kt}$, $\ln(\frac{T-A}{B-A}) = \ln(e^{-kt})$, $\ln(\frac{T-A}{B-A}) = -kt$, $t = -\frac{1}{k}(\frac{T-A}{B-A})$

63. $10^{(5x+1)} = e^{(2-3x)}$; $\ln 10^{(5x+1)} = \ln e^{(2-3x)}$, $(5x+1)\ln 10 = 2 - 3x$, $5x \ln 10 + \ln 10 = 2 - 3x$, $5x \ln 10 + 3x = 2 - \ln 10$, $x(5 \ln 10 + 3) = 2 - \ln 10$, $x = \frac{2 - \ln 10}{5 \ln 10 + 3}$, $x = -0.0208494$

65. $e^x + e^{-x} = 10$; $e^x(e^x + e^{-x} = 10)$, $e^{x^2} + 1 = 10e^x$, $e^{x^2} - 10e^x + 1 = 0$; let $e^x = u$, then $u^2 - 10u + 1 = 0$, using the quadratic formula you get $u = 5 \pm 2\sqrt{6}$. This means that $e^x = 5 \pm 2\sqrt{6}$, $\ln e^x = \ln(5 \pm 2\sqrt{6})$ so $x = \ln(5 \pm 2\sqrt{6}) = \pm 2.2924317$

67. $3x^2 + mx + 4 = 0$, find $\log_2 p + \log_2 q$; $3x^2 + mx + 4 = 0$ and $3(x^2 + \frac{m}{3}x + \frac{4}{3}) = 0$; since p and q are roots you have: $(x-p)(x-q) = x^2 + \frac{m}{3}x + \frac{4}{3}$, $x^2 - (p+q)x + pq = x^2 + \frac{m}{3}x + \frac{4}{3}$, so $pq = \frac{4}{3}$ and $\log_2 p + \log_2 q = \log_2 pq = \log_2 \frac{4}{3}$

CHAPTER 5 SUMMARY PROBLEMS

1. $y = (\frac{1}{2})^x$

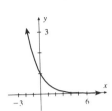

3. $y = 2^{-x}$

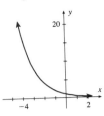

5. $\quad 10^{0.5} = \sqrt{10}$
    $.5 \log 10 = \log \sqrt{10}$
    $\quad 0.5 = \log \sqrt{10}$

7. $9^3 = 729, \ \log_9 729 = 3$

9. $\log 1 = 0, \ 10^0 = 1$

11. $\log_2 64 = 6, \quad 2^6 = 64$

13. $\log 3 \approx 0.4771212547$

15. $\log 0.0021 \approx -2.67780705$

17. $\ln 3 \approx 1.098612289$

19. $\ln 0.013 \approx -4.342805922$

21. $\log_4 15 = \dfrac{\ln 15}{\ln 4} = 1.953445298$

23. $\log_{1.06} 14650 = \dfrac{\ln 14650}{\ln 1.06} = 164.6194501$

25. $y = \log x$
    $10^y = x$

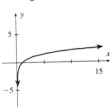

27. $y = \log_3 x$
    $3^y = x$

29. $\log_b A \cdot B = \log_b A + \log_b B$

31. $\log_b A^P = p \log_b A$

33. $\log_5 25 = x, \quad 5^x = 25,$
    $\quad 5^x = 5^2, \quad x = 2$

35. $3 \log 3 - \frac{1}{2} \log 3 = \log \sqrt{x}$
    $\log 27 - \log\sqrt{3} = \log \sqrt{x}$
    $\log \dfrac{27}{\sqrt{3}} = \log \sqrt{x},$
    $\dfrac{27}{\sqrt{x}} = \sqrt{x}$

37. $10^{(x + 2)} = 125, \ (x +2) \log 10 = \log 125, \ x + 2 = \log 125, \ x = \log 125 - 2$
    $x \approx 0.096910031$

39. $10^{-x^2} = 0.45, \ -x^2 = \log 0.45, \ -x^2 \approx -0.346787486, \quad x \approx \pm \ 0.5888866497$

41. $e^{3x} = 50, \quad 3x \ln e = \ln 50, \quad 3x = \ln 50, \quad x = \frac{\ln 50}{3}, \quad x \approx 1.304007668$

43. $e^{(1 - 2x)} = 690, \quad 1 - 2x = \ln 690, \quad -2x = \ln 690 - 1, \quad x = -\frac{\ln 690 - 1}{2}, \quad x \approx -2.768345799$

45. $5^x = 125$ can be done 3 ways: a. $5^x = 125, \quad 5^x = 5^3, \quad x = 3$
    b. $\log 5^x = \log 125, \quad x = \dfrac{\log 125}{\log 5} = 3$ \qquad c. $\ln 5^x = \ln 125, \quad x = \dfrac{\ln 125}{\ln 5} = 3$

47. $3^x = 7, \quad x \log 3 = \log 7, \quad x = \dfrac{\log 7}{\log 3}, \quad x \approx 1.771243749$

110

49. $A = A_0 e^{-kt}$, $t = 26.5$ hours; $\frac{1}{2} = e^{-k(26.5)}$, $\ln 0.5 = \ln e^{-26.5k}$, $-26.5k = \ln 0.5$,

$k = \frac{\ln 0.5}{-26.5} \approx 0.026156497$, $k \approx 0.026$

51. $A = P(1 + r)^n$, $P = 20{,}000$, $n = 15$, $r = 0.134$;

$A = 20{,}000 (1.134)^{15}$, $A \approx 20{,}000 (6.59471176)$, $A \approx \$131{,}894.24$

53. $P = 20{,}000$, $n = 15$, $r = 0.04$;

$A = 20{,}000 (1.04)^{15}$, $A \approx 20{,}000 (1.800943506)$, $A \approx \$36{,}018.87$

CUMULATIVE REVIEW I

1. a. $f(-2) = 3(-2)^2 + 2 = 12 + 2 = 14$ b. $g(t) = 2t - 1$

   c. $f(t + h) = 3(t + h)^2 + 2 = 3(t^2 + 2ht + h^2) + 2 = 3t^2 + 6ht + 3h^2 + 2$

   d. $f[g(x)] = f[2x - 1] = 3[2x - 1]^2 + 2 = 3[4x^2 - 4x + 1] + 2 =$

   $12x^2 - 12x + 3 + 2 = 12x^2 - 12x + 5$

   e. $(g \circ f)(x) = g[f(x)] = g[3x^2 + 2] = 2[3x^2 + 2] - 1 = 6x^2 + 4 - 1 = 6x^2 + 3$

   f. $\dfrac{f(x + h) - f(x)}{h} = \dfrac{[3(x + h)^2 + 2] - (3x^2 + 2)}{h} =$

   $\dfrac{[3(x^2 + 2hx + h^2) + 2] - 3x^2 - 2}{h} = \dfrac{3x^2 + 6hx + 3h^2 + 2 - 3x^2 - 2}{h} = \dfrac{6hx + 3h^2}{h} = 6x + 3h$

2. a. $f(x) = 3x^2 + 2$, not 1–1 so the inverse is not a function.

   b. $g(x) = 2x - 1$, f: $y = 2x - 1$, f^{-1}: $x = 2y - 1$ so $2y = x + 1$, $f^{-1}(x) = \dfrac{x + 1}{2}$

3. a. $y + 2 = -\frac{1}{2}(x + 3)$

   $y + 2 = -\frac{1}{2}x - \frac{3}{2}$

   $y = -\frac{1}{2}x - \frac{7}{2}$ or

   $x + 2y + 7 = 0$

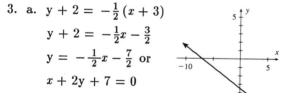

3. b. $y + 2 = -\frac{1}{2}(x + 3)^2$

   $V(-3, -2)$

   $\{(-3, -2), (-2, -\frac{5}{2}),$

   $(-4, -\frac{5}{2}), (-1, -4),$

   $(-5, -4), (0, -\frac{13}{2}),$

   $(-6, -\frac{13}{2})\}$

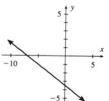

4. a. $F(-1) = 0$ b. $F(0) = 27$ c. $F(3) = -36$

$$\begin{array}{r|rrrr} -1 & 4 & -20 & 3 & 27 \\ & & -4 & 24 & -27 \\ \hline & 4 & -24 & 27 & 0 \end{array}$$

$$\begin{array}{r|rrrr} 3 & 4 & -20 & 3 & 27 \\ & & 12 & -24 & -63 \\ \hline & 4 & -8 & -21 & -36 \end{array}$$

d. $F(\frac{3}{2}) = 0$

$$\begin{array}{r|rrrr} 3/2 & 4 & -20 & 3 & 27 \\ & & 6 & -21 & -27 \\ \hline & 4 & -14 & -18 & 0 \end{array}$$

e. $(\frac{3}{2}, 0)$, then $4x^2 - 14x - 18 = 0$,

$2x^2 - 7x - 9 = (2x - 9)(x + 1) = 0$ so $x = \frac{9}{2}, -1$

Zeros: $(\frac{3}{2}, 0), (\frac{9}{2}, 0), (-1, 0)$

5. Graph: $F(x) = 4x^3 - 20x^2 + 3x + 27$

$\{(0, 27), (\frac{3}{2}, 0), (\frac{9}{2}, 0), (-1, 0), (1, 14),$

$(2, -15), (-2, -91), (3, -36),$

$(4, -25), (-\frac{1}{2}, 20)$

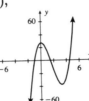

6. Given $3 + \sqrt{7}$ is a root, then $3 - \sqrt{7}$ must also be a root.

$$\begin{array}{r|ccccc} 3 + \sqrt{7} & 4 & -20 & -19 & 26 & -6 \\ & & 12 + 4\sqrt{7} & 4 + 4\sqrt{7} & -17 - 3\sqrt{7} & 6 \\ \hline & 4 & -8 + 4\sqrt{7} & -15 + 4\sqrt{7} & 9 - 3\sqrt{7} & 0 \end{array}$$

$$\begin{array}{r|cccc} 3 - \sqrt{7} & 4 & -8 + 4\sqrt{7} & -15 + 4\sqrt{7} & 9 - 3\sqrt{7} \\ & & 12 - 4\sqrt{7} & 12 - 4\sqrt{7} & -9 + 3\sqrt{7} \\ \hline & 4 & 4 & -3 & 0 \end{array}$$

$4x^2 + 4x - 3 = 0$

$(2x - 1)(2x + 3) = 0$

$x = -\frac{3}{2}, \frac{1}{2}$

$\{-\frac{3}{2}, \frac{1}{2}, 3 \pm \sqrt{7}\}$

7. $g = \dfrac{(2x - 1)(x - 1)}{(x - 2)(x + 1)}$;   a. D: $(-\infty, -1) \cup (-1, 2) \cup (2, \infty)$ since $x \neq -1$ or 2

b. $(-x, y)$, $y = \dfrac{2x^2 + 3x + 1}{x^2 + x - 2}$ none; $(x, -y)$, $y = -\dfrac{2x^2 - 3x + 1}{x^2 - x - 2}$ none;

$(-x, -y)$, $y = -\dfrac{2x^2 + 3x + 1}{x^2 + x - 2}$ none;   no symmetry

c. $x \neq -1$ or 2 so   Vertical: $x = -1, 2$;   Horizontal: $y = 2$ since $\dfrac{p}{q} = \dfrac{2}{1}$

d. $y = 0$ then $2x^2 - 3x + 1 = (2x - 1)(x - 1) = 0$, $x = \frac{1}{2}, 1$; $(\frac{1}{2}, 0)$ and $(1, 0)$

$x = 0$ then $y = -\frac{1}{2}$; $(0, -\frac{1}{2})$

e. Graph: $y = \dfrac{(2x-1)(x-1)}{(x-2)(x+1)}$

Use the information from parts
a, b, c, and d and the following
plotting points: $(2, \frac{5}{2})$, $(3, \frac{5}{2})$,
$(\frac{3}{2}, -\frac{4}{5})$

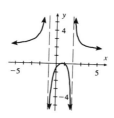

8.a. $y = -\dfrac{10}{x}$, $x \neq 0$

Vert.: $x = 0$
Horiz.: $y = 0$

b. $y = \dfrac{-3}{x+4}$, $x \neq -4$

Vert.: $x = -4$
Horiz.: $y = -3$

c. $y = \dfrac{x^2}{x(x-3)}$, $x \neq 0, 3$

Vert.: $x = 0, 3$
Horiz.: $y = 1$

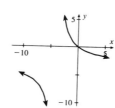

9. a. $8^x = 32$, $2^{3x} = 2^5$, $x = \frac{5}{3}$ or $\log 8^x = \log 32$, $\log 2^{3x} = \log 2^5$, $x = \frac{5}{3}$
   b. $\log_3 x = 4.12$, $3^{4.12} = x$, $x \approx 92.4$

   c. $\log_8 x = \log_8 (4 - 3x)$, $x = 4 - 3x$, $4x = 4$, $x = 1$

   d. $2 \log 3 = \frac{1}{2} \log x$, $\log 3^2 = \log \sqrt{x}$, $9 = \sqrt{x}$, $9^2 = x$, $x = 81$

10. $A = A_0 \, 2^{-kt}$

   a. $\dfrac{A}{A_0} = \dfrac{1}{2}$, $t = 5700$, $\dfrac{1}{2} = 2^{-5700k}$, $\log 0.5 = -5700k \log 2$,

   $k = \dfrac{\log 0.5}{-5700 \log 2} = \dfrac{1}{5700}$

   b. 72.4% of original carbon, $\dfrac{A}{A_0} = 0.724$, $k = \dfrac{1}{5700}$, $0.724 = 2^{(-1/5700)t}$,

   $\ln 0.724 = -\dfrac{1}{5700}t \ln 2$, $t = \dfrac{-5700 \ln 0.724}{\ln 2} \approx 2655.848866 \approx 2700$ years

113

c. $A = A_0 \, 2^{-kt}$, $\quad \dfrac{A}{A_0} = 2^{-kt}$, $\quad \log_2 \dfrac{A}{A_0} = \log_2 2^{-kt}$, $\quad \log_2 \dfrac{A}{A_0} = -kt$,

$t = -\dfrac{1}{k} \log_2 \dfrac{A}{A_0}$

EXTENDED APPLICATION: POPULATION GROWTH

1. $A = 2$ billion, $\quad A_0 = 1$ billion, $\quad t = 80$; $\quad P = P_0 \, e^{rt}$; $\quad 2 = 1 \, e^{80r}$, $\quad \dfrac{1}{2} = e^{80r}$,

$\ln 0.5 = 80r \ln e$, $\quad \ln 0.5 = 80r$, $\quad r = \dfrac{\ln 0.5}{80}$, $\quad r \approx 0.00866434$, $\quad r = .866\%$

3. $A = 4$ billion, $\quad A_0 = 3$ billion, $\quad t = 15$; $\quad 4 = 3 \, e^{15r}$, $\quad \dfrac{4}{3} = e^{15r}$, $\quad \ln \dfrac{4}{3} = 15r \ln e$,

$\ln \dfrac{4}{3} = 15r$, $\quad r = \dfrac{\ln \frac{4}{3}}{15}$, $\quad r \approx 0.019178805$, $\quad r = 1.918\%$

5. $A = 6$ billion, $\quad t = ?$, $\quad A_0 = 5$ billion, $\quad r = 2.2\%$; $\quad 6 = 5 \, e^{0.22t}$, $\quad \dfrac{6}{5} = e^{0.22t}$,

$\ln \dfrac{6}{5} = 0.22t \ln e$, $\quad t = \dfrac{\ln \frac{6}{5}}{0.22}$, $\quad t \approx 8.287343491$, $\quad 8 \text{ years} + 0.287343491(365)$

$= 8 \text{ years} + 104.8 \text{ days}$ or $8 \text{ years} + 105 \text{ days}$ (also add 2 leap year days),

The population reaches 6 billion on October 18, 1994.

7. $A = 8$ billion, $\quad t = ?$, $\quad A_0 = 5$, $\quad r = 2.2\%$; $\quad 8 = 5 \, e^{0.22t}$, $\quad \dfrac{8}{5} = e^{0.22t}$, $\quad \ln \dfrac{8}{5} = 0.22t$,

$t = \dfrac{\ln \frac{8}{5}}{0.22}$, $\quad t \approx 21.36380133$, $\quad 21 \text{ years} + 133 \text{ days and } 5 \text{ leap year days}$,

The population reaches 8 billion on November 12, 2007

9. a. $A = 2,767,000$, $\quad A_0 = 2,914,000$, $\quad t = 10$, $\quad 2,767,000 = 2,914,000 \, e^{10r}$, $\quad \dfrac{2,767,000}{2,914,000} = e^{10r}$,

$\ln \dfrac{2,767,000}{2,914,000} = 10r$, $\quad r \approx -0.051763007 \approx -0.05$, $\quad -.5\%$ so a decline

b. $A_0 = 2,767,000$, $\quad t = 10$, $\quad r \approx -0.051763007$; $\quad A = 2,767,000 \, e^{(-0.05176)10}$

$A \approx 2,767,000 \, e^{-0.05176}$, $\quad A \approx 2,767,000 \, (0.9495567331) \approx 2,627,423$

11. 1850, 1 billion;
1930, 2 billion;
1961, 3 billion;
1975, 4 billion;
1995, 10.0 bil.
1990, 5.3 billion;
2007, 8 billion

13. answers will vary
15. answers will vary

CHAPTER 6

PROBLEM SET 6.1

1. a. $\frac{\pi}{6}$ b. $\frac{\pi}{2}$ c. $\frac{3\pi}{2}$ d. $\frac{\pi}{4}$ 3. a. 180°   b. 45°   c. 60°   d. 360°

5. a.   b.   c.   d.

7. a.   b.   c.   d.

9. a.   b.   c.   d.

11. a. $400° - 360° = 40°$ b. $540° - 360° = 180°$
   c. $750° - 720° = 30°$ d. $1050° - 720° = 330°$

13. a. $-120° + 360° = 240°$ b. $500° - 360° = 140°$
   c. $-180° + 360° = 180°$ d. $1000° - 720° = 280°$

15. a. $-\frac{\pi}{4} + 2\pi = \frac{7\pi}{4}$ b. $\frac{17\pi}{4} - 4\pi = \frac{\pi}{4}$ c. $\frac{11\pi}{3} - 2\pi = \frac{5\pi}{3}$ d. $-2 + 2\pi$ or $2\pi - 2$

17. a. $9 - 6.2832 = 2.7168$ b. $-5 + 6.2832 = 1.2832$
   c. $\sqrt{50} - 6.2832 = 0.7879$ d. $-6 + 6.2832 = 0.2832$

19. a. $6.8068 - 6.2832 = 0.5236$

    b. $-0.7854 + 6.2832 = 5.4978$

    c. $150 - 46\pi = 150 - 144.5133 = 5.4867$

    d. $9.4247 - 6.2832 = 3.1415$

21. a. $60°$ b. $180° - 120° = 60°$ c. $360° - 300° = 60°$ d. $180° - 135° = 45°$

23. a. $\pi - \frac{11\pi}{12} = \frac{\pi}{12}$ b. $\pi - \frac{2\pi}{3} = \frac{\pi}{3}$ c. $2\pi - \frac{11\pi}{6} = \frac{\pi}{6}$ d. $\frac{\pi}{4}$

25. a. $\left|-\frac{\pi}{4}\right| = \frac{\pi}{4}$

    b. 0 (terminal side of angle on x - axis)

    c. $\left|-\frac{13\pi}{6}\right| - 2\pi = \frac{\pi}{6}$

    d. $-\frac{5\pi}{3} + 2\pi = \frac{\pi}{3}$

27. a. $\sqrt{50} - 2\pi \approx 0.7879$

    b. $3\sqrt{5} - 2\pi \approx 0.4250$

    c. $6.8068 - 6.2832 = 0.5236$

    d. $|-0.7854| = 0.7854$

29. $\left(\frac{\pi}{10}\right)\left(\frac{180°}{\pi}\right) = 18.00°$ 31. $\left(\frac{5\pi}{3}\right)\left(\frac{180°}{\pi}\right) = 300.00°$ 33. $\left(\frac{3\pi}{18}\right)\left(\frac{180°}{\pi}\right) = 30.00°$

35. $(-3)\left(\frac{180°}{\pi}\right) \approx -171.89°$ 37. $(-2.5)\left(\frac{180°}{\pi}\right) \approx -143.24°$ 39. $(0.51)\left(\frac{180°}{\pi}\right) \approx 29.22°$

41. $(20°)\left(\frac{\pi}{180°}\right) = \frac{\pi}{9}$ 43. $(-220°)\left(\frac{\pi}{180°}\right) = -\frac{11\pi}{9}$ 45. $(85°)\left(\frac{\pi}{180°}\right) = \frac{17\pi}{36}$

47. $(314°)\left(\frac{\pi}{180°}\right) \approx 5.48$ 49. $(350°)\left(\frac{\pi}{180°}\right) \approx 6.11$ 51. $(985°)\left(\frac{\pi}{180°}\right) \approx 17.19$

53. $s = (6 \text{ cm})(2.34) = 14.04 \text{ cm}$ 55. $s = (4 \text{ m})\left(\frac{\pi}{3}\right) \approx 4.19 \text{ m}$ 57. $s = (7 \text{ ft})(40°)\left(\frac{\pi}{180°}\right) \approx 4.89 \text{ ft}$

59. $s = (7.2 \text{ cm})(112°)\left(\frac{\pi}{180°}\right) \approx 14.07 \text{ cm}$ 61. $s = (50 \text{ cm})(100°)\left(\frac{\pi}{180°}\right) \approx 87.27 \text{ cm}$

63. $\theta = 41° - 37° = 4°$, $r = 6370 \text{ km}$, $s = (6370 \text{ km})(4°)\left(\frac{\pi}{180°}\right) \approx 440 \text{ km}$

65. $r = 384{,}417 \text{ km} + 6370 \text{ km} = 390{,}787 \text{ km}$;

    $s = (390{,}787 \text{ km})(45.75\prime)\left(\frac{1°}{60\prime}\right)\left(\frac{\pi}{180°}\right) \approx 5{,}200 \text{ km}$

PROBLEM SET 6.2

1. $\cos 50° = $ x - component on unit circle at $50° \approx 0.6$ (see figure 6.8)

3. $\sin 320° = $ y - component on unit circle at $320° \approx -0.6$ (see figure 6.8)

5. $\tan(-20°) = b/a$ where (a,b) is the point on the unit circle at $-20°$;

    $a \approx 0.9$, $b \approx -0.35$, $\tan(-20°) = b/a \approx -0.35/0.9 \approx -0.4$ (see figure 6.8)

7. $\sec 70° = 1/a$ where (a,b) is the point on the unit circle at $70°$;

    $a \approx 0.35$, $\sec 70° = 1/a \approx 1/0.35 \approx 2.9$ (see figure 6.8)

9. a. $\cos 50° \approx 0.64278761$ b. $\sin 20° \approx 0.34202014$

11. a. $\cot 250° = 1/\tan 250° \approx 0.36397023$ b. $\sec 135° = 1/\cos 135° \approx -1.4142136$

13. a. $-\cot(-18°) = -1/\tan(-18°) \approx 3.07768354$

    b. $-\sec(-213°) = -1/\cos(-213°) \approx 1.19236329$

15. a. $\sin 1 \approx 0.84147098$ (radians mode)   b. $\sin 1° \approx 0.0174524$ (degrees mode)

17. a. $\cot 2.5 = 1/\tan 2.5 \approx -1.3386481$

    b. $-\sec 1.5 = -1/\cos 1.5 \approx -14.136833$ (radians mode)

19. a. $129°09'12'' = (129 + 9/60 + 12/3600)° \approx 129.1533333°$

    $\tan 129°09'12'' \approx \tan 129.1533333° \approx -1.2281621$

    b. $240°08' = (240 + 8/3600)° \approx 240.00222222°$, $\cos 240.00222222° \approx -0.49996641$

21. a. $\frac{3}{5} \sin \frac{1}{2} \approx 0.2876553$ b. $\frac{3 \sin 2}{5} \approx 0.5455785$ (radian mode)

23. a. $2 \sec \frac{1}{3} = \dfrac{2}{\cos \frac{1}{3}} \approx 2.1164985$ b. $\dfrac{2}{\sec \frac{1}{3}} = 2 \cos \frac{1}{3} \approx 1.8899139$

25. a. $3 \cos\left(-\frac{1}{2}\right) \approx 2.6327477$ b. $\dfrac{-3}{\cos \frac{1}{2}} \approx -3.4184818$

27. a. $3 \cot\left(-\frac{1}{2}\right) = \dfrac{3}{\tan\left(-\frac{1}{2}\right)} \approx -5.4914632$ b. $\dfrac{-3}{\cot \frac{1}{2}} = -3 \tan \frac{1}{2} \approx -1.6389075$

29. cosine (x - component), Quad I $-$ positive

31. secant (r/x), Quad II $-$ negative (r is always positive)

33. cotangent (x/y), Quad IV $-$ negative (pos/neg)

35. $\cos 2$ (Quad II) $-$ negative   37. $\sec 4$ (Quad III) $-$ negative

39. $\cos(-2)$ (Quad III) $-$ negative   41. $\cos \theta < 0$, $(x < 0)$ $-$ Quad II and Quad III

43. $\sin \theta < 0$ (y < 0), $\tan \theta > 0$ (y/$x > 0$), so $x < 0$ $-$ Quad III

45. $\cos \theta < 0$ ($x < 0$), $\sin \theta < 0$ (y < 0) $-$ Quad III

47.

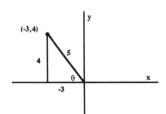

$r = \sqrt{9 + 16} = 5$

$\cos \theta = x/r = -3/5$ $\sec = r/x = -5/3$

$\sin \theta = y/r = 4/5$ $\csc \theta = r/y = 5/4$

$\tan \theta = y/x = -4/3$ $\cot \theta = x/y = -3/4$

49.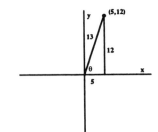

$r = \sqrt{25 + 144} = 13$

$\cos\theta = x/r = 5/13$ $\sec\theta = r/x = 13/5$

$\sin\theta = y/r = 12/13$ $\csc\theta = r/y = 13/12$

$\tan\theta = y/x = 12/5$ $\cot\theta = x/y = 5/12$

51.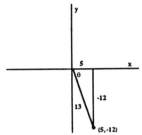

$r = \sqrt{25 + 144} = 13$

$\cos\theta = x/r = 5/13$ $\sec\theta = r/x = 13/5$

$\sin\theta = y/r = -12/13$ $\csc\theta = r/y = -13/12$

$\tan\theta = y/x = -12/5$ $\cot\theta = x/y = -5/12$

53.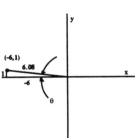

$r = \sqrt{36 + 1} = \sqrt{37}$

$\cos\theta = x/r = -6/\sqrt{37}$ $\sec\theta = r/x = -\sqrt{37}/6$

$\sin\theta = y/r = 1/\sqrt{37}$ $\csc\theta = r/y = \sqrt{37}$

$\tan\theta = y/x = -1/6$ $\cot\theta = x/y = -6$

55. $\sqrt{2x^2} = \sqrt{2}\,\sqrt{x^2} = \sqrt{2}\,|x|$ **57.** $\sqrt{2x^2} = |x|\sqrt{2} = -x\sqrt{2}$ $(x < 0$ implies $|x| = -x)$

59. $\sqrt{9x^2} = \sqrt{9}\,\sqrt{x^2} = 3\,|x| = 3x$ $(x > 0$ implies $|x| = x)$

61. $\dfrac{x}{\sqrt{4x^2}} = \dfrac{x}{2|x|} = \dfrac{x}{2x} = \dfrac{1}{2}$ $(x > 0$ implies $|x| = x)$

63. $\dfrac{\sqrt{16x^2}}{x} = \dfrac{4|x|}{x} = \dfrac{4(-x)}{x} = -4$ $(x < 0$ implies $|x| = -x)$

65. a. $\cos\theta = \dfrac{x}{r}$, $r\cos\theta = x$; $\sin\theta = \dfrac{y}{r}$, $r\sin\theta = y$; by substitution:

$P(x, y) = P(r\cos\theta, r\sin\theta)$

b. Let $A(\cos\theta, \sin\theta)$ and $B(\cos\beta, \sin\beta)$ be any two points on a unit circle.

$|AB| = \sqrt{(\cos\beta - \cos\alpha)^2 + (\sin\beta - \sin\alpha)^2}$

$ = \sqrt{\cos^2\beta - 2\cos\beta\cos\alpha + \cos^2\alpha + \sin^2\beta - 2\sin\beta\sin\alpha + \sin^2\alpha}$

$ = \sqrt{(\sin^2\beta + \cos^2\beta) + (\sin^2\alpha + \cos^2\alpha) - 2(\sin\beta\sin\alpha + \cos\beta\cos\alpha)}$

$ = \sqrt{2 - 2(\sin\beta\sin\alpha + \cos\beta\cos\alpha)}$

67. $\cos x = 1 - \dfrac{x^2}{2!} + \dfrac{x^4}{4!} - \dfrac{x^6}{6!} + \dfrac{x^8}{8!} - \cdots$

$\cos 1 = 1 - \dfrac{1}{2 \cdot 1} + \dfrac{1}{4 \cdot 3 \cdot 2 \cdot 1} - \dfrac{1}{6 \cdot 5 \cdot 4 \cdot 3 \cdot 2 \cdot 1} + \dfrac{1}{8 \cdot 7 \cdot 6 \cdot 5 \cdot 4 \cdot 3 \cdot 2 \cdot 1}$

$= 1 - \dfrac{1}{2} + \dfrac{1}{24} - \dfrac{1}{720} + \dfrac{1}{403202} = 0.5403$

PROBLEM SET 6.3

1 a. 1   b. 1   c. $\dfrac{\sqrt{3}}{2}$ d. $\dfrac{\sqrt{3}}{2}$ 3. a. 0   b. 1   c. 0   d. $\dfrac{\sqrt{2}}{2}$

5. a. $\sqrt{2}$ b. undefined   c. 2   d. $\sqrt{3}$ 7. a. 1   b. -1 c. -1 d. 0

9. a. $\cos(-300°) = \cos 60° = \dfrac{1}{2}$ b. $\sin 390° = \sin 30° = \dfrac{1}{2}$

   c. $\sin \dfrac{17\pi}{4} = \sin \dfrac{\pi}{4} = \dfrac{\sqrt{2}}{2}$ d. $\cos(-6\pi) = \cos 0 = 1$

11 a. 0.5650   b. 0.5850   13. a. 0.6428   b. 0.1763   15. a. -0.3640 b. -0.1736

17. a. 0.7337   b. -0.4791 19. a. 0.9320   b. 0.7961   21. a. 1.5574   b. 0.0709

23. a. -0.7470 b. 0.1411   25. a. 0.9004   b. 0.3555   27. see **EXAMPLE 3** in text

29. $\cos \dfrac{5\pi}{4} = \dfrac{x}{r} = \dfrac{x}{\sqrt{x^2 + y^2}} = \dfrac{x}{\sqrt{x^2 + x^2}}$ ($x = y$ since $\dfrac{5\pi}{4}$ bisects Quad III)

   $\cos \dfrac{5\pi}{4} = \dfrac{x}{\sqrt{2x^2}} = \dfrac{x}{|x|\sqrt{2}} = \dfrac{x}{-x\sqrt{2}} = (|x| = -x \text{ in Quad III}) = \dfrac{1}{-\sqrt{2}} = -\dfrac{1}{\sqrt{2}} \cdot \dfrac{\sqrt{2}}{\sqrt{2}} = -\dfrac{\sqrt{2}}{2}$

31. Choose point $(a, -a)$. $r = \sqrt{a^2 + a^2} = a\sqrt{2}$, $\sin\left(-\dfrac{\pi}{4}\right) = \dfrac{y}{r} = \dfrac{-a}{a\sqrt{2}} = \dfrac{-1}{\sqrt{2}} = -\dfrac{\sqrt{2}}{2}$

33. Choose point $(-a\sqrt{3}, -a)$ where a is any positive number. Then $r = 2a$ and

   $\cos 210° = \dfrac{x}{a} = \dfrac{-a\sqrt{3}}{2a} = -\dfrac{\sqrt{3}}{2}$ 35. $\sin \dfrac{\pi}{2} + 3 \cos \dfrac{\pi}{2} = 1 + 3(0) = 1$

37. $\cos \dfrac{2\pi}{2} = \cos \pi = -1$ 39. $2 \sin \dfrac{\pi}{4} = 2\left(\dfrac{\sqrt{2}}{2}\right) = \sqrt{2}$

41. $\cos^2 \dfrac{\pi}{4} = \left(\cos \dfrac{\pi}{4}\right)^2 = \left(\dfrac{\sqrt{2}}{2}\right)^2 = \dfrac{2}{4} = \dfrac{1}{2}$

43. $\sin^2 \dfrac{\pi}{2} + \cos^2 \dfrac{\pi}{2} = \left(\sin \dfrac{\pi}{2}\right)^2 + \left(\cos \dfrac{\pi}{2}\right)^2 = 1^2 + 0^2 = 1$

45. $\sin^2 \dfrac{\pi}{6} + \cos^2 \dfrac{\pi}{3} = \left(\sin \dfrac{\pi}{6}\right)^2 + \left(\cos \dfrac{\pi}{3}\right)^2 = \left(\dfrac{1}{2}\right)^2 + \left(\dfrac{1}{2}\right)^2 = \dfrac{1}{4} + \dfrac{1}{4} = \dfrac{1}{2}$

47. $\csc \dfrac{\pi}{2} \sin \dfrac{\pi}{2} = 1 \cdot 1 = 1$ 49. $\cos \dfrac{\pi}{4} - \cos \dfrac{\pi}{2} = \dfrac{\sqrt{2}}{2} - 0 = \dfrac{\sqrt{2}}{2}$

51. $2 \tan 30° = 2 \cdot \dfrac{\sqrt{3}}{3} = \dfrac{2\sqrt{3}}{3}$ 53. $\dfrac{\csc 60°}{2} = \dfrac{2\sqrt{3}/3}{2} = \dfrac{2\sqrt{3}}{3} \cdot \dfrac{1}{2} = \dfrac{\sqrt{3}}{3}$

55. $\sqrt{\dfrac{1 + \cos 60°}{2}} = \sqrt{\dfrac{1 + \frac{1}{2}}{2}} = \sqrt{\dfrac{\frac{3}{2}}{2}} = \sqrt{\dfrac{3}{4}} = \dfrac{\sqrt{3}}{2}$

57. $\dfrac{2 \tan 60°}{1 - \tan^2 60°} = \dfrac{2 \cdot \sqrt{3}}{1 - \left(\sqrt{3}\right)^2} = \dfrac{2\sqrt{3}}{1 - 3} = \dfrac{2\sqrt{3}}{-2} = -\sqrt{3}$

59. $\cos \dfrac{\pi}{2} \cos \dfrac{\pi}{6} + \sin \dfrac{\pi}{2} \sin \dfrac{\pi}{6} = 0 \cdot \dfrac{\sqrt{3}}{2} + 1 \cdot \dfrac{1}{2} = \dfrac{1}{2}$

61. a. Sine is negative in Quad III and positive in Quad I. By the reduction principle
   $\sin(\theta + \pi) = \pm \sin \theta$ since they share the same reference angle. Therefore
   $\sin(\theta + \pi) = - \sin \theta$.

   b. $\theta + \pi$ is in Quad IV. Sine is negative in Quad IV and positive in Quad II.

   c. $\theta + \pi$ is in Quad I. Sine is positive in Quad I and negative in Quad III.

   d. $\theta + \pi$ is in Quad II. Sine is positive in Quad II and negative in Quad IV.

   e. $\sin(\theta + \pi) = - \sin \theta$ whenever θ lies in one of the four quadrants. It is easily
   verified that $\sin(\theta + \pi) = - \sin \theta$ if θ lies on an axis. Therefore, no matter where
   θ lies $\sin(\theta + \pi) = - \sin \theta$.

63. Answers vary

PROBLEM SET 6.4

1.

$x =$ angle	$\dfrac{2\pi}{3}$	$\dfrac{3\pi}{4}$	$\dfrac{5\pi}{6}$	$\dfrac{7\pi}{6}$	$\dfrac{5\pi}{4}$	$\dfrac{4\pi}{3}$	$\dfrac{7\pi}{4}$	$\dfrac{11\pi}{6}$
Quadrant	II;$-$	II;$-$	II;$-$	III;$-$	III;$-$	III;$-$	IV;$+$	IV;$+$
$y = \cos x$	$-\dfrac{1}{2}$	$-\dfrac{\sqrt{2}}{2}$	$-\dfrac{\sqrt{3}}{2}$	$-\dfrac{\sqrt{3}}{2}$	$-\dfrac{\sqrt{2}}{2}$	$-\dfrac{1}{2}$	$\dfrac{\sqrt{2}}{2}$	$\dfrac{\sqrt{3}}{2}$
y (approx)	$-.50$	$-.71$	$-.87$	$-.87$	$-.71$	$-.50$	$.71$	$.87$

3. See graph in text

5. $y = \sin(x + \pi)$

7. $y = \tan(x + \dfrac{\pi}{3})$

9. $y = 2 \cos x$

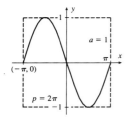

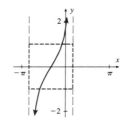

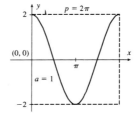

11. $y = \sin 3x$

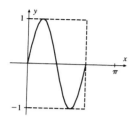

13. $y = \cos \frac{1}{2}x$

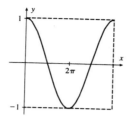

15. $y = \tan(x + \frac{\pi}{6})$

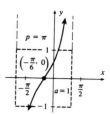

17. $y = \frac{1}{3} \tan x$

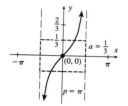

19. $y - 2 = \sin(x - \frac{\pi}{2})$

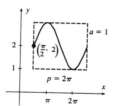

21. $y - 3 = \tan(x + \frac{\pi}{6})$

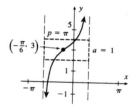

23. $y - 1 = 2 \sin x$

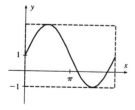

25. $y - 1 = 2 \cos(x - \frac{\pi}{4})$

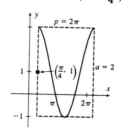

27. $y + 2 = 3 \sin(x + \frac{\pi}{6})$

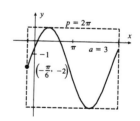

121

29. $y = 1 + \tan 2(x - \frac{\pi}{4})$

31. $y = \sin(4x + \pi)$

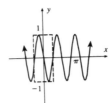

33. $y = \tan(2x - \frac{\pi}{2})$

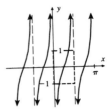

35. $y = \frac{1}{2} \cos(x + \frac{\pi}{6})$

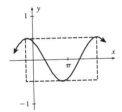

37. $y = 3 \cos(3x + 2\pi) - 2$

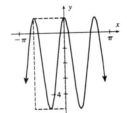

39. $y = \sqrt{2} \cos(x - \sqrt{2}) - 2$

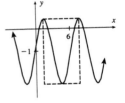

41. $y = 2 \sin 2\pi x$

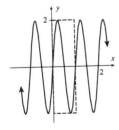

43. $y = 4 \tan \frac{\pi x}{5}$

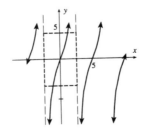

45. $y = |\sin x|$

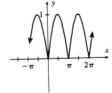

47. $y = \sin |x|$

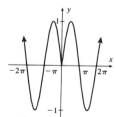

49. $y = -\sin x$

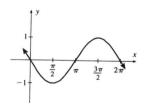

51. $y = -\tan x$

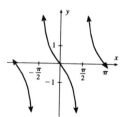

53. $y = -2\cos x$

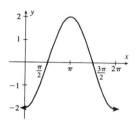

55. $y = 2\sec x$

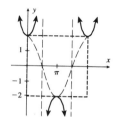

57. $y = 2\cos x - 1$

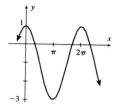

59. $y = \sin 2x + \cos x$

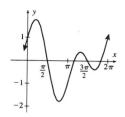

61. a. $y = x$,   b. $y = -x$,

c. $y = \dfrac{x}{\sin x}$

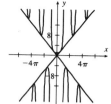

63. a. $y = x^2$,   b. $y = -x^2$

c. $y = x^2 \sin x$

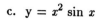
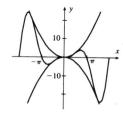

123

65. $I = 60 \cos(120\pi t - \pi)$,
    $0 \le t \le \frac{1}{30}$

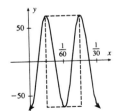

67. a. $y = 3,000 \cos(\frac{\pi}{60} t + \frac{\pi}{5})$, $0 \le t \le 120$;
    b. Maximum is at the amplitude of 3000km.
    c. 1 orbit is 120 minutes.

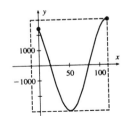

PROBLEM SET 6.5

1. a. 0   b. $\frac{\pi}{6}$ c. $\frac{\pi}{6}$ d. 0   3. a. $\frac{\pi}{3}$ b. $\frac{\pi}{4}$ c. $\frac{\pi}{4}$ d. $\frac{\pi}{4}$

5. a. $\frac{5\pi}{6}$ b. $-\frac{\pi}{4}$ c. $-\frac{\pi}{4}$ d. $\frac{2\pi}{3}$ 7. 0   b. $\frac{\pi}{3}$ c. $\frac{\pi}{3}$ d. $\frac{\pi}{3}$

9. 0.58   11. 1.65   13. 2.81   15. 1.32   17. 0.98   19. 1.34   21. 69°   23. $-28°$

25. 46°   27. 85°   29. 135°   31. 153°   33. $-72°$ 35. True statement

37. Not true. Graph is between $0 \to -\infty$ so in Quad II.   39. True statement.

41. The domain of the $\text{Cot}^{-1} x$ is different from the domain of the $\text{Tan}^{-1} x$ when $x < 0$ so this
    statement is true only for $x > 0$.

43. The reciprocal of the cosecant is the sine not the cosine. So this statement is not true.

45. $\frac{\pi}{6}$ 47. $\frac{\pi}{15}$ 49. $\frac{2\pi}{15}$ 51. 0.4163   53. 1.28   55. 0.2835   57. -0.64

59. $\frac{1}{5}\sqrt{21} \approx 0.917$ 61. $y = \text{Cos } x$; $y = \text{Cos}^{-1}x$ 63. $y + 2 = \text{Arctan } x$

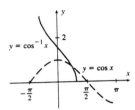

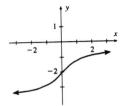

65. $y = 2 \cos^{-1} x$

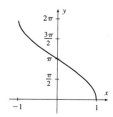

67. $y = \text{Arcsin}(x - 2)$

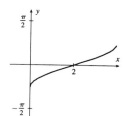

CHAPTER 6 SUMMARY PROBLEMS

1. a. lambda   b. theta   c. phi   d. alpha   e. beta

3. $x^2 + y^2 = 1$ 5. $-215° + 360° = 145°$ 7. $-\frac{5\pi}{6} + 2\pi = \frac{7\pi}{6}$

9. $180°$ 11. $-30°$ 13. $\frac{\pi}{3}$ 15. $\frac{5\pi}{6}$

17. $\left(\frac{3\pi}{2}\right)\left(\frac{180°}{\pi}\right) = 270°$ 19. $\left(-\frac{7\pi}{4}\right)\left(\frac{180°}{\pi}\right) = -315°$

21. $(300°)\left(\frac{\pi}{180°}\right) = \frac{5\pi}{3}$ 23. $(54°)\left(\frac{\pi}{180°}\right) = \frac{3\pi}{10}$ 25. $s = r\theta$, radius, angle in radians

27. $r = 15$ cm, $\theta = \left(\frac{10}{60}\right)(2\pi) = \frac{\pi}{3}$, $s = r\theta = (15 \text{ cm})\left(\frac{\pi}{3}\right) = 5\pi$ cm. ≈ 15.7 cm.

29. $360° - 300° = 60°$ 31. $|-4| - \pi = 4 - \pi$ 33. an angle in standard position

35. $\sin \theta = b$, $\csc \theta = \frac{1}{2}$ $(b \neq 0)$ 37. all   39. tangent and cotangent

41. $\tan \theta = \frac{\sin \theta}{\cos \theta} = \frac{4/5}{3/5} = \frac{4}{3}$ 43. secant   45. 1.0896   47. 1.4587

49. θ is an angle in standard position with the point (x,y) on its terminal side and
    $r = \sqrt{x^2 + y^2}$ (distance to the origin)

51. $\sin \theta = y/r$; $\csc \theta = r/y$ $(y \neq 0)$

53. $r = \sqrt{25 + 144} = 13$; $\cos \theta = \frac{5}{13}$, $\sin \theta = -\frac{12}{13}$, $\tan \theta = -\frac{12}{5}$, $\sec \theta = \frac{13}{5}$,
    $\csc \theta = -\frac{13}{12}$, $\cot \theta = -\frac{5}{12}$

55. $r = \sqrt{25 + 4} = \sqrt{29}$, $\cos \theta = -\frac{5}{\sqrt{29}} = -\frac{5}{\sqrt{29}} \cdot \frac{\sqrt{29}}{\sqrt{29}} = -\frac{5}{29}\sqrt{29}$

    $\sin \theta = \frac{2}{\sqrt{29}} = \frac{2}{29}\sqrt{29}$, $\tan \theta = -\frac{2}{5}$, $\sec \theta = -\frac{\sqrt{29}}{5}$, $\csc \theta = \frac{\sqrt{29}}{2}$, $\cot \theta = -\frac{5}{2}$

57. a. 1   b. $\frac{\sqrt{3}}{2}$ c. $\frac{\sqrt{2}}{2}$ d. $\frac{1}{2}$ e. 0   f. -1 g. 0

59. a. 0   b. $\dfrac{\sqrt{3}}{3}$ c. 1   d. $\sqrt{3}$ e. undefined   f. 0   g. undefined

61. $\cos\left(-\dfrac{5\pi}{3}\right) = \cos\dfrac{\pi}{3} = \dfrac{1}{2}$ 63. $\tan 135° = -\tan 45° = -1$ 65. 1.4587   67. 1.0896

69. $y = 2\cot x$

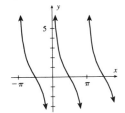

71. $y = \dfrac{1}{2}\csc x$

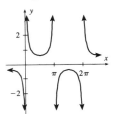

73. $y = \sin x$

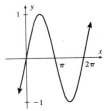

75. $y = \tan x$

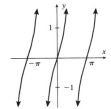

77. $y = 2\cos\dfrac{2}{3}x$

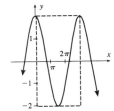

79. $y - 2 = \sin\left(x - \dfrac{\pi}{6}\right)$

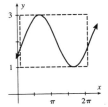

81. $-\dfrac{\pi}{2} < y < \dfrac{\pi}{2}$ 83. $0 < y < \pi$ 85. $\dfrac{\pi}{6}$ 87. $\dfrac{\pi}{6}$ 89. 0.3194   91. 1.274

93. $y = \text{Sin}^{-1}x$ 95. $y = \text{Arctan } x$

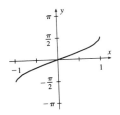

97. 1   99. -0.4292

126

CHAPTER 7

PROBLEM SET 7.1

1. $\sin x = \frac{1}{2}$; $x = \sin^{-1}\left(\frac{1}{2}\right) = \frac{\pi}{6}$

3. $\sin x = -\frac{1}{2}$; $x = \sin^{-1}\left(-\frac{1}{2}\right) = -\frac{\pi}{6}$

5. $\sin x = \frac{1}{2}$; $x' = \cos^{-1}\left(\frac{1}{2}\right) = \frac{\pi}{6}$; cosine is positive in Quad I and IV;
   $x = \frac{\pi}{6} + 2n\pi$, $\frac{5\pi}{6} + 2n\pi$

7. $\cos x = \frac{1}{2}$; $x = \cos^{-1}\left(\frac{1}{2}\right) = \frac{\pi}{3}$

9. $\cos x = -\frac{1}{2}$; $x = \cos^{-1}\left(-\frac{1}{2}\right) = 120°$

11. $\cos x = \frac{1}{2}$; $x' = \cos^{-1}\left(\frac{1}{2}\right) = \frac{\pi}{3}$; cosine is positive in Quad I and IV;
    $x = \frac{\pi}{3} + 2n\pi$, $\frac{5\pi}{3} + 2n\pi$

13. $\cos 2x = \frac{1}{2}$; $2x' = \frac{\pi}{3}$; cosine is positive in Quad I and IV; $0 \le 2x \le 4\pi$;
    $2x = \frac{\pi}{3}, \frac{5\pi}{3}, \frac{7\pi}{3}, \frac{11\pi}{3}$; $x = \frac{\pi}{6}, \frac{5\pi}{6}, \frac{7\pi}{6}, \frac{11\pi}{6}$

15. $\cos 2x = -\frac{1}{2}$; $2x' = \frac{\pi}{3}$; cosine is negative in Quad II and III; $0 \le 2x \le 4\pi$;
    $2x = \frac{2\pi}{3}, \frac{4\pi}{3}, \frac{8\pi}{3}, \frac{10\pi}{3}$; $x = \frac{\pi}{3}, \frac{2\pi}{3}, \frac{4\pi}{3}, \frac{5\pi}{3}$

17. $\sin 2x = -\frac{\sqrt{3}}{2}$; $2x' = \frac{\pi}{3}$; sine is negative in Quad III and IV; $0 \le 2x \le 4\pi$;
    $2x = \frac{4\pi}{3}, \frac{5\pi}{3}, \frac{10\pi}{3}, \frac{11\pi}{3}$; $x = \frac{2\pi}{3}, \frac{5\pi}{6}, \frac{5\pi}{3}, \frac{11\pi}{6}$

19. $\tan 3x = 1$; $3x' = \frac{\pi}{4}$; tangent is positive in Quad I and III; $0 \le 3x \le 6\pi$;
    $3x = \frac{\pi}{4}, \frac{5\pi}{4}, \frac{9\pi}{4}, \frac{13\pi}{4}, \frac{17\pi}{4}, \frac{21\pi}{4}$; $x = \frac{\pi}{12}, \frac{5\pi}{12}, \frac{3\pi}{4}, \frac{13\pi}{12}, \frac{17\pi}{12}, \frac{7\pi}{4}$

21. $\sec 2x = -\frac{2\sqrt{3}}{3}$; $2x' = \frac{\pi}{6}$; secant is negative in Quad II and III;
    $0 \le 2x \le 4\pi$; $2x = \frac{5\pi}{6}, \frac{7\pi}{6}, \frac{17\pi}{6}, \frac{19\pi}{6}$; $x = \frac{5\pi}{12}, \frac{7\pi}{12}, \frac{17\pi}{12}, \frac{19\pi}{12}$

23. $(\sec x)(\tan x) = 0$; $\sec x = 0$ (no solution) or $\tan x = 0$ at $x = 0$ and π; $x = 0, \pi$

25. $(\cot x)(\cos x) = 0$; $\cot x = 0$ at $x = \frac{\pi}{2}$ and $\frac{3\pi}{2}$ or $\cos x = 0$ at $x = \frac{\pi}{2}$ and $\frac{3\pi}{2}$; $x = \frac{\pi}{2}, \frac{3\pi}{2}$

27. $(\sec x - 2)(2\sin x - 1) = 0$; $\sec x - 2 = 0$ or $2\sin x - 1 = 0$;
    $\sec x = 2$ at $x = \frac{\pi}{3}$ and $\frac{5\pi}{3}$ or $\sin x = \frac{1}{2}$ at $x = \frac{\pi}{6}$ and $\frac{5\pi}{6}$; $x = \frac{\pi}{6}, \frac{\pi}{3}, \frac{5\pi}{6}, \frac{5\pi}{3}$

29. $\tan^2 x = \tan x$; $\tan^2 x - \tan x = 0$; $(\tan x)(\tan x - 1) = 0$;
    $\tan x = 0$ at $x = 0$ and π or $\tan x = 1$ at $x = \frac{\pi}{4}$ and $\frac{5\pi}{4}$; $x = 0, \frac{\pi}{4}, \pi, \frac{5\pi}{4}$

31. $\cos^2 x = \frac{1}{2}$; $\cos x = \pm\sqrt{\frac{1}{2}}$; $\cos x = \pm\frac{\sqrt{2}}{2}$; $x = \frac{\pi}{4}, \frac{3\pi}{4}, \frac{5\pi}{4}, \frac{7\pi}{4}$

33. $2\cos x \sin x = \sin x$; $2\cos x \sin x - \sin x = 0$; $(\sin x)(2\cos x - 1) = 0$; $\sin x = 0$ or $2\cos x - 1 = 0$; $\sin x = 0$ at $x = 0$ and π or $\cos x = \frac{1}{2}$ at $x = \frac{\pi}{3}$ and $\frac{5\pi}{3}$; $x = 0, \frac{\pi}{3}, \pi, \frac{5\pi}{3}$

35. $\cos^2 x - 1 - \cos x = 0$; $\cos^2 x - \cos x - 1 = 0$;

$\cos x = \dfrac{1 \pm \sqrt{(-1)^2 - 4(1)(-1)}}{2(1)}$; $\cos x = \dfrac{1 \pm \sqrt{5}}{2}$;

$\cos x = \dfrac{1 + \sqrt{5}}{2}$ or $\cos x = \dfrac{1 - \sqrt{5}}{2}$; $\cos x = 1.618$ (no solution)

or $\cos x = -.6180$, $x' = .9046$, x in Quad II or III;

$x = \pi - .9046$, $\pi + .9046$; $x = 2.2370, 4.0462$

37. $\tan^2 x - 3\tan x + 1 = 0$; $\tan x = \dfrac{3 \pm \sqrt{(-3)^2 - 4(1)(1)}}{2(1)}$; $\tan x = \dfrac{3 \pm \sqrt{5}}{2}$;

$\tan x = \dfrac{3 + \sqrt{5}}{2}$; $\tan x = 2.6180$, $x' = 1.2059$, x in Quad I or III,

$x = 1.2059, 4.3475$; $\tan x = \dfrac{3 - \sqrt{5}}{2}$; $\tan x = .3820$, $x' = .3649$,

x in quadrant I or III, $x = .3649, 3.5065$; $x = .3649, 1.2059, 3.5065, 4.3475$

39. $\csc^2 x - \csc x - 1 = 0$; $\csc x = \dfrac{1 \pm \sqrt{(-1)^2 - 4(1)(-1)}}{2(1)}$; $\csc x = \dfrac{1 \pm \sqrt{5}}{2}$;

$\csc x = \dfrac{1 + \sqrt{5}}{2}$; $\csc x = 1.6180$; $x' = .6662$, x in quadrant I or II;

$x = .6662, 2.4754$; $\csc x = \dfrac{1 - \sqrt{5}}{2}$; $\csc x = -.6180$ (No solution); $x = .6662, 2.4754$

41. $\sin x + 1 = \sqrt{3}$; $\sin x = \sqrt{3} - 1$; $\sin x = .7321$, $x' = .8213$,

x in Quad I or II; $x = .8213$, $\pi - .8213$; $x = .8213, 2.3203$

43. $\sin 2x + 2\cos x \sin 2x = 0$; $\sin 2x (1 + 2\cos x) = 0$;

$\sin 2x = 0$ at $2x = 0, \pi, 2\pi, 3\pi$; $x = 0, \frac{\pi}{2}, \pi, \frac{3\pi}{2}$;

$\cos x = -\frac{1}{2}$ at $x = \frac{2\pi}{3}$ and $\frac{4\pi}{3}$; $x = 0, \frac{\pi}{2}, \frac{2\pi}{3}, \pi, \frac{4\pi}{3}, \frac{3\pi}{2}$;

$x \approx .00, 1.57, 2.09, 3.14, 4.19, 4.71$

45. $\sin 2x + 1 = \sqrt{3}$; $\sin 2x = \sqrt{3} - 1$; $\sin 2x = .7321$;

$2x' = .8213$, $2x$ in Quad I or II; $2x = .8213$, $\pi - .8213$, $2\pi + .8213$,

$3\pi - .8213$; $2x = .8213, 2.3203, 7.1045, 8.6035$; $x \approx .41, 1.16, 3.55, 4.30$

128

47. $\tan 2x + 1 = \sqrt{3}$; $\tan 2x = \sqrt{3} - 1$; $\tan 2x = .7321$;

$2x' = .6319$, $2x$ in quadrant I or III; $2x = .6319$, $\pi + .6319$, $2\pi + .6319$,

$3\pi + .6319$; $2x = .6319, 3.7735, 6.9151, 10.0567$; $x \approx .32$, 1.89, 3.46, 5.03

49. $2\cos^2 x - 1 = \cos x$; $2\cos^2 x - \cos x - 1 = 0$; $(2\cos x + 1)(\cos x - 1) = 0$;

$\cos x = -\frac{1}{2}$; $x' = 1.0472$, x in Quad II or III; $x = 2.0944, 4.1888$;

$\cos x = 1$ at $x = 0$; $x \approx .00, 2.09, 4.19$

51. $1 - \sin x = 1 - 2\sin^2 x$; $2\sin^2 x - \sin x = 0$; $\sin x\,(2\sin x - 1) = 0$;

$\sin x = 0$ at $x = 0$ and π; $\sin x = \frac{1}{2}$; $x' = \frac{\pi}{6}$, x in quadrant I or II;

$x = \frac{\pi}{6}$ and $\frac{5\pi}{6}$; $x = 0, \frac{\pi}{6}, \frac{5\pi}{6}, \pi$; $x \approx .00, .52, 2.62, 3.14$

53. $\sin^2 3x + \sin 3x + 1 = 1 - \sin^2 3x$; $2\sin^2 3x + \sin 3x = 0$;

$\sin 3x\,(2\sin 3x + 1) = 0$; $\sin 3x = 0$ at $3x = 0, \pi, 2\pi, 3\pi, 4\pi, 5\pi$;

$x = 0, \frac{\pi}{3}, \frac{2\pi}{3}, \pi, \frac{4\pi}{3}, \frac{5\pi}{3}$; $2\sin 3x + 1 = 0$; $\sin 3x = -\frac{1}{2}$ at $3x' = \frac{\pi}{6}$,

$3x'$ in quad III or IV; $3x = \frac{7\pi}{6}, \frac{11\pi}{6}, \frac{19\pi}{6}, \frac{23\pi}{6}, \frac{31\pi}{6}, \frac{35\pi}{6}$;

$x = \frac{7\pi}{18}, \frac{11\pi}{18}, \frac{19\pi}{18}, \frac{23\pi}{18}, \frac{31\pi}{18}, \frac{35\pi}{18}$;

$x \approx 0, \frac{\pi}{3}, \frac{7\pi}{18}, \frac{11\pi}{18}, \frac{2\pi}{3}, \pi, \frac{19\pi}{18}, \frac{23\pi}{18}, \frac{4\pi}{3}, \frac{5\pi}{3}, \frac{31\pi}{18}, \frac{35\pi}{18}$;

$x = .00, 1.05, 1.22, 1.92, 2.09, 3.14, 3.32, 4.01, 4.19, 5.24, 5.41, 6.11$

55. $\cos x = -\frac{\sqrt{3}}{2}$; $x' = \text{Cos}^{-1}\left|-\frac{\sqrt{3}}{2}\right|$; $x' = \frac{\pi}{3}$, x in Quad II and III;
$x = \frac{5\pi}{6} + 2n\pi, \frac{7\pi}{6} + 2n\pi$

57. $\sin x = .3907$; $x' = \text{Sin}^{-1}|.3907|$; $x' = .4014$, x in Quad I or II;
$x = .4104 + 2n\pi, 2.7402 + 2n\pi$

59. $\tan x = 1.376$; $x' = \text{Tan}^{-1}|1.376|$; $x' = .9423$, x in Quad I or III; $x = .9423 + n\pi$

61. $V = \cos 2\pi t$; $.400 = \cos 2\pi t$; $\cos 2\pi t = .400$; $2\pi t = \text{Cos}^{-1}|.400|$;
$2\pi t = 1.1593$; $t = .185$

63. Graph: $y = \cos 2x - 1$

$\qquad y = \sin 2x$

Points of Intersection:

$\qquad$ .00, 2.35, 3.14, 5.50

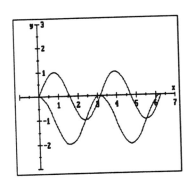

65. Graph: $y = x$

$\qquad y = -\cos x$

Since the graphs do not intersect,

there is no solution.

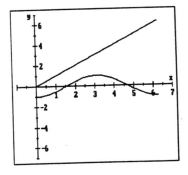

67. Graph: $y = \sin^{-1} x + 2 \cos^{-1} \frac{x}{2}$

$\qquad y = \frac{\pi}{3}$

Point of Intersection: 1.86

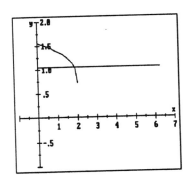

PROBLEM SET 7.2

1.   RECIPROCAL IDENTITIES:

$\quad$ 1.$\quad$ $\sec \theta = \dfrac{1}{\cos \theta}$ $\qquad$ 2.$\quad$ $\csc \theta = \dfrac{1}{\sec \theta}$ $\qquad$ 3.$\quad$ $\cot \theta = \dfrac{1}{\tan \theta}$

$\quad$ RATIO IDENTITIES:

$\quad$ 4.$\quad$ $\tan \theta = \dfrac{\sin \theta}{\cos \theta}$ $\qquad$ 5.$\quad$ $\cot \theta = \dfrac{\cos \theta}{\sin \theta}$

$\quad$ PYTHAGOREAN IDENTITIES:

$\quad$ 6.$\quad$ $\sin^2\theta + \cos^2\theta = 1$ $\quad$ 7.$\quad$ $1 + \tan^2\theta = \sec^2\theta$ $\quad$ 8.$\quad$ $\cot^2\theta + 1 = \csc^2\theta$

130

3.   sine is negative in III and IV   5.   secant is positive in I and IV

7.   cosine is negative in II and III; sine is positive is I and II; hence the quadrant is II.

9.   cosecant is positive in I and II; cosine is positive in I and IV; hence the quadrant is I.

11. $\dfrac{\cos\,(A+B)}{\sin\,(A+B)} = \cot\,(A+B)$ 13. $\dfrac{1}{\cot\left(\frac{\pi}{15}\right)} = \tan\left(\frac{\pi}{15}\right)$

15. $\cot\frac{\pi}{8}\sin\frac{\pi}{8} = \dfrac{\cos\frac{\pi}{8}}{\sin\frac{\pi}{8}} \cdot \sin\frac{\pi}{8} = \cos\frac{\pi}{8}$

17. $-\sqrt{1-\sin^2 127^\circ} = -\sqrt{\cos^2 127^\circ} = -|\cos\,127^\circ| = \cos\,127^\circ$

19. $\cos\,128^\circ\,\sec\,128^\circ = \cos\,128^\circ \cdot \dfrac{1}{\cos\,128^\circ} = 1$

21. $\sec^2\frac{\pi}{6} - \tan^2\frac{\pi}{6} = \left(\tan^2\frac{\pi}{6}+1\right) - \tan^2\frac{\pi}{6} = 1$

23. $\tan^2 135^\circ - \sec^2 135^\circ = \tan^2 135^\circ - (\tan^2 135^\circ + 1) = -1$

25. $\csc\theta = \dfrac{r}{y} = \dfrac{1}{y/r} = \dfrac{1}{\sin\theta}$

27. $\cot\theta = \dfrac{x}{y} = \dfrac{x}{y}\cdot\dfrac{1/r}{1/r} = \dfrac{x/r}{y/r} = \dfrac{\cos\theta}{\sin\theta}$

29. $1 + \tan^2\theta = 1 + \left(\dfrac{y}{x}\right)^2 = 1 + \dfrac{y^2}{x^2} = \dfrac{x^2+y^2}{x^2} = \dfrac{r^2}{x^2} = \left(\dfrac{r}{x}\right)^2 = \sec^2\theta$

31. $\cot\theta = \cot\theta;$ $\tan\theta = \dfrac{1}{\cot\theta};$ $1+\cot^2\theta = \csc^2\theta;$ $\csc\theta = \pm\sqrt{1+\cot^2\theta};$

    $\sin\theta = \dfrac{\pm 1}{\sqrt{1+\cot^2\theta}};$ $\tan^2\theta + 1 = \sec^2\theta;$ $\dfrac{1}{\cot^2\theta} + 1 = \sec^2\theta;$

    $\dfrac{1+\cot^2\theta}{\cot^2\theta} = \sec^2\theta;$ $\sec\theta = \dfrac{\pm\sqrt{1+\cot^2\theta}}{\cot\theta};$ $\cos\theta = \dfrac{\pm\cot\theta}{\sqrt{1+\cot^2\theta}}$

33. $\csc\theta = \csc\theta;$ $\sin\theta = \dfrac{1}{\csc\theta};$ $1+\cot^2\theta = \csc^2\theta;$ $\cot^2\theta = \csc^2\theta - 1;$

    $\cot\theta = \pm\sqrt{\csc^2\theta - 1};$ $\tan\theta = \dfrac{\pm 1}{\sqrt{\csc^2\theta - 1}};$ $\sin^2\theta + \cos^2\theta = 1;$

    $\cos^2\theta = 1 - \sin^2\theta;$ $\cos^2\theta = 1 - \dfrac{1}{\csc^2\theta};$ $\cos^2\theta = \dfrac{\csc^2\theta - 1}{\csc^2\theta};$

    $\cos\theta = \dfrac{\pm\sqrt{\csc^2\theta - 1}}{\csc\theta};$ $\sec\theta = \dfrac{\pm\csc\theta}{\sqrt{\csc^2\theta - 1}}$

35. $\cos \theta = \frac{3}{5}$; $\csc \theta < 0$; Since cosine > 0 and cosecant < 0, the quadrant is IV.

Thus, $\sin \theta = -\sqrt{1-\cos^2\theta} = -\sqrt{1-\left(\frac{3}{5}\right)^2} = -\sqrt{1-\frac{9}{25}} = -\sqrt{\frac{16}{25}} = -\frac{4}{5}$;

$\tan \theta = \frac{\sin \theta}{\cos \theta} = \frac{-4/5}{3/5} = -\frac{4}{3}$

Using the reciprocal identities, $\sec \theta = \frac{5}{3}$, $\csc \theta = -\frac{5}{4}$, and $\cot \theta = -\frac{3}{4}$

37. $\cos \theta = \frac{5}{13}$; $\tan \theta > 0$; Since cosine > 0 and tangent > 0, the quadrant is I.

Thus, $\sin \theta = \sqrt{1-\cos^2\theta} = \sqrt{1-\left(\frac{5}{13}\right)^2} = \sqrt{1-\frac{25}{169}} = \sqrt{\frac{144}{169}} = \frac{12}{13}$;

$\tan \theta = \frac{\sin \theta}{\cos \theta} = \frac{12/13}{5/13} = \frac{12}{5}$;

Using the reciprocal identities, $\sec \theta = \frac{13}{5}$, $\csc \theta = \frac{13}{12}$, and $\cot \theta = \frac{5}{12}$

39. $\tan \theta = \frac{5}{12}$; $\sin \theta < 0$; Since tangent > 0 and sine < 0, the quadrant is III.

Thus, $\sec \theta = -\sqrt{\tan^2\theta + 1} = -\sqrt{\left(\frac{5}{12}\right)^2 + 1} = -\sqrt{\frac{25}{144} + 1} = -\sqrt{\frac{169}{144}} = -\frac{13}{12}$;

Using the reciprocal identities, $\cot \theta = \frac{12}{5}$ and $\cos \theta = -\frac{12}{13}$

$\sin \theta = -\sqrt{1-\cos^2\theta} = -\sqrt{1-\left(-\frac{12}{13}\right)^2} = -\sqrt{1-\frac{144}{169}} = -\sqrt{\frac{25}{169}} = -\frac{5}{13}$;

Using the reciprocal identity, $\csc \theta = -\frac{13}{5}$

41. $\sin \theta = \frac{2}{3}$; $\sec \theta < 0$; Since sine > 0 and secant < 0, the quadrant is II;

Thus, $\cos \theta = -\sqrt{1-\sin^2\theta} = -\sqrt{1-\left(\frac{2}{3}\right)^2} = -\sqrt{1-\frac{4}{9}} = -\sqrt{\frac{5}{9}} = -\frac{\sqrt{5}}{3}$;

$\tan \theta = \frac{\sin \theta}{\cos \theta} = \frac{2/3}{-\sqrt{5}/3} = -\frac{2}{\sqrt{5}}$;

Using the reciprocal identities, $\csc \theta = \frac{3}{2}$, $\sec \theta = -\frac{3}{\sqrt{5}}$, and $\cot \theta = -\frac{\sqrt{5}}{2}$

43. $\sec \theta = \frac{\sqrt{34}}{5}$; $\tan \theta > 0$; Since secant > 0 and tangent > 0, the quadrant is I.

Thus, $\tan \theta = \sqrt{\sec^2\theta - 1} = \sqrt{\left(\frac{\sqrt{34}}{5}\right)^2 - 1} = \sqrt{\frac{34}{25} - 1} = \sqrt{\frac{9}{25}} = \frac{3}{5}$;

Using the reciprocal identities, $\cos \theta = \frac{5}{\sqrt{34}}$ and $\cot \theta = \frac{5}{3}$;

$\sin \theta = \sqrt{1-\cos^2\theta} = \sqrt{1-\left(\frac{5}{\sqrt{34}}\right)^2} = \sqrt{1-\frac{25}{34}} = \sqrt{\frac{9}{34}} = \frac{3}{\sqrt{34}}$;

Using the reciprocal identity, $\csc \theta = \frac{\sqrt{34}}{3}$

45. $\csc \theta = -\dfrac{\sqrt{10}}{3}$; $\cos \theta < 0$; Since cosecant < 0, and cosine < 0, the quadrant is III.

   Thus, $\cot \theta = \sqrt{\csc^2\theta - 1} = \sqrt{\left(-\dfrac{\sqrt{10}}{3}\right)^2 - 1} = \sqrt{\dfrac{10}{9} - 1} = \sqrt{\dfrac{1}{9}} = \dfrac{1}{3}$;

   Using the reciprocal identities, $\sin \theta = -\dfrac{3}{\sqrt{10}}$, and $\tan \theta = 3$;

   $\cos \theta = -\sqrt{1 - \sin^2\theta} = -\sqrt{1 - \left(-\dfrac{3}{\sqrt{10}}\right)^2} = -\sqrt{1 - \dfrac{9}{10}} = -\sqrt{\dfrac{1}{10}} = -\dfrac{1}{\sqrt{10}}$;

   Using the reciprocal identity, $\sec \theta = -\sqrt{10}$

47. $\dfrac{1 - \cos^2\theta}{\sin \theta} = \dfrac{\sin^2\theta}{\sin \theta} = \sin \theta$

49. $\dfrac{1}{1 + \cos \theta} + \dfrac{1}{1 - \cos \theta} = \dfrac{(1 - \cos \theta) + (1 + \cos \theta)}{(1 + \cos \theta)(1 - \cos \theta)} = \dfrac{2}{1 - \cos^2\theta} = \dfrac{2}{\sin^2\theta}$

51. $\sin \theta + \dfrac{\cos^2\theta}{\sin \theta} = \dfrac{\sin^2\theta + \cos^2\theta}{\sin \theta} = \dfrac{1}{\sin \theta}$

53. $\dfrac{\sin \theta - \dfrac{\cos^2\theta}{\sin \theta}}{\cos \theta} = \dfrac{\dfrac{\sin^2\theta - \cos^2\theta}{\sin \theta}}{\cos \theta} = \dfrac{\sin^2\theta - \cos^2\theta}{\sin \theta \cos \theta}$

55. $\dfrac{\dfrac{\sin \theta}{\cos \theta} + \dfrac{\cos \theta}{\sin \theta}}{\dfrac{1}{\sin \theta}} = \dfrac{\dfrac{\sin^2\theta + \cos^2\theta}{\sin \theta \cos \theta}}{\dfrac{1}{\sin \theta}} = \dfrac{\dfrac{1}{\sin \theta \cos \theta}}{\dfrac{1}{\sin \theta}} = \dfrac{1}{\cos \theta}$

57. $\sec \theta + \tan \theta = \dfrac{1}{\cos \theta} + \dfrac{\sin \theta}{\cos \theta} = \dfrac{1 + \sin \theta}{\cos \theta}$

59. $\dfrac{\sec \theta + \csc \theta}{\tan \theta \cot \theta} = \dfrac{\dfrac{1}{\cos \theta} + \dfrac{1}{\sin \theta}}{\dfrac{\sin \theta}{\cos \theta} \cdot \dfrac{\cos \theta}{\sin \theta}} = \dfrac{\dfrac{1}{\cos \theta} + \dfrac{1}{\sin \theta}}{1} = \dfrac{\sin \theta + \cos \theta}{\cos \theta \sin \theta}$

61. $\csc^2\theta + \cot^2\theta = \dfrac{1}{\sin^2\theta} + \dfrac{\cos^2\theta}{\sin^2\theta} = \dfrac{1 + \cos^2\theta}{\sin^2\theta}$

63. $(\tan \theta - \csc \theta)(\cos \theta \sin \theta) = \left(\dfrac{\sin \theta}{\cos \theta} - \dfrac{1}{\sin \theta}\right)(\cos \theta \sin \theta) = \sin^2\theta - \cos \theta$

PROBLEM SET 7.3

1. $\sin^3\theta + \cos^2\theta \sin \theta = \sin \theta \, (\sin^2\theta + \cos^2\theta)$

   $= \sin \theta \, (1)$

   $= \sin \theta$

3. $\cot \theta \ \tan^2\theta = \dfrac{1}{\tan \theta} \cdot \tan^2\theta$

$$= \tan \theta$$

5. $\tan^2\theta \ \sin^2\theta = \tan^2\theta \ (1 - \cos^2\theta)$

$$= \tan^2\theta - \tan^2\theta \ \cos^2\theta$$

$$= \tan^2\theta - \dfrac{\sin^2\theta}{\cos^2\theta} \cdot \cos^2\theta$$

$$= \tan^2\theta - \sin^2\theta$$

7. $\tan A + \cot A = \dfrac{\sin A}{\cos A} + \dfrac{\cos A}{\sin A}$

$$= \dfrac{\sin^2 A + \cos^2 A}{\cos A \ \sin A}$$

$$= \dfrac{1}{\cos A \ \sin A}$$

$$= \dfrac{1}{\cos A} \cdot \dfrac{1}{\sin A}$$

$$= \sec A \ \csc A$$

9. $\dfrac{\sec x + \csc x}{\csc x \ \sec x} = \dfrac{\sec x}{\csc x \ \sec x} + \dfrac{\csc x}{\csc x \ \sec x}$

$$= \dfrac{1}{\csc x} + \dfrac{1}{\sec x}$$

$$= \sin x + \cos x$$

11. $\dfrac{1 - \sec^2 t}{\sec^2 t} = \dfrac{1}{\sec^2 t} - \dfrac{\sec^2 t}{\sec^2 t}$

$$= \cos^2 t - 1$$

$$= -(1 - \cos^2 t)$$

$$= -\sin^2 t$$

13. $(\sec \theta - \cos \theta)^2 = \sec^2\theta - 2 \ \sec \theta \ \cos \theta + \cos^2\theta$

$$= \sec^2\theta - 2 \cdot \dfrac{1}{\cos \theta} \cdot \cos \theta + \cos^2\theta$$

$$= \sec^2\theta - 2 + \cos^2\theta$$

$$= \sec^2\theta - 1 - 1 + \cos^2\theta$$

$$= (\sec^2\theta - 1) - (1 - \cos^2\theta)$$

$$= \tan^2\theta - \sin^2\theta$$

15. $\dfrac{1 - \sin^2 2\theta}{1 + \sin 2\theta} = \dfrac{(1 + \sin 2\theta)(1 - \sin 2\theta)}{1 + \sin 2\theta}$

$$= 1 - \sin 2\theta$$

134

17.
$$\frac{\sin^2\gamma + \sin\gamma\cos\gamma + \sin\gamma}{\sin\gamma + \cos\gamma + 1} = \frac{\sin\gamma(\sin\gamma + \cos\gamma + 1)}{(\sin\gamma + \cos\gamma + 1)}$$
$$= \sin\gamma$$

19.
$$\sin 2\alpha \cos 2\alpha (\tan 2\alpha + \cot 2\alpha) = \sin 2\alpha \cos 2\alpha \left(\frac{\sin 2\alpha}{\cos 2\alpha} + \frac{\cos 2\alpha}{\sin 2\alpha}\right)$$
$$= \sin^2 2\alpha + \cos^2 2\alpha$$
$$= 1$$

21.
$$\csc 3\beta - \cos 3\beta \cot 3\beta = \frac{1}{\sin 3\beta} - \cos 3\beta \cdot \frac{\cos 3\beta}{\sin 3\beta}$$
$$= \frac{1 - \cos^2 3\beta}{\sin 3\beta}$$
$$= \frac{\sin^2 3\beta}{\sin 3\beta}$$
$$= \sin 3\beta$$

23.
$$\frac{\sin^2 B - \cos^2 B}{\sin B + \cos B} = \frac{(\sin B + \cos B)(\sin B - \cos B)}{\sin B + \cos B}$$
$$= \sin B - \cos B$$

25.
$$\tan^2 2\gamma + \sin^2 2\gamma + \cos^2 2\gamma = \tan^2 2\gamma + 1$$
$$= \sec^2 2\gamma$$

27.
$$\frac{\tan\theta + \cot\theta}{\sec\theta\csc\theta} = \frac{\tan\theta}{\sec\theta\csc\theta} + \frac{\cot\theta}{\sec\theta\csc\theta}$$
$$= \tan\theta\cos\theta\sin\theta + \cot\theta\cos\theta\sin\theta$$
$$= \frac{\sin\theta}{\cos\theta}\cdot\cos\theta\sin\theta + \frac{\cos\theta}{\sin\theta}\cdot\cos\theta\sin\theta$$
$$= \sin^2\theta + \cos^2\theta$$
$$= 1$$

29.
$$1 + \sin^2\gamma = 1 + (1 - \cos^2\gamma)$$
$$= 2 - \cos^2\gamma$$

31.
$$\frac{\sin\alpha}{\tan\alpha} + \frac{\cos\alpha}{\cot\alpha} = \sin\alpha\cot\alpha + \cos\alpha\tan\alpha$$
$$= \sin\alpha\cdot\frac{\cos\alpha}{\sin\alpha} + \cos\alpha\cdot\frac{\sin\alpha}{\cos\alpha}$$
$$= \cos\alpha + \sin\alpha$$

33. $\sec\,\beta + \cos\,\beta = \dfrac{1}{\cos\,\beta} + \cos\,\beta$

$$= \dfrac{1 + \cos^2\beta}{\cos\,\beta}$$

$$= \dfrac{1 + (1 - \sin^2\beta)}{\cos\,\beta}$$

$$= \dfrac{2 - \sin^2\beta}{\cos\,\beta}$$

35. $\dfrac{\tan\,2\theta + \cot\,2\theta}{\sec\,2\theta} = \dfrac{\tan\,2\theta}{\sec\,2\theta} + \dfrac{\cot\,2\theta}{\sec\,2\theta}$

$$= \tan\,2\theta\,\cos\,2\theta + \cot\,2\theta\,\cos\,2\theta$$

$$= \dfrac{\sin\,2\theta}{\cos\,2\theta} \cdot \cos\,2\theta + \dfrac{\cos\,2\theta}{\sin\,2\theta} \cdot \cos\,2\theta$$

$$= \sin\,2\theta + \dfrac{\cos^2 2\theta}{\sin\,2\theta}$$

$$= \dfrac{\sin^2 2\theta + \cos^2 2\theta}{\sin\,2\theta}$$

$$= \dfrac{1}{\sin\,2\theta}$$

$$= \csc\,2\theta$$

37. $\dfrac{\sec\,\gamma + \tan^2\gamma}{\sec\,\gamma} = \dfrac{\sec\,\gamma}{\sec\,\gamma} + \dfrac{\tan^2\gamma}{\sec\,\gamma}$

$$= 1 + \dfrac{\sec^2\gamma - 1}{\sec\,\gamma}$$

$$= 1 + \dfrac{\sec^2\gamma}{\sec\,\gamma} - \dfrac{1}{\sec\,\gamma}$$

$$= 1 + \sec\,\gamma - \cos\,\gamma$$

39. $\dfrac{1 + \tan\,C}{1 - \tan\,C} = \dfrac{1 + \tan\,C}{1 - \tan\,C} \cdot \dfrac{1 + \tan\,C}{1 + \tan\,C}$

$$= \dfrac{1 + 2\,\tan\,C + \tan^2 C}{1 - \tan^2 C}$$

$$= \dfrac{1 + \tan^2 C + 2\,\tan\,C}{1 - (\sec^2 C - 1)}$$

$$= \dfrac{\sec^2 C + 2\,\tan\,C}{2 - \sec^2 C}$$

41.

$$\frac{\sin^3 x - \cos^3 x}{\sin x - \cos x} = \frac{(\sin x - \cos x)(\sin^2 x + \sin x \cos x + \cos^2 x)}{\sin x - \cos x}$$

$$= \sin^2 x + \sin x \cos x + \cos^2 x$$

$$= \sin^2 x + \cos^2 x + \sin x \cos x$$

$$= 1 + \sin x \cos x$$

43.

$$\left(\frac{1 - \cos \theta}{\sin \theta}\right)^2 = \frac{(1 - \cos \theta)^2}{\sin^2 \theta}$$

$$= \frac{(1 - \cos \theta)^2}{1 - \cos^2 \theta}$$

$$= \frac{(1 - \cos \theta)^2}{(1 + \cos \theta)(1 - \cos \theta)}$$

$$= \frac{1 - \cos \theta}{1 + \cos \theta}$$

45.

$$\frac{(\cos^2 \gamma - \sin^2 \gamma)^2}{\cos^4 \gamma - \sin^4 \gamma} = \frac{(\cos^2 \gamma - \sin^2 \gamma)^2}{(\cos^2 \gamma + \sin^2 \gamma)(\cos^2 \gamma - \sin^2 \gamma)}$$

$$= \frac{(\cos^2 \gamma - \sin^2 \gamma)^2}{(1)(\cos^2 \gamma - \sin^2 \gamma)}$$

$$= \cos^2 \gamma - \sin^2 \gamma$$

$$= \cos^2 \gamma - (1 - \cos^2 \gamma)$$

$$= 2 \cos^2 \gamma - 1$$

47.

$$\frac{1}{\sec \theta + \tan \theta} = \frac{1}{\sec \theta + \tan \theta} \cdot \frac{\sec \theta - \tan \theta}{\sec \theta - \tan \theta}$$

$$= \frac{\sec \theta - \tan \theta}{\sec^2 \theta - \tan^2 \theta}$$

$$= \frac{\sec \theta - \tan \theta}{(\tan^2 \theta + 1) - \tan^2 \theta}$$

$$= \sec \theta - \tan \theta$$

49. $\sec^2 2\gamma + \csc^2 2\gamma = \dfrac{1}{\cos^2 2\gamma} + \dfrac{1}{\sin^2 2\gamma}$

$$= \frac{\sin^2 2\gamma + \cos^2 2\gamma}{\cos^2 2\gamma \, \sin^2 2\gamma}$$

$$= \frac{1}{\cos^2 2\gamma \, \sin^2 2\gamma}$$

$$= \frac{1}{\cos^2 2\gamma} \cdot \frac{1}{\sin^2 2\gamma}$$

$$= \sec^2 2\gamma \, \csc^2 2\gamma$$

$$= \csc^2 2\gamma \, \sec^2 2\gamma$$

51. $\dfrac{1 - \sec^3\theta}{1 - \sec\theta} = \dfrac{(1 - \sec\theta)(1 + \sec\theta + \sec^2\theta)}{1 - \sec\theta}$

$$= 1 + \sec\theta + \sec^2\theta$$

$$= 1 + \sec\theta + (\tan^2\theta + 1)$$

$$= \tan^2\theta + \sec\theta + 2$$

53. $\dfrac{\tan^2\theta - 2\tan\theta}{2\tan\theta - 4} = \dfrac{\tan\theta\,(\tan\theta - 2)}{2(\tan\theta - 2)}$

$$= \tfrac{1}{2}\tan\theta$$

55. $\sqrt{(3\cos\theta - 4\sin\theta)^2 + (3\sin\theta + 4\cos\theta)^2}$

$$= \sqrt{9\cos^2\theta - 24\cos\theta\,\sin\theta + 16\sin^2\theta + 9\sin^2\theta + 24\sin\theta\,\cos\theta + 16\cos^2\theta}$$

$$= \sqrt{9\cos^2\theta + 9\sin^2\theta + 16\sin^2\theta + 16\cos^2\theta}$$

$$= \sqrt{9(\cos^2\theta + \sin^2\theta) + 16(\sin^2\theta + \cos^2\theta)}$$

$$= \sqrt{9 + 16}$$

$$= \sqrt{25}$$

$$= 5$$

57.
$$(2\sin^2\gamma - 1)(\sec^2\gamma + \csc^2\gamma) = \left(\sin^2\gamma + (\sin^2\gamma - 1)\right)\left(\frac{1}{\cos^2\gamma} + \frac{1}{\sin^2\gamma}\right)$$

$$= \left(\sin^2\gamma - \cos^2\gamma\right)\left(\frac{\sin^2\gamma + \cos^2\gamma}{\cos^2\gamma\,\sin^2\gamma}\right)$$

$$= \left(\sin^2\gamma - \cos^2\gamma\right)\left(\frac{1}{\cos^2\gamma\,\sin^2\gamma}\right)$$

$$= \frac{\sin^2\gamma - \cos^2\gamma}{\cos^2\gamma\,\sin^2\gamma}$$

$$= \frac{1}{\cos^2\gamma} - \frac{1}{\sin^2\gamma}$$

$$= \sec^2\gamma - \csc^2\gamma$$

59.
$$\frac{\csc\theta + 1}{\cot^2\theta + \csc\theta + 1} = \frac{\dfrac{1}{\sin\theta} + 1}{\dfrac{\cos^2\theta}{\sin^2\theta} + \dfrac{1}{\sin\theta} + 1}$$

$$= \frac{\dfrac{1 + \sin\theta}{\sin\theta}}{\dfrac{\cos^2\theta + \sin\theta + \sin^2\theta}{\sin^2\theta}}$$

$$= \frac{1 + \sin\theta}{\sin\theta} \cdot \frac{\sin^2\theta}{\cos^2\theta + \sin^2\theta + \sin\theta}$$

$$= \frac{1 + \sin\theta}{\sin\theta} \cdot \frac{\sin^2\theta}{1 + \sin\theta}$$

$$= \sin\theta$$

$$= \sin\theta \cdot \frac{\sin\theta + \cos\theta}{\sin\theta + \cos\theta}$$

$$= \frac{\sin^2\theta + \sin\theta\,\cos\theta}{\sin\theta + \cos\theta}$$

61.
$$(\cos\alpha - \cos\beta)^2 + (\sin\alpha - \sin\beta)^2$$

$$= \cos^2\alpha - 2\cos\alpha\cos\beta + \cos^2\beta + \sin^2\alpha - 2\sin\alpha\sin\beta + \sin^2\beta$$

$$= \cos^2\alpha + \sin^2\alpha + \cos^2\beta + \sin^2\beta - 2\cos\alpha\cos\beta - 2\sin\alpha\sin\beta$$

$$= 1 + 1 - 2(\cos\alpha\cos\beta - \sin\alpha\sin\beta)$$

$$= 2 - 2\,(\cos\alpha\cos\beta + \sin\alpha\sin\beta)$$

63. $(\sin A \sin B + \cos A \cos B) \sec A \csc B$

$$= (\sin A \sin B + \cos A \cos B) \frac{1}{\cos A} \cdot \frac{1}{\sin B}$$

$$= \frac{\sin A}{\cos B} + \frac{\cos B}{\sin B}$$

$$= \tan A + \cot B$$

65. $(\sin A \cos A \cos B + \sin B \cos B \cos A) \sec A \sec B$

$$= (\sin A \cos A \cos B + \sin B \cos B \cos A) \cdot \frac{1}{\cos A} \cdot \frac{1}{\cos B}$$

$$= \sin A + \sin B$$

67. $\sin \theta + \cos \theta + 1 = (\sin \theta + \cos \theta + 1) \cdot \left(\frac{\sin \theta + \cos \theta - 1}{\sin \theta + \cos \theta - 1} \right)$

$$= \frac{(\sin \theta + \cos \theta)^2 - 1}{\sin \theta + \cos \theta - 1}$$

$$= \frac{\sin^2\theta + 2 \sin \theta \cos \theta + \cos^2\theta - 1}{\sin \theta + \cos \theta - 1}$$

$$= \frac{\sin^2\theta + \cos^2\theta + 2 \sin \theta \cos \theta - 1}{\sin \theta + \cos \theta - 1}$$

$$= \frac{1 + 2 \sin \theta \cos \theta - 1}{\sin \theta + \cos \theta - 1}$$

$$= \frac{2 \sin \theta \cos \theta}{\sin \theta + \cos \theta - 1}$$

69. $\dfrac{\csc \theta + 1}{\csc \theta - 1} - \dfrac{\sec \theta - \tan \theta}{\sec \theta + \tan \theta} = \dfrac{\frac{1}{\sin \theta} + 1}{\frac{1}{\sin \theta} - 1} - \dfrac{\frac{1}{\cos \theta} - \frac{\sin \theta}{\cos \theta}}{\frac{1}{\cos \theta} + \frac{\sin \theta}{\cos \theta}}$

$$= \frac{\frac{1 + \sin \theta}{\sin \theta}}{\frac{1 - \sin \theta}{\sin \theta}} - \frac{\frac{1 - \sin \theta}{\cos \theta}}{\frac{1 + \sin \theta}{\cos \theta}}$$

$$= \frac{1 + \sin \theta}{1 - \sin \theta} - \frac{1 - \sin \theta}{1 + \sin \theta}$$

$$= \frac{(1 + \sin \theta)^2 - (1 - \sin \theta)^2}{(1 - \sin \theta)(1 + \sin \theta)}$$

$$= \frac{1 + 2 \sin \theta + \sin^2\theta - 1 + 2 \sin \theta - \sin^2\theta}{1 - \sin^2\theta}$$

$$= \frac{4 \sin \theta}{\cos^2\theta}$$

$$= \frac{4 \sin \theta}{\cos \theta \cos \theta}$$

$$= 4 \cdot \frac{\sin \theta}{\cos \theta} \cdot \frac{1}{\cos \theta}$$

$$= 4 \tan \theta \sec \theta$$

71. $$\frac{\cos\theta+1}{\cos\theta-1}+\frac{1-\sec\theta}{1+\sec\theta}=\frac{\cos\theta+1}{\cos\theta-1}+\frac{1-\dfrac{1}{\cos\theta}}{1+\dfrac{1}{\cos\theta}}$$

$$=\frac{\cos\theta+1}{\cos\theta-1}+\frac{\dfrac{\cos\theta-1}{\cos\theta}}{\dfrac{\cos\theta+1}{\cos\theta}}$$

$$=\frac{\cos\theta+1}{\cos\theta-1}+\frac{\cos\theta-1}{\cos\theta+1}$$

$$=\frac{(\cos\theta+1)^2+(\cos\theta-1)^2}{(\cos\theta-1)(\cos\theta+1)}$$

$$=\frac{\cos^2\theta+2\cos\theta+1+\cos^2\theta-2\cos\theta+1}{\cos^2\theta-1}$$

$$=\frac{2\cos^2\theta+2}{-\sin^2\theta}$$

$$=-2\cot^2\theta-2\csc^2\theta$$

PROBLEM SET 7.4

1. $\cos(30°+\theta)=\cos30°\cos\theta-\sin30°\sin\theta=\dfrac{\sqrt{3}}{2}\cos\theta-\dfrac{1}{2}\sin\theta=\dfrac{\sqrt{3}\cos\theta-\sin\theta}{2}$

3. $\cos\left(\theta-\dfrac{\pi}{4}\right)=\cos\theta\cos\dfrac{\pi}{4}+\sin\theta\sin\dfrac{\pi}{4}=(\cos\theta)\!\left(\dfrac{\sqrt{2}}{2}\right)+(\sin\theta)\!\left(\dfrac{\sqrt{2}}{2}\right)=\dfrac{\sqrt{2}}{2}\left(\cos\theta+\sin\theta\right)$

5. $\tan(45°+\theta)=\dfrac{\tan45°+\tan\theta}{1-\tan45°\tan\theta}=\dfrac{1+\tan\theta}{1-\tan\theta}$

7. $\tan(\theta+\theta)=\dfrac{\tan\theta+\tan\theta}{1-\tan\theta\tan\theta}=\dfrac{2\tan\theta}{1-\tan^2\theta}$

9. $\cos(\theta+\theta)=\cos\theta\cos\theta-\sin\theta\sin\theta=\cos^2\theta-\sin^2\theta$

11. $\sin38°=\cos(90°-38°)=\cos52°$

13. $\sin\dfrac{\pi}{6}=\cos\left(\dfrac{\pi}{2}-\dfrac{\pi}{6}\right)=\cos\left(\dfrac{3\pi}{6}-\dfrac{\pi}{6}\right)=\cos\left(\dfrac{2\pi}{6}\right)=\cos\dfrac{\pi}{3}$

15. $\cot\dfrac{2\pi}{3}=\tan\left(\dfrac{\pi}{2}-\dfrac{2\pi}{3}\right)=\tan\left(\dfrac{3\pi}{6}-\dfrac{4\pi}{6}\right)=\tan\left(-\dfrac{\pi}{6}\right)=-\tan\dfrac{\pi}{6}$

17. $\tan(-49°)=-\tan49°$

19. $\sin(-31°) = -\sin 31°$

21. $\tan(-24°) = -\tan 24°$

23. $\cos 114° \cos 85° + \sin 114° \sin 85° = \cos(114° - 85°) = \cos 29° = .8746$

25. $\sin 18° \cos 23° + \cos 18° \sin 23° = \sin(18° + 23°) = \sin 41° = .6561$

27. $\dfrac{\tan 59° - \tan 25°}{1 + \tan 59° \tan 25°} = \tan(59° - 25°) = \tan 34° = .6745$

29. $\sin(-15°) = \sin(30° - 45°) = \sin 30° \cos 45° - \cos 30° \sin 45°$

$$= \frac{1}{2} \cdot \frac{\sqrt{2}}{2} - \frac{\sqrt{3}}{2} \cdot \frac{\sqrt{2}}{2} = \frac{\sqrt{2} - \sqrt{6}}{4}$$

$\cos(-15°) = \cos(30° - 45°) = \cos 30° \cos 45° + \sin 30° \sin 45°$

$$= \frac{\sqrt{3}}{2} \cdot \frac{\sqrt{2}}{2} + \frac{1}{2} \cdot \frac{\sqrt{2}}{2} = \frac{\sqrt{6} + \sqrt{2}}{4}$$

$\tan(-15°) = \tan(30° - 45°)$

$$= \frac{\tan 30° - \tan 45°}{1 + \tan 30° \tan 45°}$$

$$= \frac{\dfrac{\sqrt{3}}{3} - 1}{1 + \dfrac{\sqrt{3}}{3} \cdot 1}$$

$$= \frac{\sqrt{3} - 3}{3 + \sqrt{3}}$$

$$= \frac{\sqrt{3} - 3}{3 + \sqrt{3}} \cdot \frac{3 - \sqrt{3}}{3 - \sqrt{3}}$$

$$= \frac{3\sqrt{3} - 3 - 9 + 3\sqrt{3}}{9 - 3}$$

$$= \frac{-12 + 6\sqrt{3}}{6}$$

$$= -2 + \sqrt{3}$$

31. $\sin 75° = \sin(45° + 30°) = \sin 45° \cos 30° + \cos 45° \sin 30°$

$\quad = \dfrac{\sqrt{2}}{2} \cdot \dfrac{\sqrt{3}}{2} + \dfrac{\sqrt{2}}{2} \cdot \dfrac{1}{2} = \dfrac{\sqrt{6} + \sqrt{2}}{4}$

$\cos 75° = \cos(45° + 30°) = \cos 45° \cos 30° - \sin 45° \sin 30°$

$\quad = \dfrac{\sqrt{2}}{2} \cdot \dfrac{\sqrt{3}}{2} - \dfrac{\sqrt{2}}{2} \cdot \dfrac{1}{2} = \dfrac{\sqrt{6} - \sqrt{2}}{4}$

$\quad\quad \tan 75° = \tan(45° + 30°)$

$$= \frac{\tan 45° + \tan 30°}{1 - \tan 45° \tan 30°}$$

$$= \frac{1 + \dfrac{\sqrt{3}}{3}}{1 - 1 \cdot \dfrac{\sqrt{3}}{3}}$$

$$= \frac{3 + \sqrt{3}}{3 - \sqrt{3}}$$

$$= \frac{3 + \sqrt{3}}{3 - \sqrt{3}} \cdot \frac{3 + \sqrt{3}}{3 + \sqrt{3}}$$

$$= \frac{9 + 6\sqrt{3} + 3}{9 - 3}$$

$$= \frac{12 + 6\sqrt{3}}{6}$$

$$= 2 + \sqrt{3}$$

33. $\sin 105° = \sin(60° + 45°) = \sin 60° \cos 45° + \cos 60° \sin 45°$

$\quad = \dfrac{\sqrt{3}}{2} \cdot \dfrac{\sqrt{2}}{2} + \dfrac{1}{2} \cdot \dfrac{\sqrt{2}}{2} = \dfrac{\sqrt{6} + \sqrt{2}}{4}$

$\cos 105° = \cos(60° + 45°) = \cos 60° \cos 45° - \sin 60° \sin 45°$

$\quad = \dfrac{1}{2} \cdot \dfrac{\sqrt{2}}{2} - \dfrac{\sqrt{3}}{2} \cdot \dfrac{\sqrt{2}}{2} = \dfrac{\sqrt{2} - \sqrt{6}}{4}$

$\quad\quad \tan 105° = \tan(60° + 45°)$

$$= \frac{\tan 60° + \tan 45°}{1 - \tan 60° \tan 45°}$$

$$= \frac{\sqrt{3} + 1}{1 - \sqrt{3} \cdot 1}$$

$$= \frac{\sqrt{3} + 1}{1 - \sqrt{3}} \cdot \frac{1 + \sqrt{3}}{1 + \sqrt{3}}$$

$$= \frac{\sqrt{3} + 3 + 1 + \sqrt{3}}{1 - 3}$$

$$= \frac{4 + 2\sqrt{3}}{-2}$$

$$= -2 - \sqrt{3}$$

35. **a.** $\cos\left(\frac{2\pi}{3} - \theta\right) = \cos\left[-\left(\theta - \frac{2\pi}{3}\right)\right] = \cos\left(\theta - \frac{2\pi}{3}\right)$

   b. $\sin\left(\frac{2\pi}{3} - \theta\right) = \sin\left[-\left(\theta - \frac{2\pi}{3}\right)\right] = -\sin\left(\theta - \frac{2\pi}{3}\right)$

   c. $\tan\left(\frac{2\pi}{3} - \theta\right) = \tan\left[-\left(\theta - \frac{2\pi}{3}\right)\right] = -\tan\left(\theta - \frac{2\pi}{3}\right)$

37.

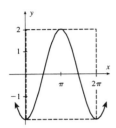

39.

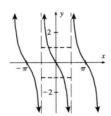

41.

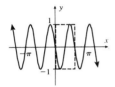

43.

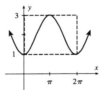

45.

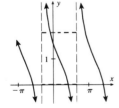

144

47. a. $s(-x) = \sin(-x)$

$\qquad = -\sin x$

$\qquad = -s(x)$

$\qquad s(x)$ is odd

b. $c(-x) = \cos(-x)$

$\qquad = \cos x$

$\qquad = c(x)$

$\qquad c(x)$ is even

49. a. $f(-x) = \dfrac{\sin(-x)}{-x}$

$\qquad = \dfrac{-\sin x}{-x}$

$\qquad = \dfrac{\sin x}{x}$

$\qquad = f(x)$

$\qquad f(x)$ is even

b. $g(-x) = \dfrac{-x}{\cos(-x)}$

$\qquad = \dfrac{-x}{\cos x}$

$\qquad = -\dfrac{x}{\cos x}$

$\qquad = -g(x)$

$\qquad g(x)$ is odd

51. $\quad \sin(\alpha - \beta) = \sin[\alpha + (-\beta)]$

$\qquad = \sin \alpha \cos(-\beta) + \cos \alpha \sin(-\beta)$

$\qquad = \sin \alpha \cos \beta + \cos \alpha (-\sin \beta)$

$\qquad = \sin \alpha \cos \beta - \cos \alpha \sin \beta$

53. $\quad \cot(\alpha + \beta) = \dfrac{\cos(\alpha + \beta)}{\sin(\alpha + \beta)}$

$\qquad = \dfrac{\cos \alpha \cos \beta - \sin \alpha \sin \beta}{\sin \alpha \cos \beta + \cos \alpha \sin \beta}$

$\qquad = \dfrac{\cos \alpha \cos \beta - \sin \alpha \sin \beta}{\sin \alpha \cos \beta + \cos \alpha \sin \beta} \cdot \dfrac{\frac{1}{\sin \alpha \sin \beta}}{\frac{1}{\sin \alpha \sin \beta}}$

$\qquad = \dfrac{\frac{\cos \alpha \cos \beta}{\sin \alpha \sin \beta} - 1}{\frac{\cos \beta}{\sin \beta} + \frac{\cos \alpha}{\sin \alpha}}$

$\qquad = \dfrac{\cot \alpha \cot \beta - 1}{\cot \beta + \cot \alpha}$

55. $\quad \dfrac{\cos 5\theta}{\sin \theta} - \dfrac{\sin 5\theta}{\cos \theta} = \dfrac{\cos 5\theta \cos \theta - \sin 5\theta \sin \theta}{\sin \theta \cos \theta}$

$\qquad = \dfrac{\cos(5\theta + \theta)}{\sin \theta \cos \theta}$

$\qquad = \dfrac{\cos 6\theta}{\sin \theta \cos \theta}$

57.
$$\sec\left(\frac{\pi}{2}-\theta\right)=\frac{1}{\cos\left(\frac{\pi}{2}-\theta\right)}$$
$$=\frac{1}{\sin\theta}$$
$$=\csc\theta$$

59.
$$\csc\left(\frac{\pi}{2}-\theta\right)=\frac{1}{\sin\left(\frac{\pi}{2}-\theta\right)}$$
$$=\frac{1}{\cos\theta}$$
$$=\sec\theta$$

61.
$$\sec(-\theta)=\frac{1}{\cos(-\theta)}$$
$$=\frac{1}{\cos\theta}$$
$$=\sec\theta$$

63.
$$\sin(\alpha+\beta)\cos\beta-\cos(\alpha+\beta)\sin\beta$$
$$=(\sin\alpha\cos\beta+\cos\alpha\sin\beta)\cos\beta-(\cos\alpha\cos\beta-\sin\alpha\sin\beta)\sin\beta$$
$$=\sin\alpha\cos^2\beta+\cos\alpha\sin\beta\cos\beta-\cos\alpha\cos\beta\sin\beta+\sin\alpha\sin^2\beta$$
$$=\sin\alpha\cos^2\beta+\sin\alpha\sin^2\beta$$
$$=\sin\alpha\left(\cos^2\beta+\sin^2\beta\right)$$
$$=\sin\alpha$$

65.
$$\tan\alpha=\tan[(\alpha+\beta)-\beta]$$
$$=\frac{\tan(\alpha+\beta)-\tan\beta}{1+\tan(\alpha+\beta)\tan\beta}$$

67.
$$\frac{\cos(\theta+h)-\cos\theta}{h}=\frac{\cos\theta\cos h-\sin\theta\sin h-\cos\theta}{h}$$
$$=-\frac{\sin\theta\sin h}{h}+\frac{\cos\theta\cos h-\cos\theta}{h}$$
$$=-\sin\theta\left(\frac{\sin h}{h}\right)-\cos\theta\left(\frac{1-\cos h}{h}\right)$$

69.
$$\cos(\alpha+\beta+\gamma)=\cos[(\alpha+\beta)+\gamma]$$
$$=\cos(\alpha+\beta)\cos\gamma-\sin(\alpha+\beta)\sin\gamma$$
$$=(\cos\alpha\cos\beta-\sin\alpha\sin\beta)\cos\gamma-(\sin\alpha\cos\beta+\cos\alpha\sin\beta)\sin\gamma$$
$$=\cos\alpha\cos\beta\cos\gamma-\sin\alpha\sin\beta\cos\gamma-\sin\alpha\cos\beta\sin\gamma-\cos\alpha\sin\beta\sin\gamma$$
$$=\cos\alpha\cos\beta\cos\gamma-\cos\alpha\sin\beta\sin\gamma-\sin\alpha\cos\beta\sin\gamma-\sin\alpha\sin\beta\cos\gamma$$

71.

$$\cot(\alpha + \beta + \gamma) = \cot[(\alpha + \beta) + \gamma]$$

$$= \frac{\cot(\alpha + \beta) \cot \gamma - 1}{\cot \gamma + \cot(\alpha + \beta)}$$

$$= \frac{\left(\dfrac{\cot \alpha \cot \beta - 1}{\cot \beta + \cot \alpha}\right) \cdot \cot \gamma - 1}{\cot \gamma + \left(\dfrac{\cot \alpha \cot \beta - 1}{\cot \beta + \cot \alpha}\right)}$$

$$= \frac{(\cot \alpha \cot \beta - 1) \cot \gamma - (\cot \beta + \cot \alpha)}{\cot \gamma (\cot \beta + \cot \alpha) + (\cot \alpha \cot \beta - 1)}$$

$$= \frac{\cot \alpha \cot \beta \cot \gamma - \cot \gamma - \cot \beta - \cot \alpha}{\cot \gamma \cot \beta + \cot \gamma \cot \alpha + \cot \alpha \cot \beta - 1}$$

$$= \frac{\cot \alpha \cot \beta \cot \gamma - \cot \alpha - \cot \beta - \cot \gamma}{\cot \beta \cot \gamma + \cot \alpha \cot \gamma + \cot \alpha \cot \beta - 1}$$

PROBLEM SET 7.5

1. $2 \cos^2 22.5° = \cos 2(22.5°) = \cos 45° = \dfrac{\sqrt{2}}{2}$

3. $\sqrt{\dfrac{1 - \cos 60°}{2}} = \sin \frac{1}{2}(60°) = \sin 30° = \dfrac{1}{2}$

5. $\cos\left(\dfrac{\pi}{8}\right) = \cos \frac{1}{2}\left(\dfrac{\pi}{4}\right) = \sqrt{\dfrac{1 + \cos \frac{\pi}{4}}{2}} = \sqrt{\dfrac{1 + \frac{\sqrt{2}}{2}}{2}}$

$$= \sqrt{\dfrac{\frac{2 + \sqrt{2}}{2}}{2}} = \sqrt{\dfrac{2 + \sqrt{2}}{4}} = \frac{1}{2}\sqrt{2 + \sqrt{2}}$$

7. $\cos(\theta + \phi) = \cos \theta \cos \phi - \sin \theta \sin \phi$

9. $\dfrac{\sin 2\theta}{2} = \dfrac{2 \sin \theta \cos \theta}{2} = \sin \theta \cos \theta$

11. $\cos \dfrac{\theta}{2} = \pm \sqrt{\dfrac{1 + \cos \theta}{2}}$; $\pm$ is missing; also sinus is minus

13. $\sin \theta = \dfrac{3}{5}$; θ in Quadrant I

$$\cos \theta = \sqrt{1 - \sin^2\theta} = \sqrt{1 - \left(\dfrac{3}{5}\right)^2} = \sqrt{1 - \dfrac{9}{25}} = \sqrt{\dfrac{16}{25}} = \dfrac{4}{5}$$

$$\cos 2\theta = \cos^2\theta - \sin^2\theta = \left(\dfrac{4}{5}\right)^2 - \left(\dfrac{3}{5}\right)^2 = \dfrac{16}{25} - \dfrac{9}{25} = \dfrac{7}{25}$$

$$\sin 2\theta = 2 \sin \theta \cos \theta = 2 \cdot \dfrac{3}{5} \cdot \dfrac{4}{5} = \dfrac{24}{25}$$

$$\tan 2\theta = \dfrac{\sin 2\theta}{\cos 2\theta} = \dfrac{24/25}{7/25} = \dfrac{24}{7}$$

15. $\tan \theta = -\frac{5}{12}$; θ in Quadrant IV

$\sec \theta = \sqrt{\tan^2\theta + 1} = \sqrt{\left(-\frac{5}{12}\right)^2 + 1} = \sqrt{\frac{25}{144} + 1} = \sqrt{\frac{169}{144}} = \frac{13}{12}$

$\cos \theta = \frac{1}{\sec \theta} = \frac{12}{13}$

$\sin \theta = -\sqrt{1 - \cos^2\theta} = -\sqrt{1 - \left(\frac{12}{13}\right)^2} = -\sqrt{\frac{25}{169}} = -\frac{5}{13}$

$\cos 2\theta = \cos^2\theta - \sin^2\theta = \left(\frac{12}{13}\right)^2 - \left(-\frac{5}{13}\right)^2 = \frac{144}{169} - \frac{25}{169} = \frac{119}{169}$

$\sin 2\theta = 2 \sin \theta \cos \theta = 2 \cdot \left(-\frac{5}{13}\right) \cdot \frac{12}{13} = -\frac{120}{169}$

$\tan 2\theta = \frac{\sin 2\theta}{\cos 2\theta} = \frac{-120/169}{119/169} = -\frac{120}{119}$

17. $\cos \theta = \frac{5}{9}$; θ in Quadrant I

$\sin \theta = \sqrt{1 - \cos^2\theta} = \sqrt{1 - \left(\frac{5}{9}\right)^2} = \sqrt{1 - \frac{25}{81}} = \sqrt{\frac{56}{81}} = \frac{2\sqrt{14}}{9}$

$\cos 2\theta = \cos^2\theta - \sin^2\theta = \left(\frac{5}{9}\right)^2 - \left(\frac{2\sqrt{14}}{9}\right)^2 = \frac{25}{81} - \frac{56}{81} = -\frac{31}{81}$

$\sin 2\theta = 2 \sin \theta \cos \theta = 2 \cdot \left(\frac{2\sqrt{14}}{9}\right) \cdot \frac{5}{9} = \frac{20\sqrt{14}}{81}$

$\tan 2\theta = \frac{\sin 2\theta}{\cos 2\theta} = \frac{20\sqrt{14}/81}{-31/81} = -\frac{20\sqrt{14}}{31}$

19. $\sin \theta = \frac{3}{5}$; θ in Quadrant I

$\cos \theta = \sqrt{1 - \sin^2\theta} = \sqrt{1 - \left(\frac{3}{5}\right)^2} = \sqrt{1 - \frac{9}{25}} = \sqrt{\frac{16}{25}} = \frac{4}{5}$

$\cos \tfrac{1}{2}\theta = \sqrt{\frac{1 + \cos \theta}{2}} = \sqrt{\frac{1 + \frac{4}{5}}{2}} = \sqrt{\frac{9}{10}} = \frac{3}{\sqrt{10}}$

$\sin \tfrac{1}{2}\theta = \sqrt{\frac{1 - \cos \theta}{2}} = \sqrt{\frac{1 - \frac{4}{5}}{2}} = \sqrt{\frac{1}{10}} = \frac{1}{\sqrt{10}}$

$\tan \tfrac{1}{2}\theta = \frac{\sin \tfrac{1}{2}\theta}{\cos \tfrac{1}{2}\theta} = \frac{\frac{1}{\sqrt{10}}}{\frac{3}{\sqrt{10}}} = \frac{1}{3}$

21. $\tan \theta = -\frac{5}{12}$; θ in Quadrant IV

$$\sec \theta = \sqrt{1 + \tan^2\theta} = \sqrt{1 + \left(-\frac{5}{12}\right)^2} = \sqrt{1 + \frac{25}{144}} = \sqrt{\frac{169}{144}} = \frac{13}{12}$$

$$\cos \theta = \frac{1}{\sec \theta} = \frac{12}{13}$$

$$\sin \theta = -\sqrt{1 - \cos^2\theta} = -\sqrt{1 - \left(\frac{12}{13}\right)^2} = -\sqrt{\frac{25}{169}} = -\frac{5}{13}$$

$$\cos \tfrac{1}{2}\theta = -\sqrt{\frac{1 + \cos \theta}{2}} = -\sqrt{\frac{1 + \frac{12}{13}}{2}} = -\sqrt{\frac{25}{26}} = -\frac{5}{\sqrt{26}}$$

$$\sin \tfrac{1}{2}\theta = \sqrt{\frac{1 - \cos \theta}{2}} = \sqrt{\frac{1 - \frac{12}{13}}{2}} = \sqrt{\frac{1}{26}} = \frac{1}{\sqrt{26}}$$

$$\tan \tfrac{1}{2}\theta = \frac{\sin \frac{1}{2}\theta}{\cos \frac{1}{2}\theta} = \frac{1/\sqrt{26}}{-5/\sqrt{26}} = -\frac{1}{5}$$

23. $\cos \theta = \frac{5}{9}$; θ in Quadrant I

$$\cos \tfrac{1}{2}\theta = \sqrt{\frac{1 + \cos \theta}{2}} = \sqrt{\frac{1 + \frac{5}{9}}{2}} = \sqrt{\frac{14}{18}} = \sqrt{\frac{7}{9}} = \frac{\sqrt{7}}{3}$$

$$\sin \tfrac{1}{2}\theta = \sqrt{\frac{1 - \cos \theta}{2}} = \sqrt{\frac{1 - \frac{5}{9}}{2}} = \sqrt{\frac{4}{18}} = \sqrt{\frac{2}{9}} = \frac{\sqrt{2}}{3}$$

$$\tan \tfrac{1}{2}\theta = \frac{\sin \frac{1}{2}\theta}{\cos \frac{1}{2}\theta} = \frac{\sqrt{2}/3}{\sqrt{7}/3} = \frac{\sqrt{2}}{\sqrt{7}} = \frac{\sqrt{2}}{\sqrt{7}} \cdot \frac{\sqrt{7}}{\sqrt{7}} = \frac{\sqrt{14}}{7}$$

25. $2 \sin 35° \sin 24° = \cos(35° - 24°) - \cos(35° + 24°) = \cos 11° - \cos 59°$

27. $\sin 225° \sin 300° = \frac{1}{2}[\cos(225° - 300°) - \cos(225° + 300°)]$

$$= \tfrac{1}{2}[\cos(-75°) - \cos 525°] = \tfrac{1}{2}(\cos 75° - \cos 165°) = \tfrac{1}{2}\cos 75° - \tfrac{1}{2}\cos 165°$$

29. $\sin 2\theta \sin 5\theta = \frac{1}{2}[\cos(2\theta - 5\theta) - \cos(2\theta + 5\theta)] = \frac{1}{2}[\cos(-3\theta) - \cos 7\theta]$

$$= \tfrac{1}{2}\cos 3\theta - \tfrac{1}{2}\cos 7\theta$$

31. $\sin 43° + \sin 63° = 2 \sin\left(\frac{43 + 63}{2}\right) \cos\left(\frac{43 - 63}{2}\right) = 2 \sin 53° \cos(-10°)$

$$= 2 \sin 53° \cos 10°$$

33. $\sin 215° + \sin 300° = 2 \sin\left(\dfrac{215° + 300°}{2}\right) \cos\left(\dfrac{215° - 300°}{2}\right)$

$= 2 \sin 257.5° \cos(-42.5°) = 2 \sin 257.5° \cos 42.5°$

35. $\cos 6x + \cos 2x = 2 \cos\left(\dfrac{6x + 2x}{2}\right) \cos\left(\dfrac{6x - 2x}{2}\right) = 2 \cos 4x \cos 2x$

37. $\sin x - \sin 3x = 0$; $\ 2 \sin\left(\dfrac{x - 3x}{2}\right) \cos\left(\dfrac{x + 3x}{2}\right) = 0$; $\ 2 \sin(-x) \cos 2x = 0$;

$-2 \sin x \cos 2x = 0$; $\ \sin x = 0$ at $x = 0$ and π; $\ \cos 2x = 0$ at $2x' = \dfrac{\pi}{2}$;

$2x = \dfrac{\pi}{2}, \dfrac{3\pi}{2}, \dfrac{5\pi}{2}, \dfrac{7\pi}{2}$; $\ x = \dfrac{\pi}{4}, \dfrac{3\pi}{4}, \dfrac{5\pi}{4}, \dfrac{7\pi}{4}$; $\ x = 0, \dfrac{\pi}{4}, \dfrac{3\pi}{4}, \pi, \dfrac{5\pi}{4}, \dfrac{7\pi}{4}$

39. $\cos 5y + \cos 3y = 0$; $\ 2 \cos\left(\dfrac{5y + 3y}{2}\right) \cos\left(\dfrac{5y - 3y}{2}\right) = 0$;

$2 \cos 4y \cos y = 0$; $\ \cos 4y = 0$ at $4y' = \dfrac{\pi}{2}$;

$4y = \dfrac{\pi}{2}, \dfrac{3\pi}{2}, \dfrac{5\pi}{2}, \dfrac{7\pi}{2}, \dfrac{9\pi}{2}, \dfrac{11\pi}{2}, \dfrac{13\pi}{2}, \dfrac{15\pi}{2}$;

$y = \dfrac{\pi}{8}, \dfrac{3\pi}{8}, \dfrac{5\pi}{8}, \dfrac{7\pi}{8}, \dfrac{9\pi}{8}, \dfrac{11\pi}{8}, \dfrac{13\pi}{8}, \dfrac{15\pi}{8}$; $\ \cos y = 0$ at $y = \dfrac{\pi}{2}$ and $\dfrac{3\pi}{2}$;

$y = \dfrac{\pi}{8}, \dfrac{3\pi}{8}, \dfrac{\pi}{2}, \dfrac{5\pi}{8}, \dfrac{7\pi}{8}, \dfrac{9\pi}{8}, \dfrac{11\pi}{8}, \dfrac{3\pi}{2}, \dfrac{13\pi}{8}, \dfrac{15\pi}{8}$

41. $\cos 5x - \cos 3x = 0$; $\ -2 \sin\left(\dfrac{5x + 3x}{2}\right) \sin\left(\dfrac{5x - 3x}{2}\right) = 0$;

$-2 \sin 4x \sin x = 0$; $\ \sin 4x = 0$ at $4x' = 0$;

$4x = 0, \pi, 2\pi, 3\pi, 4\pi, 5\pi, 6\pi, 7\pi$; $\ 4x = 0, \dfrac{\pi}{4}, \dfrac{\pi}{2}, \dfrac{3\pi}{4}, \pi, \dfrac{5\pi}{4}, \dfrac{3\pi}{2}, \dfrac{7\pi}{4}$;

$\sin x = 0$ at $x = 0$ and π; $\ y = 0, \dfrac{\pi}{4}, \dfrac{\pi}{2}, \dfrac{3\pi}{4}, \pi, \dfrac{5\pi}{4}, \dfrac{3\pi}{2}, \dfrac{7\pi}{4}$

43. $\sin 3z - \sin 5z = 0$; $\ 2 \sin\left(\dfrac{3z - 5z}{2}\right) \cos\left(\dfrac{3z + 5z}{2}\right) = 0$;

$2 \sin(-z) \cos 4z = 0$; $\ -2 \sin z \cos 4z = 0$; $\ \sin z = 0$ at $z = 0$ and π;

$\cos 4z = 0$ at $4z = \dfrac{\pi}{2}, \dfrac{3\pi}{2}, \dfrac{5\pi}{2}, \dfrac{7\pi}{2}, \dfrac{9\pi}{2}, \dfrac{11\pi}{2}, \dfrac{13\pi}{2}, \dfrac{15\pi}{2}$;

$z = \dfrac{\pi}{8}, \dfrac{3\pi}{8}, \dfrac{5\pi}{8}, \dfrac{7\pi}{8}, \dfrac{9\pi}{8}, \dfrac{11\pi}{8}, \dfrac{13\pi}{8}, \dfrac{15\pi}{8}$;

$y = 0, \dfrac{\pi}{8}, \dfrac{3\pi}{8}, \dfrac{5\pi}{8}, \dfrac{7\pi}{8}, \pi, \dfrac{9\pi}{8}, \dfrac{11\pi}{8}, \dfrac{13\pi}{8}, \dfrac{15\pi}{8}$

45. $\cot 2\theta = -\frac{3}{4}$

$\tan 2\theta = \dfrac{1}{\cot 2\theta} = -\dfrac{4}{3}$

$\sec 2\theta = -\sqrt{\tan^2 2\theta + 1} = -\sqrt{\left(-\dfrac{4}{3}\right)^2 + 1} = -\sqrt{\dfrac{25}{9}} = -\dfrac{5}{3}$

$\cos 2\theta = \dfrac{1}{\sec 2\theta} = -\dfrac{3}{5}$

$\sin \theta = \sqrt{\dfrac{1 - \cos 2\theta}{2}} = \sqrt{\dfrac{1 - \left(-\dfrac{3}{5}\right)}{2}} = \sqrt{\dfrac{8}{10}} = \sqrt{\dfrac{4}{5}} = \dfrac{2}{\sqrt{5}}$

$\cos \theta = \sqrt{\dfrac{1 + \cos 2\theta}{2}} = \sqrt{\dfrac{1 + \left(-\dfrac{3}{5}\right)}{2}} = \sqrt{\dfrac{2}{10}} = \sqrt{\dfrac{1}{5}} = \dfrac{1}{\sqrt{5}}$

$\tan \theta = \dfrac{\sin \theta}{\cos \theta} = \dfrac{2/\sqrt{5}}{1/\sqrt{5}} = 2$

47. $\cot 2\theta = -\frac{4}{3}$

$\tan 2\theta = \dfrac{1}{\cot 2\theta} = -\dfrac{3}{4}$

$\sec 2\theta = \sqrt{\tan^2 2\theta + 1} = -\sqrt{\left(-\dfrac{3}{4}\right)^2 + 1} = -\sqrt{\dfrac{25}{16}} = -\dfrac{5}{4}$

$\cos 2\theta = \dfrac{1}{\sec 2\theta} = -\dfrac{4}{5}$

$\sin \theta = \sqrt{\dfrac{1 - \cos 2\theta}{2}} = \sqrt{\dfrac{1 - \left(-\dfrac{4}{5}\right)}{2}} = \sqrt{\dfrac{9}{10}} = \dfrac{3}{\sqrt{10}}$

$\cos \theta = \sqrt{\dfrac{1 + \cos 2\theta}{2}} = \sqrt{\dfrac{1 + \left(-\dfrac{4}{5}\right)}{2}} = \sqrt{\dfrac{1}{10}} = \dfrac{1}{\sqrt{10}}$

$\tan \theta = \dfrac{\sin \theta}{\cos \theta} = \dfrac{3/\sqrt{10}}{1/\sqrt{10}} = 3$

49. $\cot 2\theta = -\dfrac{1}{\sqrt{3}}$

$\tan 2\theta = \dfrac{1}{\cot 2\theta} = -\sqrt{3}$

$\sec 2\theta = \sqrt{\tan^2 2\theta + 1} = -\sqrt{\left(-\sqrt{3}\right)^2 + 1} = -\sqrt{4} = -2$

$\cos 2\theta = \dfrac{1}{\sec 2\theta} = -\dfrac{1}{2}$

$\sin \theta = \sqrt{\dfrac{1 - \cos 2\theta}{2}} = \sqrt{\dfrac{1 - \left(-\dfrac{1}{2}\right)}{2}} = \sqrt{\dfrac{3}{4}} = \dfrac{\sqrt{3}}{2}$

$\cos \theta = \sqrt{\dfrac{1 + \cos 2\theta}{2}} = \sqrt{\dfrac{1 + \left(-\dfrac{1}{2}\right)}{2}} = \sqrt{\dfrac{1}{4}} = \dfrac{1}{2}$

$\tan \theta = \dfrac{\sin \theta}{\cos \theta} = \dfrac{\sqrt{3}/2}{1/2} = \sqrt{3}$

51. a.

$$\sin \frac{\theta}{2} = \frac{1}{M}$$

$$\sin \tfrac{1}{2}\left(\tfrac{\pi}{6}\right) = \frac{1}{M}$$

$$\sqrt{\frac{1 - \cos \frac{\pi}{6}}{2}} = \frac{1}{M}$$

$$\sqrt{\frac{1 - \frac{\sqrt{3}}{2}}{2}} = \frac{1}{M}$$

$$\sqrt{\frac{2 - \sqrt{3}}{4}} = \frac{1}{M}$$

$$\frac{\sqrt{2 - \sqrt{3}}}{2} = \frac{1}{M}$$

$$M = \frac{2}{\sqrt{2 - \sqrt{3}}}$$

$$M \approx 3.9$$

b. From part (a)

$$\frac{1}{M} = \sin \tfrac{1}{2}\left(\tfrac{\pi}{6}\right)$$

$$\frac{1}{M} = \sin \frac{\pi}{12}$$

$$\frac{1}{M} = \sin \left(\tfrac{\pi}{3} - \tfrac{\pi}{4}\right)$$

$$\frac{1}{M} = \sin \tfrac{\pi}{3} \cos \tfrac{\pi}{4} - \cos \tfrac{\pi}{3} \sin \tfrac{\pi}{4}$$

$$\frac{1}{M} = \frac{\sqrt{3}}{2} \cdot \frac{\sqrt{2}}{2} - \frac{1}{2} \cdot \frac{\sqrt{2}}{2}$$

$$\frac{1}{M} = \frac{\sqrt{2}(\sqrt{3} - 1)}{4}$$

$$M = \frac{4}{\sqrt{2}(\sqrt{3} - 1)}$$

$$M = \frac{4}{\sqrt{2}(\sqrt{3} - 1)} \cdot \frac{\sqrt{3} + 1}{\sqrt{3} + 1}$$

$$M = \frac{4(\sqrt{3} + 1)}{\sqrt{2}(3 - 1)} \cdot \frac{\sqrt{2}}{\sqrt{2}}$$

$$M = \frac{4\sqrt{2}(\sqrt{3} + 1)}{4}$$

$$M = \sqrt{2}(\sqrt{3} + 1)$$

53.

$$\cos 4\theta = \cos 2(2\theta)$$
$$= \cos^2 2\theta - \sin^2 2\theta$$

55.

$$\tan \tfrac{3}{2}\beta = \tan 2\left(\tfrac{3}{4}\beta\right)$$

$$= \frac{2 \tan \tfrac{3}{4}\beta}{1 - \tan^2 \tfrac{3}{4}\beta}$$

57.

$$\tan \tfrac{1}{2}\theta = \frac{\sin \tfrac{1}{2}\theta}{\cos \tfrac{1}{2}\theta}$$

$$= \frac{\pm\sqrt{\dfrac{1-\cos \theta}{2}}}{\pm\sqrt{\dfrac{1+\cos \theta}{2}}}$$

$$= \sqrt{\frac{1-\cos \theta}{1+\cos \theta}}$$

$$= \sqrt{\frac{1-\cos \theta}{1+\cos \theta}} \cdot \sqrt{\frac{1+\cos \theta}{1+\cos \theta}}$$

$$= \sqrt{\frac{1-\cos^2\theta}{(1+\cos \theta)^2}}$$

$$= \sqrt{\frac{\sin^2\theta}{(1+\cos \theta)^2}}$$

$$= \frac{\sin \theta}{1+\cos \theta}$$

59.

$$\frac{\cos 5w + \cos w}{\cos w - \cos 5w} = \frac{2 \cos\left(\dfrac{5w + w}{2}\right) \cos\left(\dfrac{5w - w}{2}\right)}{-2 \sin\left(\dfrac{w + 5w}{2}\right) \sin\left(\dfrac{w - 5w}{2}\right)}$$

$$= -\frac{\cos 3w \cos 2w}{\sin 3w \sin(-2w)}$$

$$= \frac{\cos 2w/\sin 2w}{\sin 3w/\cos 3w}$$

$$= \frac{\cot 2w}{\tan 3w}$$

61.

$$\sin 3\theta = \sin(2\theta + \theta)$$

$$= \sin 2\theta \cos \theta + \cos 2\theta \sin \theta$$

$$= 2 \sin \theta \cos \theta \cos \theta + (1 - 2 \sin^2\theta)\sin \theta$$

$$= 2 \sin \theta \cos^2\theta + \sin \theta - 2 \sin^3\theta$$

$$= 2 \sin \theta (1 - \sin^2\theta) + \sin \theta - 2 \sin^3\theta$$

$$= 2 \sin \theta - 2 \sin^3\theta + \sin\theta - 2 \sin^3\theta$$

$$= 3 \sin\theta - 4 \sin^3\theta$$

63. $\sin 4\theta = \sin 2(2\theta)$

$= 2 \sin 2\theta \cos 2\theta$

$= 2(2 \sin \theta \cos \theta)(1 - 2 \sin^2\theta)$

$= 4 \sin \theta \cos \theta - 8 \sin^3\theta \cos \theta$

65. $\frac{1}{8}(3 + 4 \cos 2\theta + \cos 4\theta) = \frac{1}{8}[3 + 4(2 \cos^2\theta - 1) + (2 \cos^2 2\theta - 1)]$

$= \frac{1}{8}[3 + 8 \cos^2\theta - 4 + 2(2 \cos^2\theta - 1)^2 - 1]$

$= \frac{1}{8}(3 + 8 \cos^2\theta - 4 + 8 \cos^4\theta - 8 \cos^2\theta + 2 - 1)$

$= \frac{1}{8}(8 \cos^4\theta)$

$= \cos^2\theta$

PROBLEM SET 7.6

1.   cis 90° is on the positive imaginary axis   **3.**   True

5.   Quadrant IV

7.   a. $|3 + i| = \sqrt{3^2 + 1^2} = \sqrt{10}$

b. $|7 - i| = \sqrt{7^2 + (-1)^2} = \sqrt{50} = 5\sqrt{2}$

c. $|3 + 2i| = \sqrt{3^2 + 2^2} = \sqrt{13}$

d. $|-3 - 2i| = \sqrt{(-3)^2 + (-2)^2} = \sqrt{13}$

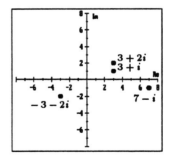

9.   a. $|-2 + 5i| = \sqrt{(-2)^2 + 5^2} = \sqrt{29}$

b. $|-5 + 4i| = \sqrt{(-5)^2 + 4^2} = \sqrt{41}$

c. $|4 - i| = \sqrt{4^2 + (-1)^2} = \sqrt{17}$

d. $|-1 + i| = \sqrt{(-1)^2 + 1^2} = \sqrt{2}$

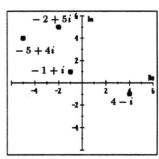

11. a. $\sqrt{3}-i$; $r=\sqrt{(\sqrt{3})^2+(-1)^2}=2$; $\tan\theta'=\left|\dfrac{-1}{\sqrt{3}}\right|$; $\theta'=30°$;

θ in Quadrant IV, $\theta=330°$; $\sqrt{3}-i=2$ cis $330°$

   b. $\sqrt{3}+i$; $r=\sqrt{(\sqrt{3})^2+1^2}=2$; $\tan\theta'=\left|\dfrac{1}{\sqrt{3}}\right|$; $\theta'=30°$;

θ in Quadrant I, $\theta=30°$; $\sqrt{3}+i=2$ cis $30°$

11.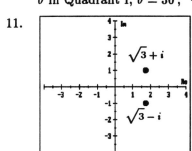

13.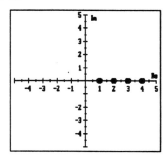

13. a.   1; $r=\sqrt{1^2+0^2}=1$; $\tan\theta'=\left|\dfrac{0}{1}\right|$; $\theta'=0°$; θ on positive real axis;

$\theta=0°$;   1 = cis $0°$

   b.   2; $r=\sqrt{2^2+0^2}=2$; $\tan\theta'=\left|\dfrac{0}{2}\right|$; $\theta'=0°$; θ on positive real axis;

$\theta=0°$;   2 = 2 cis $0°$

   c.   3; $r=\sqrt{3^2+0^2}=3$; $\tan\theta'=\left|\dfrac{0}{3}\right|$; $\theta'=0°$; θ on positive real axis;

$\theta=0°$;   3 = 3 cis $0°$

   d.   4; $r=\sqrt{4^2+0^2}=4$; $\tan\theta'=\left|\dfrac{0}{4}\right|$; $\theta'=0°$; θ on positive real axis;

$\theta=0°$;   4 = 4 cis $0°$

15. a. $-5i$; $r=\sqrt{0^2+(-5)^2}=5$;

θ on negative imaginary axis;

$\theta=270°$; $-5i=5$ cis $270°$

   b. $-6i$; $r=\sqrt{0^2+(-6)^2}=6$;

θ on negative imaginary axis;

$\theta=270°$; $-6i=6$ cis $270°$

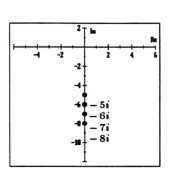

   c. $-7i$; $r=\sqrt{0^2+(-7)^2}=3$;

θ on negative imaginary axis;

$\theta=270°$; $-7i=7$ cis $270°$

   d. $-8i$; $r=\sqrt{0^2+(-8)^2}=4$;

θ on negative imaginary axis; $\theta=270°$; $-8i=8$ cis $270°$

155

17. $5.7956 - 1.5529i$

$r = \sqrt{(5.7956)^2 + (-1.5529)^2} \approx \sqrt{36} = 6$

$\tan \theta' = \left| \dfrac{-1.5529}{5.7956} \right|$

$\theta' \approx 15°$

θ in Quadrant IV, $\theta = 345°$

$5.7956 - 1.5529i \approx 6 \text{ cis } 345°$

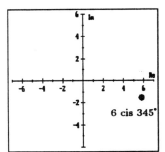

19. $-.6946 + 3.9392i$

$r = \sqrt{(-.6946)^2 + (3.9392)^2} \approx \sqrt{16} = 4$

$\tan \theta' = \left| \dfrac{3.9392}{-.6946} \right|$

$\theta' = 80°$

θ in Quadrant II, $\theta = 100°$

$-.6946 + 3.9392i \approx 4 \text{ cis } 100°$

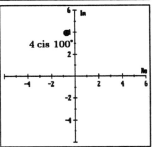

21. $3(\cos 60° + i \sin 60°)$

$= 3\left(\dfrac{1}{2} + i \dfrac{\sqrt{3}}{2} \right)$

$= \dfrac{3}{2} + \dfrac{3\sqrt{3}}{2}\, i$

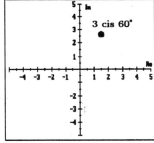

23. $5 \text{ cis}\left(\dfrac{3\pi}{2} \right)$

$= 5\left(\cos \dfrac{3\pi}{2} + i \sin \dfrac{3\pi}{2} \right)$

$= 5[0 + i(-1)]$

$= -5i$

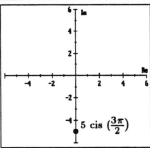

25. $9 \text{ cis } 190°$

$= 9(\cos 190° + i \sin 190°)$

$= -8.8633 - 1.5628\, i$

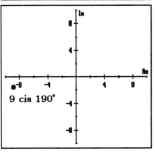

156

27. $3 \text{ cis } 48° \cdot 5 \text{ cis } 92° = 15 \text{ cis}(48° + 92°) = 15 \text{ cis } 140°$

29. $\dfrac{5(\cos 315° + i \sin 315°)}{2(\cos 48° + i \sin 48°)} = \dfrac{5}{2}\left[\text{cis}(315° - 48°)\right] = \dfrac{5}{2} \text{ cis } 267°$

31. $\dfrac{12 \text{ cis } 250°}{4 \text{ cis } 120°} = 3 \text{ cis}(250° - 120°) = 3 \text{ cis } 130°$

33. $(3 \text{ cis } 60°)^4 = 3^4 \text{ cis}(4 \cdot 60°) = 81 \text{ cis } 240°$

35. $(2 - 2i)^4$; $a = 2$ and $b = -2$; $r = \sqrt{2^2 + (-2)^2} = \sqrt{8} = 2\sqrt{2}$;
$\tan \theta' = \left|\dfrac{-2}{2}\right|$; $\theta' = 45°$; θ in Quadrant IV, $\theta = 315°$;
$2 - 2i = 2\sqrt{2} \text{ cis } 315°$
$(2 - 2i)^4 = \left(2\sqrt{2} \text{ cis } 315°\right)^4 = (2\sqrt{2})^4 \text{ cis }(4 \cdot 315°) = 64 \text{ cis } 1260°$
$\qquad = 64 \text{ cis } 180° = 64(\cos 180° + i \sin 180°) = 64(-1 + 0) = -64$

37. $(\sqrt{3} - i)^8$; $a = \sqrt{3}$ and $b = -1$; $r = \sqrt{(\sqrt{3})^2 + (-1)^2} = \sqrt{4} = 2$;
$\tan \theta' = \left|\dfrac{-1}{\sqrt{3}}\right|$; $\theta' = 30°$; θ in Quadrant IV, $\theta = 330°$; $\sqrt{3} - i = 2 \text{ cis } 330°$;
$(\sqrt{3} - i)^8 = \left(2 \text{ cis } 330°\right)^8 = (2)^8 \text{ cis}(8 \cdot 330°) = 256 \text{ cis } 2640°$
$\qquad = 256 \text{ cis } 120° = 256(\cos 120° + i \sin 120°)$
$\qquad = 256\left(-\dfrac{1}{2} + i \dfrac{\sqrt{3}}{2}\right) = -128 + 128\sqrt{3}\, i$

39. $\sqrt[4]{81 \text{ cis } 88°} = 81^{1/4} \text{ cis}\left(\dfrac{88° + 360°k}{4}\right) = 3 \text{ cis}\left(\dfrac{88° + 360°k}{4}\right)$

$\qquad k = 0: \quad 3 \text{ cis}\left(\dfrac{88° + 360° \cdot 0}{4}\right) = 3 \text{ cis } 22°$

$\qquad k = 1: \quad 3 \text{ cis}\left(\dfrac{88° + 360° \cdot 1}{4}\right) = 3 \text{ cis } 112°$

$\qquad k = 2: \quad 3 \text{ cis}\left(\dfrac{88° + 360° \cdot 2}{4}\right) = 3 \text{ cis } 202°$

$\qquad k = 3: \quad 3 \text{ cis}\left(\dfrac{88° + 360° \cdot 3}{4}\right) = 3 \text{ cis } 292°$

$\sqrt[4]{81 \text{ cis } 88°} = 3 \text{ cis } 22°, \ 3 \text{ cis } 112°, \ 3 \text{ cis } 202°, \ 3 \text{ cis } 292°$

41. $\sqrt[3]{-1} = \sqrt[3]{-1 + 0i} = \sqrt[3]{\text{cis } 180°} = \text{cis}\left(\dfrac{180° + 360°k}{3}\right)$

$\quad\quad k = 0: \quad \text{cis}\left(\dfrac{180° + 360° \cdot 0}{3}\right) = \text{cis } 60°$

$\quad\quad k = 1: \quad \text{cis}\left(\dfrac{180° + 360° \cdot 1}{3}\right) = \text{cis } 180°$

$\quad\quad k = 2: \quad \text{cis}\left(\dfrac{180° + 360° \cdot 2}{3}\right) = \text{cis } 300°$

$\sqrt[3]{-1} = \text{cis } 60°, \text{cis } 180°, \text{cis } 300°$

43. $\sqrt[4]{1 + i} = \sqrt[4]{\sqrt{2}\text{ cis } 45°} = (\sqrt{2})^4\ \text{cis}\left(\dfrac{45° + 360°k}{4}\right) = 4\ \text{cis}\left(\dfrac{45° + 360°k}{4}\right)$

$\quad\quad k = 0: \quad \sqrt[8]{2}\ \text{cis}\left(\dfrac{45° + 360° \cdot 0}{4}\right) = \sqrt[8]{2}\ \text{cis } 11.25°$

$\quad\quad k = 1: \quad \sqrt[8]{2}\ \text{cis}\left(\dfrac{45° + 360° \cdot 1}{4}\right) = \sqrt[8]{2}\ \text{cis } 101.25°$

$\quad\quad k = 2: \quad \sqrt[8]{2}\ \text{cis}\left(\dfrac{45° + 360° \cdot 2}{4}\right) = \sqrt[8]{2}\ \text{cis } 191.25°$

$\quad\quad k = 3: \quad \sqrt[8]{2}\ \text{cis}\left(\dfrac{45° + 360° \cdot 3}{4}\right) = \sqrt[8]{2}\ \text{cis } 281.25°$

$\sqrt[4]{1 + i} = \sqrt[8]{2}\ \text{cis } 11.25°, \ \sqrt[8]{2}\ \text{cis } 101.25°, \ \sqrt[8]{2}\ \text{cis } 191.25°, \ \sqrt[8]{2}\ \text{cis } 281.25°$

45. $\sqrt[5]{32 \text{ cis } 200°} = 32^{1/5}\ \text{cis}\left(\dfrac{200° + 360°k}{5}\right) = 2\ \text{cis}\left(\dfrac{200° + 360°k}{5}\right)$

$\quad\quad k = 0: \quad 2\ \text{cis}\left(\dfrac{200° + 360° \cdot 0}{5}\right) = 2 \text{ cis } 40°$

$\quad\quad k = 1: \quad 2\ \text{cis}\left(\dfrac{200° + 360° \cdot 1}{5}\right) = 2 \text{ cis } 112°$

$\quad\quad k = 2: \quad 2\ \text{cis}\left(\dfrac{200° + 360° \cdot 2}{5}\right) = 2 \text{ cis } 184°$

$\quad\quad k = 3: \quad 2\ \text{cis}\left(\dfrac{200° + 360° \cdot 3}{5}\right) = 2 \text{ cis } 256°$

$\quad\quad k = 4: \quad 2\ \text{cis}\left(\dfrac{200° + 360° \cdot 4}{5}\right) = 2 \text{ cis } 328°$

$\sqrt[5]{32 \text{ cis } 200°} = 2 \text{ cis } 40°, \ 2 \text{ cis } 112°, \ 2 \text{ cis } 184°, \ 2 \text{ cis } 256°, \ 2 \text{ cis } 328°$

47. $\sqrt[3]{27} = \sqrt[3]{27 + 0\,i} = \sqrt[3]{27 \text{ cis } 0°} = 27^{1/3}\ \text{cis}\left(\dfrac{0° + 360°k}{3}\right) = 3 \text{ cis } 120°k$

$\quad\quad k = 0: \quad 3 \text{ cis } 120° \cdot 0 = 3 \text{ cis } 0°$

$\quad\quad k = 1: \quad 3 \text{ cis } 120° \cdot 1 = 3 \text{ cis } 120°$

$\quad\quad k = 2: \quad 3 \text{ cis } 120° \cdot 2 = 3 \text{ cis } 240°$

$\sqrt[3]{27} = 3 \text{ cis } 0°, \ 3 \text{ cis } 120°, \ 3 \text{ cis } 240°$

49. $\sqrt[4]{-1-i} = \sqrt[4]{\sqrt{2} \text{ cis } 225°} = (\sqrt{2})^{1/4} \text{ cis}\left(\frac{225° + 360°k}{4}\right) = \sqrt[8]{2} \text{ cis}\left(\frac{225° + 360°k}{4}\right)$

$\quad k = 0: \qquad \sqrt[8]{2} \text{ cis}\left(\frac{225° + 360°\cdot 0}{4}\right) = \sqrt[8]{2} \text{ cis } 56.25°$

$\quad k = 1: \qquad \sqrt[8]{2} \text{ cis}\left(\frac{225° + 360°\cdot 1}{4}\right) = \sqrt[8]{2} \text{ cis } 146.25°$

$\quad k = 2: \qquad \sqrt[8]{2} \text{ cis}\left(\frac{225° + 360°\cdot 2}{4}\right) = \sqrt[8]{2} \text{ cis } 236.25°$

$\quad k = 3: \qquad \sqrt[8]{2} \text{ cis}\left(\frac{225° + 360°\cdot 3}{4}\right) = \sqrt[8]{2} \text{ cis } 326.25°$

$\sqrt[4]{-1-i} = \sqrt[8]{2} \text{ cis } 56.25°, \ \sqrt[8]{2} \text{ cis } 146.25°, \ \sqrt[8]{2} \text{ cis } 236.25°, \ \sqrt[8]{2} \text{ cis } 326.25°$

51. $\sqrt[9]{-1+i} = \sqrt[9]{\sqrt{2} \text{ cis } 135°} = 2^{1/18} \text{ cis}\left(\frac{135° + 360°k}{9}\right)$

$\quad k = 0: \qquad 2^{1/18} \text{ cis}\left(\frac{135° + 360°\cdot 0}{9}\right) = 2^{1/18} \text{ cis } 15°$

$\quad k = 1: \qquad 2^{1/18} \text{ cis}\left(\frac{135° + 360°\cdot 1}{9}\right) = 2^{1/18} \text{ cis } 55°$

$\quad k = 2: \qquad 2^{1/18} \text{ cis}\left(\frac{135° + 360°\cdot 2}{9}\right) = 2^{1/18} \text{ cis } 95°$

$\quad k = 3: \qquad 2^{1/18} \text{ cis}\left(\frac{135° + 360°\cdot 3}{9}\right) = 2^{1/18} \text{ cis } 135°$

$\quad k = 4: \qquad 2^{1/18} \text{ cis}\left(\frac{135° + 360°\cdot 4}{9}\right) = 2^{1/18} \text{ cis } 175°$

$\quad k = 5: \qquad 2^{1/18} \text{ cis}\left(\frac{135° + 360°\cdot 5}{9}\right) = 2^{1/18} \text{ cis } 215°$

$\quad k = 6: \qquad 2^{1/18} \text{ cis}\left(\frac{135° + 360°\cdot 6}{9}\right) = 2^{1/18} \text{ cis } 255°$

$\quad k = 7: \qquad 2^{1/18} \text{ cis}\left(\frac{135° + 360°\cdot 7}{9}\right) = 2^{1/18} \text{ cis } 295°$

$\quad k = 8: \qquad 2^{1/18} \text{ cis}\left(\frac{135° + 360°\cdot 8}{9}\right) = 2^{1/18} \text{ cis } 335°$

$\sqrt[9]{-1+i} = 2^{1/18} \text{ cis } 15°, \ 2^{1/18} \text{ cis } 55°, \ 2^{1/18} \text{ cis } 95°, \ 2^{1/18} \text{ cis } 135°,$
$\qquad 2^{1/18} \text{ cis } 175°, \ 2^{1/18} \text{ cis } 215°, \ 2^{1/18} \text{ cis } 255°, \ 2^{1/18} \text{ cis } 295°,$
$\qquad 2^{1/18} \text{ cis } 335°$

53. $\sqrt[6]{-64} = \sqrt[6]{64 \text{ cis } 180°} = 64^{1/6} \text{ cis}\left(\dfrac{180° + 360°k}{6}\right) = 2 \text{ cis}(30° + 60°k)$

$k = 0:$ $2 \text{ cis}(30° + 60° \cdot 0) = 2 \text{ cis } 30°$

$k = 1:$ $2 \text{ cis}(30° + 60° \cdot 1) = 2 \text{ cis } 90°$

$k = 2:$ $2 \text{ cis}(30° + 60° \cdot 2) = 2 \text{ cis } 150°$

$k = 3:$ $2 \text{ cis}(30° + 60° \cdot 3) = 2 \text{ cis } 210°$

$k = 4:$ $2 \text{ cis}(30° + 60° \cdot 4) = 2 \text{ cis } 270°$

$k = 5:$ $2 \text{ cis}(30° + 60° \cdot 5) = 2 \text{ cis } 330°$

$\sqrt[6]{-64} = 2 \text{ cis } 30°,\ 2 \text{ cis } 90°,\ 2 \text{ cis } 150°,\ 2 \text{ cis } 210°,\ 2 \text{ cis } 270°,\ 2 \text{ cis } 330°$

55. $\sqrt[10]{i} = \sqrt[10]{\text{cis } 90°} = \text{cis}\left(\dfrac{90° + 360°k}{10}\right) = \text{cis}(9° + 36°k)$

$k = 0:$ $\text{cis}(9° + 36° \cdot 0) = \text{cis } 9°$

$k = 1:$ $\text{cis}(9° + 36° \cdot 1) = \text{cis } 45°$

$k = 2:$ $\text{cis}(9° + 36° \cdot 2) = \text{cis } 81°$

$k = 3:$ $\text{cis}(9° + 36° \cdot 3) = \text{cis } 117°$

$k = 4:$ $\text{cis}(9° + 36° \cdot 4) = \text{cis } 153°$

$k = 5:$ $\text{cis}(9° + 36° \cdot 5) = \text{cis } 189°$

$k = 6:$ $\text{cis}(9° + 36° \cdot 6) = \text{cis } 225°$

$k = 7:$ $\text{cis}(9° + 36° \cdot 7) = \text{cis } 261°$

$k = 8:$ $\text{cis}(9° + 36° \cdot 8) = \text{cis } 297°$

$k = 9:$ $\text{cis}(9° + 36° \cdot 9) = \text{cis } 333°$

$\sqrt[10]{i} = \text{cis } 9°,\ \text{cis } 45°,\ \text{cis } 81°,\ \text{cis } 117°,\ \text{cis } 153°,\ \text{cis } 189°,\ \text{cis } 225°,\ \text{cis } 261°,\ \text{cis } 297°,\ \text{cis } 333°$

57. $\sqrt[3]{-8} = \sqrt[3]{8 \text{ cis } 180°} = 8^{1/3} \text{ cis}\left(\dfrac{180° + 360°k}{3}\right) = 2 \text{ cis}(60° + 120°k)$

$\quad k = 0: \quad 2 \text{ cis}(60° + 120° \cdot 0) = 2 \text{ cis } 60° = 2(\cos 60° + i \sin 60°)$

$\qquad\qquad = 2\left(\dfrac{1}{2} + i\dfrac{\sqrt{3}}{2}\right) = 1 + \sqrt{3}\,i$

$\quad k = 1: \quad 2 \text{ cis}(60° + 120° \cdot 1) = 2 \text{ cis } 180° = 2\,(\cos 180° + i \sin 180°)$

$\qquad\qquad = 2(-1 + i \cdot 0) = -2$

$\quad k = 2: \quad 2 \text{ cis}(60° + 120° \cdot 2) = 2 \text{ cis } 300° = 2\,(\cos 300° + i \sin 300°)$

$\qquad\qquad = 2\left[\dfrac{1}{2} + i \cdot \left(-\dfrac{\sqrt{3}}{2}\right)\right] = 1 - \sqrt{3}\,i$

$\sqrt[3]{-8} = 1 + \sqrt{3}\,i, \; -2, \; 1 - \sqrt{3}\,i$

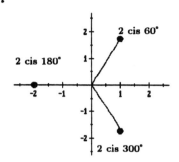

59. $\sqrt[3]{4\sqrt{3} - 4i} = \sqrt[3]{8 \text{ cis } 330°} = 2 \text{ cis}\left(\dfrac{330° + 360°k}{3}\right) = 2 \text{ cis}(110° + 120°k)$

$\quad k = 0: \quad 2 \text{ cis}(110° + 120° \cdot 0) = 2 \text{ cis } 110° = 2(\cos 110° + i \sin 110°)$

$\qquad\qquad = -.6840 + 1.8794\,i$

$\quad k = 1: \quad 2 \text{ cis}(110° + 120° \cdot 1) = 2 \text{ cis } 230° = 2(\cos 110° + i \sin 110°)$

$\qquad\qquad = -1.2856 - 1.5321\,i$

$\quad k = 2: \quad 2 \text{ cis}(110° + 120° \cdot 2) = 2 \text{ cis } 350° = 2(\cos 110° + i \sin 110°)$

$\qquad\qquad = 1.9696 - .3473\,i$

$\sqrt[3]{4\sqrt{3} - 4i} = -.6840 + 1.8794\,i, \; -1.2856 - 1.5321\,i, \; 1.9696 - .3473\,i$

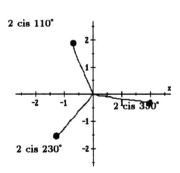

61. $\sqrt[4]{12.2567 + 10.2846\,i}$; $r = \sqrt{(12.2567)^2 + (10.2846)^2} \approx 16$;

$\tan\theta = \dfrac{10.2846}{12.2567}$; $\theta \approx 40°$; $\sqrt[4]{12.2567 + 10.2846\,i} \approx \sqrt[4]{16 \text{ cis } 40°}$

$= 2 \text{ cis}\left(\dfrac{40° + 360°k}{4}\right) = 2 \text{ cis}(10° + 90°k)$

$k = 0$: $2 \text{ cis}(10° + 90°\cdot 0) = 2 \text{ cis } 10° = 2(\cos 10° + i \sin 10°)$

$= 1.9696 + .3473\,i$

$k = 1$: $2 \text{ cis}(10° + 90°\cdot 1) = 2 \text{ cis } 100° = 2(\cos 10° + i \sin 10°)$

$= -.3473 + 1.9696\,i$

$k = 2$: $2 \text{ cis}(10° + 90°\cdot 2) = 2 \text{ cis } 190° = 2(\cos 10° + i \sin 10°)$

$= -1.9696 - .3473\,i$

$k = 3$: $2 \text{ cis}(10° + 90°\cdot 3) = 2 \text{ cis } 280° = 2(\cos 10° + i \sin 10°)$

$= .3473 - 1.9696\,i$

$\sqrt[4]{12.2567 + 10.2846\,i} = 1.9696 + .3473\,i$

$-.3473 + 1.9696\,i,$

$-1.9696 - .3473\,i,$

$.3473 - 1.9696\,i$

63. $\sqrt[5]{(-16 + 16\sqrt{3}\,i)^3}$; $r = \sqrt{(-16)^2 + (16\sqrt{3})^2} = \sqrt{256 + 768} = 32$;

$\tan\theta' = \left|\dfrac{16\sqrt{3}}{16}\right|$; $\theta' = 60°$; θ in Quadrant II, $\theta = 120°$

$\sqrt[5]{(-16 + 16\sqrt{3}\,i)^3} = \sqrt[5]{(32 \text{ cis } 120°)^3} = \sqrt[5]{32^3 \text{ cis}(3 \cdot 120°)}$

$= \sqrt[5]{32768 \text{ cis } 360°} = 32768^{1/5}\text{cis}\left(\dfrac{360°k}{5}\right) = 8 \text{ cis } 72°k$

$k = 0$: $8 \text{ cis } 72°\cdot 0 = 8 \text{ cis } 0°$

$k = 1$: $8 \text{ cis } 72°\cdot 1 = 8 \text{ cis } 72°$

$k = 2$: $8 \text{ cis } 72°\cdot 2 = 8 \text{ cis } 144°$

$k = 3$: $8 \text{ cis } 72°\cdot 3 = 8 \text{ cis } 216°$

$k = 4$: $8 \text{ cis } 72°\cdot 4 = 8 \text{ cis } 288°$

$\sqrt[5]{(-16 + 16\sqrt{3}\,i)^3} = 8 \text{ cis } 0°,\ 8 \text{ cis } 72°,\ 8 \text{ cis } 144°,\ 8 \text{ cis } 216°,\ 8 \text{ cis } 288°$

65. Prove: $\sqrt[n]{r}\ \text{cis}\left(\dfrac{\theta+360°k}{n}\right)=(r\ \text{cis}\ \theta)^{1/n}$

Suppose z_1 and z_2 are complex numbers such that z_1 is an nth root of z_2 and $z_1^n = z_2$. Also let $z_1 = r_1\ \text{cis}\ \theta_1$ and $z_2 = r_2\ \text{cis}\ \theta_2$. By DeMoivre's Theorem, $r_1^n\ \text{cis}\ n\theta_1 = r_2\ \text{cis}\ \theta_2$. If two complex numbers are equal then their absolute values are also equal which means $r_1^n = r_2$. Since r_1 and r_2 are nonnegative, $r_1 = \sqrt[n]{r_2}$. Substituting r_1^n for r_2 we find $r_1^n\ \text{cis}\ n\theta_1 = r_1^n\ \text{cis}\ \theta_2$, so $\text{cis}\ n\theta_1 = \text{cis}\ \theta_2$. It follows that $\cos n\theta_1 = \cos \theta_2$ and $\sin n\theta_1 = \sin \theta_2$. Since both the sine and cosine function have a period of $360°$, these equations are <u>both</u> true if and only if $n\theta_1$ and θ_2 differ by a multiple of $360°$. Thus for some integer k, $n\theta_1 = \theta_2 + 360°k$ or, equivilently, $\theta_1 = \dfrac{\theta_2 + 360°k}{n}$. Substitute this into the trigonometric form for z_1 to get $z_1 = \sqrt[n]{r_2}\ \text{cis}\left(\dfrac{\theta_2 + 360°k}{n}\right)$.

Thus $r_2\ \text{cis}\ \theta_2 = z_2 = z_1^n = \left[\sqrt[n]{r_2}\ \text{cis}\left(\dfrac{\theta_2 + 360°k}{n}\right)\right]^n$

67. If
$$\cos \theta = 1 - \frac{\theta^2}{2!} + \frac{\theta^4}{4!} - \frac{\theta^6}{6!} + \cdots + \frac{(-1)^n\theta^{2n}}{(2n)!} + \cdots$$
$$\sin \theta = \theta - \frac{\theta^3}{3!} + \frac{\theta^5}{5!} - \frac{\theta^7}{7!} + \cdots + \frac{(-1)^n\theta^{2n+1}}{(2n+1)!} + \cdots$$

and $e^{i\theta} = 1 + (i\theta) + \dfrac{(i\theta)^2}{2!} + \dfrac{(i\theta)^3}{3!} + \dfrac{(i\theta)^4}{4!} + \cdots + \dfrac{(i\theta)^n}{n!} + \cdots$

Prove: $e^{i\theta} = \cos \theta + i \sin \theta$

$e^{i\theta} = 1 + (i\theta) + \dfrac{(i\theta)^2}{2!} + \dfrac{(i\theta)^3}{3!} + \dfrac{(i\theta)^4}{4!} + \cdots + \dfrac{(i\theta)^n}{n!} + \cdots$

$= 1 + (i\theta) + \dfrac{i^2\theta^2}{2!} + \dfrac{i^3\theta^3}{3!} + \dfrac{i^4\theta^4}{4!} + \cdots + \dfrac{i^n\theta^n}{n!} + \cdots$

$= 1 + (i\theta) - \dfrac{\theta^2}{2!} - \dfrac{i\theta^3}{3!} + \dfrac{\theta^4}{4!} + \dfrac{i\theta^5}{5!} - \dfrac{\theta^6}{6!} \cdots + \dfrac{i^n\theta^n}{n!} + \cdots$

$= \left(1 - \dfrac{\theta^2}{2!} + \dfrac{\theta^4}{4!} - \dfrac{\theta^6}{6!} + \cdots + \dfrac{(-1)^n(\theta)^{2n}}{(2n)!} + \cdots\right) +$

$\left(i\theta - \dfrac{i\theta^3}{3!} + \dfrac{i\theta^5}{5!} + \cdots + \dfrac{(-1)^n(\theta)^{2n+1}}{(2n+1)!} + \cdots\right)$

$= \cos \theta + i \sin \theta$

CHAPTER 7 SUMMARY

1. $3 \tan 2\theta - \sqrt{3} = 0$; $\tan 2\theta = \dfrac{\sqrt{3}}{3}$; $2\theta' = \dfrac{\pi}{6}$; since $\tan 2\theta > 0$,

   2θ in Quad I or III; $2\theta = \dfrac{\pi}{6}, \dfrac{7\pi}{6}, \dfrac{13\pi}{6}, \dfrac{19\pi}{6}$; $\theta = \dfrac{\pi}{12}, \dfrac{7\pi}{12}, \dfrac{13\pi}{12}, \dfrac{19\pi}{12}$

3. $2 \sin 3\theta = \sqrt{2}$; $\sin 3\theta = \dfrac{\sqrt{2}}{2}$; $3\theta' = \dfrac{\pi}{4}$; since $\sin 3\theta > 0$,

   3θ in Quad I or II; $3\theta = \dfrac{\pi}{4}, \dfrac{3\pi}{4}, \dfrac{9\pi}{4}, \dfrac{11\pi}{4}, \dfrac{17\pi}{4}, \dfrac{19\pi}{4}$;

   $\theta = \dfrac{\pi}{12}, \dfrac{\pi}{4}, \dfrac{3\pi}{4}, \dfrac{11\pi}{12}, \dfrac{17\pi}{12}, \dfrac{19\pi}{12}$

5. $4 \cos^2\theta = 1$; $4 \cos^2\theta - 1 = 0$; $(2 \cos\theta + 1)(2 \cos\theta - 1) = 0$

   $2 \cos\theta + 1 = 0$ or $2 \cos\theta - 1 = 0$; $\cos\theta = -\dfrac{1}{2}$ or $\cos\theta = \dfrac{1}{2}$

   $\theta' = \dfrac{\pi}{3}$; since θ is positive or negative, θ can be in all four quadrants;

   $\theta = \dfrac{\pi}{3}, \dfrac{2\pi}{3}, \dfrac{4\pi}{3}, \dfrac{5\pi}{3}$

7. $\dfrac{1}{2} \cos^2\theta = 1$; $\cos^2\theta = 2$; $\cos\theta = \pm\sqrt{2}$;

   Since $-1 \le \cos\theta \le 1$, there is no solution.

9. 1. $\sec\theta = \dfrac{1}{\cos\theta}$ 2. $\csc\theta = \dfrac{1}{\sin\theta}$ 3. $\cot\theta = \dfrac{1}{\tan\theta}$

11. 1. $\sin^2\theta + \cos^2\theta = 1$ 2. $1 + \tan^2\theta = \sec^2\theta$ 3. $\cot^2\theta + 1 = \csc^2\theta$

13. $\sin\delta = \dfrac{3}{5}$, $\tan\delta < 0$; δ in Quadrant II;

   $\cos\delta = -\sqrt{1 - \sin^2\delta} = -\sqrt{1 - \left(\dfrac{3}{5}\right)^2} = -\sqrt{\dfrac{16}{25}} = -\dfrac{4}{5}$

   $\tan\delta = \dfrac{\sin\delta}{\cos\delta} = \dfrac{3/5}{-4/5} = -\dfrac{3}{4}$

   Using reciprocal identities, $\sec\delta = -\dfrac{5}{4}$, $\csc\delta = \dfrac{5}{3}$, and $\cot\delta = -\dfrac{4}{3}$

15. $\sin\omega = -\dfrac{3}{5}$, $\tan\omega > 0$; ω in Quadrant III;

   $\cos\delta = -\sqrt{1 - \sin^2\delta} = -\sqrt{1 - \left(-\dfrac{3}{5}\right)^2} = -\sqrt{\dfrac{16}{25}} = -\dfrac{4}{5}$

   $\tan\delta = \dfrac{\sin\delta}{\cos\delta} = \dfrac{-3/5}{-4/5} = \dfrac{3}{4}$

   Using reciprocal identities, $\sec\delta = -\dfrac{5}{4}$, $\csc\delta = -\dfrac{5}{3}$, and $\cot\delta = \dfrac{4}{3}$

17. $\dfrac{\sin\theta}{\cos\theta} + \dfrac{1}{\sin\theta} = \dfrac{\sin^2\theta + \cos\theta}{\cos\theta \sin\theta}$

19. $\tan^2\theta + \sec^2\theta = \dfrac{\sin^2\theta}{\cos^2\theta} + \dfrac{1}{\cos^2\theta} = \dfrac{\sin^2\theta + 1}{\cos^2\theta}$

21. $\dfrac{\csc^2\alpha}{1+\cot^2\alpha} = \dfrac{\csc^2\alpha}{\csc^2\alpha} = 1$

23. $\dfrac{\cos\theta \cot\theta - \tan\theta}{\csc\theta} = \dfrac{\cos\theta \cdot \dfrac{\cos\theta}{\sin\theta} - \dfrac{\sin\theta}{\cos\theta}}{\dfrac{1}{\sin\theta}}$

$$= \left(\dfrac{\cos^3\theta - \sin^2\theta}{\sin\theta \cos\theta}\right) \cdot \dfrac{\sin\theta}{1}$$

$$= \dfrac{\cos^3\theta - \sin^2\theta}{\cos\theta}$$

$$= \cos^2\theta - \dfrac{\sin^2\theta}{\cos\theta}$$

$$= \cos\theta \cdot \dfrac{1}{\sec\theta} - \sin\theta \cdot \dfrac{\sin\theta}{\cos\theta}$$

$$= \dfrac{\cos\theta}{\sec\theta} - \sin\theta \tan\theta$$

$$= \dfrac{\cos\theta}{\sec\theta} - \sin\theta \cdot \dfrac{1}{\cot\theta}$$

$$= \dfrac{\cos\theta}{\sec\theta} - \dfrac{\sin\theta}{\cot\theta}$$

25. $\dfrac{\sin^2\theta - \cos^2\theta}{\sin\theta + \cos\theta} = \dfrac{(\sin\theta + \cos\theta)(\sin\theta - \cos\theta)}{\sin\theta + \cos\theta} = \sin\theta - \cos\theta$

27. $\cos^2 x \tan x \csc x \sec x = \cos^2 x \cdot \dfrac{\sin x}{\cos x} \cdot \dfrac{1}{\sin x} \cdot \dfrac{1}{\cos x} = 1$

29. $\sin 38° = \cos(90° - 38°) = \cos 52°$

31. $\tan \dfrac{\pi}{8} = \cot\left(\dfrac{\pi}{2} - \dfrac{\pi}{8}\right) = \cot \dfrac{3\pi}{8}$

33. $\cos\left(\dfrac{\pi}{6} - \theta\right) = \cos\left[-\left(\theta - \dfrac{\pi}{6}\right)\right] = \cos\left(\theta - \dfrac{\pi}{6}\right)$

35.

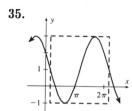

37. $\cos(\theta - 30°) = \cos\theta\cos 30° + \sin\theta\sin 30°$

$$= \cos\theta \cdot \frac{\sqrt{3}}{2} + \sin\theta \cdot \frac{1}{2}$$

$$= \tfrac{1}{2}(\sqrt{3}\cos\theta + \sin\theta)$$

39. $\dfrac{\tan 23° - \tan 85°}{1 + \tan 23° \tan 85°} = \tan(23° - 85°) = \tan(-62°) \approx -1.881$

41. $\dfrac{2\tan\frac{\pi}{6}}{1 - \tan^2\frac{\pi}{6}} = \tan 2(\frac{\pi}{6}) = \tan\frac{\pi}{3} = \sqrt{3}$

43. $\cos\theta = \frac{4}{5}$, 2θ in Quad IV; $\sin\theta = \pm\sqrt{1 - \cos^2\theta} = \pm\sqrt{1 - \left(\frac{4}{5}\right)^2}$

$$= \pm\sqrt{1 - \tfrac{16}{25}} = \pm\sqrt{\tfrac{9}{25}} = \pm\tfrac{3}{5}; \quad \sin 2\theta = 2\sin\theta\cos\theta = 2\left(\pm\tfrac{3}{5}\right)\left(\tfrac{4}{5}\right)$$

$$= \pm\tfrac{24}{25}; \text{ in Quad IV } \sin 2\theta \text{ is negative; thus } \sin 2\theta = -\tfrac{24}{25}$$

45. $-\sqrt{\dfrac{1 + \cos 240°}{2}} = \cos\tfrac{1}{2}(240°) = \cos 120° = -\tfrac{1}{2}$

47. $\cot 2\theta = \frac{4}{3}$, θ in Quad I; using the reciprocal identity, $\tan 2\theta = \frac{3}{4}$;

$$\sec 2\theta = \sqrt{1 + \tan^2 2\theta} = \sqrt{1 + \left(\tfrac{3}{4}\right)^2} = \sqrt{1 + \tfrac{9}{16}} = \sqrt{\tfrac{25}{16}} = \tfrac{5}{4};$$

using the reciprocal identity, $\cos 2\theta = \frac{4}{5}$;

$$\sin\theta = \sin\tfrac{1}{2}(2\theta) = \sqrt{\dfrac{1 - \cos 2\theta}{2}} = \sqrt{\dfrac{1 - \frac{4}{5}}{2}} = \sqrt{\tfrac{1}{10}} = \tfrac{1}{\sqrt{10}}$$

49. $\sin 3\theta\cos\theta = \tfrac{1}{2}\sin(3\theta + \theta) + \tfrac{1}{2}\sin(3\theta - \theta) = \tfrac{1}{2}\sin 4\theta + \tfrac{1}{2}\sin 2\theta$

51. $\sin x - \sin 3x = 0$; $2\sin\left(\dfrac{x - 3x}{2}\right)\cos\left(\dfrac{x + 3x}{2}\right) = 0$; $2\sin(-x)\cos 2x = 0$;

$-2\sin x\cos 2x = 0$; $\sin x = 0$ when $x = 0$ and π; $\cos 2x = 0$ when

$2x = \frac{\pi}{2}, \frac{3\pi}{2}, \frac{5\pi}{2}, \frac{7\pi}{2}$; $x = 0, \frac{\pi}{4}, \frac{3\pi}{4}, \pi, \frac{5\pi}{4}, \frac{7\pi}{4}$

53. $7 - 7i$; $a = 7$ and $b = -7$; $r = \sqrt{7^2 + (-7)^2} = \sqrt{49 + 49} = \sqrt{98} = 7\sqrt{2}$;

θ in Quad IV; $\theta' = \tan^{-1}\left|\frac{-7}{7}\right| = \tan^{-1} 1 = 45°$; $\theta = 360° - 45° = 315°$;

so $7 - 7i = 7\sqrt{2}$ cis $315°$

55. $\frac{7}{2}\sqrt{3} - \frac{7}{2}i$; $a = \frac{7}{2}\sqrt{3}$ and $b = -\frac{7}{2}$; θ in Quad IV;

$$r = \sqrt{\left(\frac{7}{2}\sqrt{3}\right)^2 + \left(\frac{7}{2}\right)^2} = \sqrt{\frac{147}{4} + \frac{49}{4}} = \sqrt{\frac{196}{4}} = \sqrt{49} = 7$$

$$\theta' = \tan^{-1}\left|\frac{-7/2}{7\sqrt{3}/2}\right| = \tan^{-1}\frac{1}{\sqrt{3}} = 30°; \ \theta = 360° - 30° = 330°;$$

so $\frac{7}{2}\sqrt{3} - \frac{7}{2}i = 7 \text{ cis } 330°$

57. $4\left(\cos\frac{7\pi}{4} + i\sin\frac{7\pi}{4}\right) = 4\left(\frac{\sqrt{2}}{2} - \frac{\sqrt{2}}{2}i\right) = 2\sqrt{2} - 2\sqrt{2}\,i$

59. $5 \text{ cis } 270° = 5(\cos 270° + i\sin 270°) = 5(0 - i) = -5i$

61. $(\sqrt{12} - 2i)^4 = (2\sqrt{3} - 2i)^4$; $r = \sqrt{(2\sqrt{3})^2 + (-2)^2} = \sqrt{12 + 4} = \sqrt{16} = 4$

θ in Quad IV; $\theta' = \tan^{-1}\left|\frac{-2}{2\sqrt{3}}\right| = \tan^{-1}\frac{1}{\sqrt{3}} = 30°; \ \theta = 360° - 30° = 330°;$

so $2\sqrt{3} - 2i = 4 \text{ cis } 330°$; $(2\sqrt{3} - 2i)^4 = (4 \text{ cis } 330°)^4 = 4^4 \text{ cis}(4 \cdot 330°)$

$= 256 \text{ cis } 1320° = 256 \text{ cis } 240° = 256(\cos 240° + i\sin 240°) = 256\left(-\frac{1}{2} - \frac{\sqrt{3}}{2}i\right)$

$= -128 - 128\sqrt{3}\,i$

63. $\dfrac{2 \text{ cis } 158° \cdot 4 \text{ cis } 212°}{(2 \text{ cis } 312°)^3} = \dfrac{2 \cdot 4 \text{ cis}(158° + 212°)}{2^3 \text{ cis}(3 \cdot 312°)} = \dfrac{8 \text{ cis } 370°}{8 \text{ cis } 936°} = \dfrac{\text{cis } 370°}{\text{cis } 216°}$

$= \text{cis}(370° - 216°) = \text{cis } 154°$

65. If n is a natural number, then $(r \text{ cis } \theta)^n = r^n \text{ cis } n\theta$
    for a complex number $r \text{ cis } \theta = r(\cos\theta + i\sin\theta)$.

67. $\frac{7}{2}\sqrt{3} - \frac{7}{2}i = 7 \text{ cis } 330°$ (see problem 55) ;

$$\sqrt{\frac{7}{2}\sqrt{3} - \frac{1}{2}i} = \sqrt{7 \text{ cis } 330°} = \sqrt{7} \text{ cis}\left(\frac{330° + 360°k}{2}\right);$$

$k = 0$: $\quad \sqrt{7} \text{ cis}\left(\frac{330° + 360° \cdot 0}{2}\right) = \sqrt{7} \text{ cis } 165°$

$\qquad\qquad\qquad\qquad = \sqrt{7}(\cos 165° + i\sin 165°)$

$\qquad\qquad\qquad\qquad \approx -2.5556 + .6848\,i$

$k = 1$: $\quad \sqrt{7} \text{ cis}\left(\frac{330° + 360° \cdot 1}{2}\right) = \sqrt{7} \text{ cis } 345°$

$\qquad\qquad\qquad\qquad = \sqrt{7}(\cos 345° + i\sin 345°)$

$\qquad\qquad\qquad\qquad \approx 2.5556 - .6848\,i$

$\sqrt{\frac{7}{2}\sqrt{3} - \frac{1}{2}i} \approx -2.5556 + .6848\,i, \ 2.5556 - .6848\,i$

CHAPTER 8

PROBLEM SET 8.1

1. Given: $a = 80$; $\beta = 60°$; $\gamma = 90°$; $\alpha = 90° - 60° = 30°$; $\sin 30° = \frac{80}{c}$;

   $c = \frac{80}{\sin 30°} = 160$; $\tan 30° = \frac{80}{b}$; $b = \frac{80}{\tan 30°} \approx 140$

3. Given: $a = 9.0$; $\beta = 45°$; $\gamma = 90°$; $\alpha = 90° - 45° = 45°$; $\tan 45° = \frac{9.0}{b}$;

   $b = \frac{9.0}{\tan 45°} = 9.0$; $\sin 45° = \frac{9.0}{c}$; $c = \frac{9.0}{\sin 45°} \approx 13$

5. Given: $b = 15$; $\alpha = 37°$; $\gamma = 90°$; $\beta = 90° - 37° = 53°$; $\tan 37° = \frac{a}{15}$;

   $a = 15 \tan 37° \approx 11$; $\cos 37° = \frac{15}{c}$; $c = \frac{15}{\cos 37°} \approx 19$

7. Given: $a = 69$; $c = 73$; $\gamma = 90°$; $\sin \alpha = \frac{69}{73}$; $\alpha = \sin^{-1} \frac{69}{73} \approx 71°$;

   $\beta \approx 90° - 71° = 19°$; $b = \sqrt{73^2 - 69^2} = \sqrt{568} \approx 24$

9. Given: $b = 13$; $c = 22$; $\gamma = 90°$; $\sin \beta = \frac{13}{22}$; $\beta = \sin^{-1} \frac{13}{22} \approx 36°$;

   $\alpha \approx 90° - 36° = 54°$; $a = \sqrt{22^2 - 13^2} = \sqrt{315} \approx 18$

11. Given: $a = 24$; $b = 29$; $\gamma = 90°$; $\tan \alpha = \frac{24}{29}$; $\alpha = \tan^{-1} \frac{24}{29} \approx 40°$;

    $\beta \approx 90° - 40° = 50°$; $c = \sqrt{24^2 + 29^2} = \sqrt{1417} \approx 38$

13. Given: $a = 29$; $\alpha = 76°$; $\gamma = 90°$; $\beta = 90° - 76° = 14°$; $\sin 76° = \frac{29}{c}$;

    $c = \frac{29}{\sin 76°} \approx 30$; $\tan 76° = \frac{29}{b}$; $b = \frac{29}{\tan 76°} \approx 7.2$

15. Given: $a = 49$; $\beta = 45°$; $\gamma = 90°$; $\alpha = 90° - 45° = 45°$; $\tan 45° = \frac{b}{49}$;

    $b = 49 \tan 45° = 49$; $\sin 45° = \frac{49}{c}$; $c = \frac{49}{\sin 45°} \approx 69$

17. Given: $b = 90$; $\beta = 13°$; $\gamma = 90°$; $\alpha = 90° - 13° = 77°$; $\tan 77° = \frac{a}{90}$;

    $a = 90 \tan 77° \approx 390$; $\sin 13° = \frac{90}{c}$; $c = \frac{90}{\sin 13°} \approx 400$

19. Given: $b = 82$; $\alpha = 50°$; $\gamma = 90°$; $\beta = 90° - 50° = 40°$; $\tan 50° = \frac{a}{82}$;

    $a = 82 \tan 50° \approx 98$; $\sin 40° = \frac{82}{c}$; $c = \frac{82}{\sin 40°} \approx 130$

21. Given: $c = 36$; $\alpha = 6°$; $\gamma = 90°$; $\beta = 90° - 6° = 84°$; $\sin 6° = \frac{a}{36}$;

    $a = 36 \sin 6° \approx 3.8$; $\cos 6° = \frac{b}{36}$; $b = 36 \cos 6° \approx 36$

23. Given: $\alpha = 56.00°$; $b = 2,350$; $\gamma = 90.00°$; $\beta = 90.00° - 56.00° = 34.00°$;

$\tan 56.00° = \dfrac{a}{2,350}$; $a = 2,350 \tan 56.00° \approx 3,484$; $\cos 56.00° = \dfrac{2,350}{c}$;

$c = \dfrac{2,350}{\cos 56.00°} \approx 4,202$

25. Given: $b = 3,100$; $c = 3,500$; $\gamma = 90.00°$; $\sin \beta = \dfrac{3,100}{3,500}$;

$\beta = \sin^{-1} \dfrac{3,100}{3,500} \approx 62.34°$; $\alpha \approx 90.00° - 62.34° = 27.66°$;

$a = \sqrt{3,500^2 - 3,100^2} = \sqrt{2,640,000} \approx 1,625$

27. Given: $\alpha = 42°$; $b = 350$; $\gamma = 90°$; $\beta = 90° - 42° = 48°$; $\tan 42° = \dfrac{a}{350}$;

$a = 350 \tan 42° \approx 320$; $\cos 42° = \dfrac{350}{c}$; $c = \dfrac{350}{\cos 42°} \approx 470$

29. Given: $b = 4,100$; $c = 4,300$; $\gamma = 90.00°$; $\sin \beta = \dfrac{4,100}{4,300}$;

$\beta = \sin^{-1} \dfrac{4,100}{4,300} \approx 72.46°$; $\alpha \approx 90.00° - 72.46° = 17.54°$;

$a = \sqrt{4,300^2 - 4,100^2} = \sqrt{1,680,000} \approx 1,296$

31. $\tan 38° = \dfrac{h}{30}$

$h = 30 \tan 38°$

$h \approx 23$ m tall

33. $\tan 53° = \dfrac{d}{150}$

$d = 150 \tan 53°$

$d \approx 200$ m

35. $\tan 19° = \dfrac{d}{1,000}$

$d = 1,000 \tan 19°$

$d \approx 340$ ft

37. $\tan 49° = \dfrac{|BC|}{300}$

$|BC| = 300 \tan 49°$

$|BC| \approx 350$ ft

39. $\sin 52° = \dfrac{h}{16}$

$h = 16 \sin 52°$

$h \approx 13$ ft

41. $\cos \alpha = \dfrac{9}{16}$

$\alpha = \cos^{-1} \dfrac{9}{16}$

$\alpha = 56° \approx 60°$

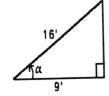

43. $\tan 51.36° = \dfrac{h}{1,000}$

$h = 1,000 \tan 51.36°$

$h \approx 1,251$ ft

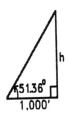

45. $\tan 14.8° = \dfrac{h}{x}$, so $x = \dfrac{h}{\tan 14.8°}$

$\tan 13.5° = \dfrac{h}{x + 100} = \dfrac{h}{\dfrac{h}{\tan 14.8°} + 100}$

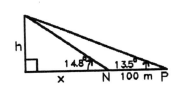

$\left(\dfrac{\tan 13.5°}{\tan 14.8°}\right) \cdot h + 100 \tan 13.5° = h$

$100 \tan 13.5° = h \cdot \left(1 - \dfrac{\tan 13.5°}{\tan 14.8°}\right);$ $h = \dfrac{100 \tan 13.5°}{1 - \dfrac{\tan 13.5°}{\tan 14.8°}} \approx 263$ m

47. $\alpha = 35°$ (alternate interior angles)

so $\angle QRP = 35° + 25° = 60°$

$\tan 60° = \dfrac{|PQ|}{100}$

$|PQ| = 100 \tan 60°$

$|PQ| \approx 170$ m

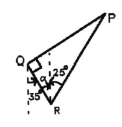

49. $\tan 55.81° = \dfrac{y}{1,000}$

$$y = 1,000 \tan 55.81°$$

$\tan 51.34° = \dfrac{x}{1,000}$

$$x = 1,000 \tan 51.34°$$

$$l = y - x = 1,000(\tan 55.81° - \tan 51.34°)$$

$$l \approx 222.0 \text{ ft}$$

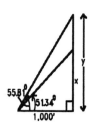

51. The circumference of the wheel is $2\pi(2.50) \approx 15.7$ ft

After three revolutions the center of the wheel has

travelled $3 \cdot 15.7 = 47.1$ feet. In the diagram,

$\sin 15° = \dfrac{d}{4.71}$, so $d = 4.71 \sin 15°$, and

$\sin 75° = \dfrac{x}{2.50}$, so $x = 2.50 \sin 75°$.

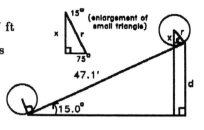

The final height is: $h = d + x = 47.1 \sin 15° + 2.50 \sin 75° \approx 14.6$ ft

53. $\cos 47° = \dfrac{|EV|}{92,900,000}$; $|EV| = 92,9000,000 \cos 47° \approx 63,400,000 = 6.34 \times 10^7$ miles

55. $\angle ACB = \frac{1}{5} \cdot 360° = 72°$; $\angle ACM = \frac{1}{2} \cdot \angle ACB = 36°$

$\cos 36° = \dfrac{|CM|}{783.5}$

$|CM| = 783.5 \cos 36° \approx 633.9$ feet

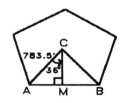

57. $\tan \alpha = \dfrac{h}{|PC|}$ so $|PC| = \dfrac{h}{\tan \alpha}$. Also, $\tan \beta = \dfrac{h}{d + |PC|} = \dfrac{h}{d + \dfrac{h}{\tan \alpha}}$.

So, $(\tan \beta) \cdot \left(d + \dfrac{h}{\tan \alpha} \right) = h$; $d \tan \beta + \dfrac{\tan \beta}{\tan \alpha} \cdot h = h$; $d \tan \beta = h \cdot \left(1 - \dfrac{\tan \beta}{\tan \alpha} \right)$;

$$h = \dfrac{d \tan \beta}{1 - \dfrac{\tan \beta}{\tan \alpha}} \cdot \dfrac{\tan \alpha}{\tan \alpha} = \dfrac{d \tan \alpha \tan \beta}{\tan \alpha - \tan \beta} = \dfrac{d \cdot \dfrac{\sin \alpha}{\cos \alpha} \cdot \dfrac{\sin \beta}{\cos \beta}}{\dfrac{\sin \alpha}{\cos \alpha} - \dfrac{\sin \beta}{\cos \beta}} \cdot \dfrac{\cos \alpha \cos \beta}{\cos \alpha \cos \beta}$$

$$= \dfrac{d \sin \alpha \sin \beta}{\sin \alpha \cos \beta - \cos \alpha \sin \beta} = \dfrac{d \sin \alpha \sin \beta}{\sin(\alpha - \beta)}$$

59. 1.75 hours have passed since the sun rose. It takes 6 hours for the sun to rise from the horizon to directly overhead. Hence the present angle of elevation of the sun is $\left(\frac{1.75}{6}\right) \cdot 90° = 26.25°$

$\tan 26.25° = \frac{6}{d}$; $\quad d = \frac{6}{\tan 26.25°} \approx 12$ feet

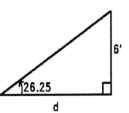

61. The sum of any two sides of a triangle is always greater than the length of the third side, hence $a + b > c$. So we must now show that $c > \frac{a+b}{\sqrt{2}}$.

Since c is positive, our assertion will be proven if we demonstrate that $\sqrt{2} > \frac{a+b}{c}$

$$\frac{a+b}{c} = \frac{a}{c} + \frac{b}{c}$$

$$= \sin A + \cos B$$

$$= \sin A + \sin(90° - A)$$

$$= 2 \sin\left(\frac{A + (90° - A)}{2}\right) \cos\left(\frac{A - (90° - A)}{2}\right)$$

$$= 2 \sin 45° \cos(2A - 90°)$$

$$= 2 \cdot \frac{\sqrt{2}}{2} \cdot (\cos 2A \cos 90° + \sin 2A \sin 90°)$$

$$= \sqrt{2} \sin 2A$$

$$< \sqrt{2} \quad \text{(since } \sin 2A < 1 \text{ for all angles } 2A\text{)}$$

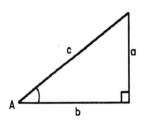

PROBLEM SET 8.2

1. $\cos \alpha = \frac{b^2 + c^2 - a^2}{2bc}$; $\quad \alpha = \cos^{-1}\left(\frac{8.0^2 + 2.0^2 - 7.0^2}{2(8.0)(2.0)}\right) \approx 54°$

3. $\cos \alpha = \frac{b^2 + c^2 - a^2}{2bc}$; $\quad \alpha = \cos^{-1}\left(\frac{9.0^2 + 8.0^2 - 11^2}{2(9.0)(8.0)}\right) \approx 80°$

5. $\cos \beta = \frac{a^2 + c^2 - b^2}{2ac}$; $\quad \beta = \cos^{-1}\left(\frac{15^2 + 20^2 - 8^2}{2(15)(20)}\right) \approx 21°$

7. $c^2 = a^2 + b^2 - 2ab \cos \gamma$; $\quad c = \sqrt{18^2 + 25^2 - 2(18)(25)\cos 30°} \approx 13$

9. $c^2 = a^2 + b^2 - 2ab \cos \gamma$; $\quad c = \sqrt{15^2 + 8.0^2 - 2(15)(8.0)\cos 38°} \approx 10$

11. $a^2 = b^2 + c^2 - 2bc \cos \alpha$; $a = \sqrt{14^2 + 12^2 - 2(14)(12)\cos 82°} \approx 17$

13. $a^2 = b^2 + c^2 - 2bc \cos \alpha$; $a = \sqrt{18^2 + 15^2 - 2(18)(15)\cos 50°} \approx 14$

15. $\cos \beta = \dfrac{a^2 + c^2 - b^2}{bac}$; $\beta = \cos^{-1}\left(\dfrac{241^2 + 100^2 - 187^2}{2(241)(100)}\right) \approx 46.6°$

17. $a^2 = b^2 + c^2 - 2bc \cos \alpha$; $a = \sqrt{123^2 + 485^2 - 2(123)(485)\cos 163.0°} \approx 604$

19. $\cos \alpha = \dfrac{b^2 + c^2 - a^2}{2bc}$; $\alpha = \cos^{-1}\left(\dfrac{9.0^2 + 8.0^2 - 11^2}{2(9.0)(8.0)}\right) \approx 80°$

21. $\cos \alpha = \dfrac{b^2 + c^2 - a^2}{2bc}$; $\alpha = \cos^{-1}\left(\dfrac{310^2 + 250^2 - 123^2}{2(310)(250)}\right) \approx 22.2°$

23. $\cos \beta = \dfrac{a^2 + c^2 - b^2}{2ac}$; $\beta = \cos^{-1}\left(\dfrac{38^2 + 25^2 - 41^2}{2(38)(25)}\right) \approx 78°$

25. Given: $b = 5.2$; $c = 3.4$; $\alpha = 54°$ (2 significant digits)

$a = \sqrt{5.2^2 + 3.4^2 - 2(5.2)(3.4)\cos 54.4°} = 4.244541409 \approx 4.2$

$\beta = \cos^{-1}\left(\dfrac{a^2 + c^2 - b^2}{2ac}\right) \approx \cos^{-1}\left(\dfrac{4.244541409^2 + 3.4^2 - 5.2^2}{2(4.244541409)(3.4)}\right) \approx 84.8° \approx 85°$

$\gamma = 180° - (\alpha + \beta) \approx 180° - (54.4° + 84.8°) = 40.8° \approx 41°$

27. Given: $a = 214$; $b = 320$; $\gamma = 14.8°$

$c = \sqrt{214^2 + 320^2 - 2(214)(320)\cos 14.8°} = 125.6179474 \approx 126$

$\beta = \cos^{-1}\left(\dfrac{a^2 + c^2 - b^2}{2ac}\right) \approx \cos^{-1}\left(\dfrac{214^2 + 125.6179474^2 - 320^2}{2(214)(125.6179474)}\right) \approx 139.4°$

$\alpha = 180° - (\beta + \gamma) \approx 180° - (139.4° + 14.8°) = 25.8°$

29. Given: $a = 140$; $b = 85.0$; $c = 105$

$\alpha = \cos^{-1}\left(\dfrac{b^2 + c^2 - a^2}{2bc}\right) \approx \cos^{-1}\left(\dfrac{85.0^2 + 105^2 - 140^2}{2(85.0)(105)}\right) \approx 94.3°$

$\beta = \cos^{-1}\left(\dfrac{a^2 + c^2 - b^2}{2ac}\right) \approx \cos^{-1}\left(\dfrac{140^2 + 105^2 - 85.0^2}{2(140)(105)}\right) \approx 37.3°$

$\gamma = 180° - (\alpha + \beta) \approx 180° - (94.3° + 37.3°) = 48.4°$

173

31. Triangle not unique. Must be given at least one side.

33. Given: $b = 520$; $c = 235$; $\alpha = 110.5°$

$$a = \sqrt{520^2 + 235^2 - 2(520)(235)\cos 110.5°} = 641.2610108 \approx 641$$

$$\beta = \cos^{-1}\left(\frac{a^2 + c^2 - b^2}{2ac}\right) \approx \cos^{-1}\left(\frac{641.2610108^2 + 235^2 - 520^2}{2(641.2610108)(235)}\right) \approx 49.4°$$

$$\gamma = 180° - (\alpha + \beta) \approx 180° - (110.5° + 49.4°) = 20.1°$$

35. Given: $a = 341$; $b = 340$; $\gamma = 23.4°$

$$c = \sqrt{341^2 + 340^2 - 2(341)(340)\cos 23.4°} = 138.1016198 \approx 138$$

$$\beta = \cos^{-1}\left(\frac{a^2 + c^2 - b^2}{2ac}\right) \approx \cos^{-1}\left(\frac{341^2 + 138.1016198^2 - 340^2}{2(341)(138.1016198)}\right) \approx 77.9°$$

$$\alpha = 180° - (\beta + \gamma) \approx 180° - (77.9° + 23.4°) = 78.7°$$

37. Use $\triangle ACH$: $\sin 45° = \dfrac{380}{|AC|}$; $|AC| = \dfrac{380}{\sin 45°} \approx 537$ ft

39. Use $\triangle BFH$: $|BF|^2 = |BH|^2 + |HF|^2 - 2 \cdot |BH| \cdot |HF| \cdot \cos 45°$;

$$|BF| = \sqrt{420^2 + 90^2 - 2(420)(90)\cos 45°} \approx 362 \text{ ft}$$

41. From $\triangle SFH$, $|SH| = \sqrt{90^2 + 90^2} = 90\sqrt{2}$;

$$|SB| = |BH| - |SH| = 420 - 90\sqrt{2} \approx 293 \text{ ft}$$

43. $\sin \alpha = \dfrac{y}{b}$ and $\cos \alpha = \dfrac{x}{b}$, so $C = (b \cos \alpha, \ b \sin \alpha)$

From the distance formula:

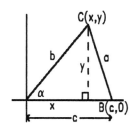

$$a = |BC| = \sqrt{(b \cos \alpha - c)^2 + (b \sin \alpha - 0)^2}; \text{ so}$$

$$a^2 = (b \cos \alpha - c)^2 + (b \sin \alpha)^2$$

$$= b^2 \cos^2\alpha - 2bc \cos \alpha + c^2 + b^2 \sin^2\alpha$$

$$= b^2(\sin^2\alpha + \cos^2\alpha) + c^2 - 2bc \cos \alpha$$

$$= b^2 + c^2 - 2bc \cos \alpha$$

45. See proof in text (page 345).

47. From problem 44: $b^2 = a^2 + c^2 - 2ac \cos \beta$; so $2ac \cos \beta = a^2 + c^2 - b^2$;

thus $\cos \beta = \dfrac{a^2 + c^2 - b^2}{2ac}$

49. $\angle BWN = 49° + 9° = 58°$

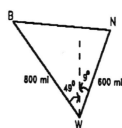

$|BN| = \sqrt{600^2 + 800^2 - 2(600)(800)\cos 58°}$

$|BN| \approx 700$ miles

51. $\alpha = 51°$ (alternate interior angles), so

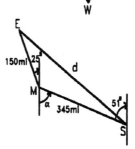

$\angle SME = 25° + (180° - 51°) = 154°$

$|SE| = \sqrt{345^2 + 150^2 - 2(345)(150)\cos 154°}$

$|SE| \approx 484$ miles

53. $q = |DP| = \frac{1}{2}(1.75) + \frac{1}{2}(2.00) = 1.88$

$p = |DQ| = \frac{1}{2}(1.75) + \frac{1}{2}(2.50) = 2.13$

$d = |PQ| = \frac{1}{2}(2.00) + \frac{1}{2}(2.50) = 2.25;$

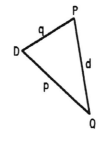

$\angle D = \cos^{-1}\left(\dfrac{1.88^2 + 2.13^2 - 2.25^2}{2(1.88)(2.13)}\right) \approx 68.1°$

$\angle P = \cos^{-1}\left(\dfrac{1.88^2 + 2.25^2 - 2.13^2}{2(1.88)(2.25)}\right) \approx 61.2°$

$\angle Q \approx 180° - (68.1 + 61.2°) = 50.7°$

55. $\angle ACB = \frac{1}{3}(360°) = 120°;$

$|AB| = \sqrt{3^2 + 3^2 - 2(3)(3)\cos 120°} \approx 5.19615242$

perimeter $= 3 \cdot |AB| \approx 16$ in (to two significant digits)

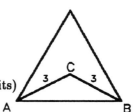

57. $\angle ACB = \frac{1}{4}(360°) = 90°$

$|AB| = \sqrt{15^2 + 15^2} = 15\sqrt{2}$ (Pythagorean Theorem)

Area $= |AB|^2 = (15\sqrt{2})^2 = 225(2) = 450$ in^2

59. $a \cos \gamma + c \cos \alpha = a \cdot \dfrac{a^2 + b^2 - c^2}{2ab} + c \cdot \dfrac{b^2 + c^2 - a^2}{2bc}$ (law of cosines)

$= \dfrac{a^2 + b^2 - c^2}{2b} + \dfrac{b^2 + c^2 - a^2}{2b}$

$= \dfrac{2b^2}{2b}$

$= b$

61. $\dfrac{(b + c - a)(a + b + c)}{2bc} = \dfrac{ab + b^2 + bc + ac + bc + c^2 - a^2 - ab - ac}{2bc}$

$= \dfrac{b^2 + c^2 - a^2 + 2bc}{2bc}$

$= \dfrac{b^2 + c^2 - a^2}{2bc} + \dfrac{2bc}{2bc}$

$= \cos \alpha + 1$ (using law of cosines)

63. $\dfrac{(a + b - c)(a + b + c)}{2ab} = \dfrac{a^2 + ab + ac + ab + b^2 + bc - ac - bc - c^2}{2ab}$

$= \dfrac{a^2 + b^2 - c^2 + 2ab}{2ab}$

$= \dfrac{a^2 + b^2 - c^2}{2ab} + \dfrac{2ab}{2ab}$

$= \cos \gamma + 1$ (using law of cosines)

PROBLEM SET 8.3

1. Given: $a = 10$; $\alpha = 48°$; $\beta = 62°$

$\gamma = 180° - (\alpha + \beta) = 180° - (48° + 62°) = 70°$

$\dfrac{b}{\sin 62°} = \dfrac{c}{\sin 70°} = \dfrac{10}{\sin 48°}$; $b = \dfrac{10 \sin 62°}{\sin 48°} \approx 12$; $c = \dfrac{10 \sin 70°}{\sin 48°} \approx 13$

3. Given: $a = 30$; $\beta = 50°$; $\gamma = 100°$

$\alpha = 180° - (\beta + \gamma) = 180° - (50° + 100°) = 30°$

$\dfrac{b}{\sin 50°} = \dfrac{c}{\sin 100°} = \dfrac{30}{\sin 30°}$; $\quad b = \dfrac{30 \sin 50°}{\sin 30°} \approx 46$; $\quad c = \dfrac{30 \sin 100°}{\sin 30°} \approx 59$

5. Given: $b = 40$; $\alpha = 50°$; $\gamma = 60°$

$\beta = 180° - (\alpha + \gamma) = 180° - (50° + 60°) = 70°$

$\dfrac{a}{\sin 50°} = \dfrac{c}{\sin 60°} = \dfrac{40}{\sin 70°}$; $\quad a = \dfrac{40 \sin 50°}{\sin 70°} \approx 33$; $\quad c = \dfrac{40 \sin 60°}{\sin 70°} \approx 37$

7. Given: $c = 43$; $\alpha = 120°$; $\gamma = 7°$

$\beta = 180° - (\alpha + \gamma) = 180° - (120° + 7°) = 53°$

$\dfrac{a}{\sin 120°} = \dfrac{b}{\sin 53°} = \dfrac{43}{\sin 7°}$; $\quad a = \dfrac{43 \sin 120°}{\sin 7°} \approx 310$; $\quad b = \dfrac{43 \sin 53°}{\sin 7°} \approx 280$

9. Given: $a = 107$; $\alpha = 18.3°$; $\beta = 54.0°$

$\gamma = 180° - (\alpha + \beta) = 180° - (18.3° + 54.0°) = 107.7°$

$\dfrac{b}{\sin 54.0°} = \dfrac{c}{\sin 107.7°} = \dfrac{107}{\sin 18.3°}$; $\quad b = \dfrac{107 \sin 54.0°}{\sin 18.3°} \approx 276$; $\quad c = \dfrac{107 \sin 107.7°}{\sin 18.3°} \approx 325$

11. Given: $a = 85$; $\alpha = 48.5°$; $\gamma = 72.4°$; To two significant digits: $\alpha = 49°$; $\gamma = 72°$

$\beta = 180° - (\alpha + \gamma) = 180° - (48.5° + 72.4°) = 59.1° \approx 59°$

$\dfrac{b}{\sin 59.1°} = \dfrac{c}{\sin 72.4°} = \dfrac{85}{\sin 48.5°}$; $\quad b = \dfrac{85 \sin 59.1°}{\sin 48.5°} \approx 97$; $\quad c = \dfrac{85 \sin 72.4°}{\sin 48.5°} \approx 110$

13. Given: $a = 105$; $\alpha = 48.5°$; $\beta = 62.7°$

$\gamma = 180° - (\alpha + \beta) = 180° - (48.5° + 62.7°) = 68.8°$

$\dfrac{b}{\sin 62.7°} = \dfrac{c}{\sin 68.8°} = \dfrac{105}{\sin 48.5°}$; $\quad b = \dfrac{105 \sin 62.7°}{\sin 48.5°} \approx 125$; $\quad c = \dfrac{105 \sin 68.8°}{\sin 48.5°} \approx 131$

15. Given: $b = 10.3$; $\alpha = 78°$; $\beta = 102°$;

$\gamma = 180° - (\alpha + \beta) = 180° - (78° + 102°) = 0°$; No triangle formed. $\alpha + \beta + \gamma > 180°$

17. Given: $c = 105$; $\beta = 148.0°$; $\gamma = 22.5°$

$\alpha = 180° - (\beta + \gamma) = 180° - (148.0° + 22.5°) = 9.5°$

$\dfrac{a}{\sin 9.5°} = \dfrac{b}{\sin 148.0°} = \dfrac{105}{\sin 22.5°}$; $\quad a = \dfrac{104 \sin 9.5°}{\sin 22.5°} \approx 45.3$; $\quad b = \dfrac{105 \sin 148.0°}{\sin 22.5°} \approx 145$

177

19. Given: $a = 41.0$; $\alpha = 45.2°$; $\beta = 21.5°$

$\gamma = 180° - (\alpha + \beta) = 180° - (45.2° + 21.5°) = 113.3°$

$\dfrac{b}{\sin 21.5°} = \dfrac{c}{\sin 113.3°} = \dfrac{41.0}{\sin 45.2°}$; $b = \dfrac{41.0 \sin 21.5°}{\sin 45.2°} \approx 21.2$;

$c = \dfrac{41.0 \sin 113.3°}{\sin 45.2°} \approx 53.1$

21. Given: $b = 58.3$; $\alpha = 120°$; $\gamma = 68.0°$

$\beta = 180° - (\alpha + \gamma) = 180° - (120° + 68.0°) = -8°$; No such triangle; $\alpha + \beta + \gamma > 180°$

23. Given: $a = 26$; $b = 71$; $c = 88$

$\gamma = \cos^{-1}\left(\dfrac{26^2 + 71^2 - 88^2}{2(26)(71)}\right) = 123.3001440° \approx 123°$

$\dfrac{71}{\sin \beta} = \dfrac{88}{\sin 123.3001440°}$; $\beta = \sin^{-1}\left(\dfrac{71 \sin 123.3001441°}{88}\right) \approx 42°$;

$\alpha = 180° - (\beta + \gamma) \approx 180° - (42° + 123°) = 15°$

25. Given: $\alpha = 48°$; $\beta = 105°$; $\gamma = 27°$

No unique solution; Must be given at least one side.

27. Given: $a = 80.6$; $b = 23.2$; $\gamma = 89.2°$

$c = \sqrt{23.2^2 + 80.6^2 - 2(23.2)(80.6)\cos 89.2°} = 83.56065868 \approx 83.6$

$\dfrac{\sin \alpha}{80.6} = \dfrac{\sin 89.2°}{83.56065868}$; $\alpha = \sin^{-1}\left(\dfrac{80.6 \sin 89.2°}{83.56065868}\right) \approx 74.7°$

$\beta \approx 180° - (\alpha + \gamma) = 180° - (74.7° + 89.2°) = 16.1°$

29. Given: $c = 28.36$; $\beta = 42.10°$; $\gamma = 102.30°$

$\alpha = 180° - (\beta + \gamma) = 180° - (42.10° + 102.30°) = 35.60°$

$\dfrac{a}{\sin 35.60°} = \dfrac{b}{\sin 42.10°} = \dfrac{28.36}{\sin 102.30°}$;

$a = \dfrac{28.36 \sin 35.50°}{\sin 102.30°} \approx 16.90$; $b = \dfrac{38.36 \sin 42.10°}{\sin 102.30°} \approx 19.46$

31. Given: $a = 10$; $b = 48$; $c = 52$

$$\gamma = \cos^{-1}\left(\frac{10^2 + 48^2 - 52^2}{2(10)(48)}\right) = 108.2099569° \approx 108°$$

$$\frac{\sin \beta}{48} = \frac{\sin 108.2099569°}{52}; \quad \beta = \sin^{-1}\left(\frac{48 \sin 108.2099569°}{52}\right) \approx 61°$$

$$\alpha = 180° - (\beta + \gamma) \approx 180° - (61° + 108°) = 11°$$

33. Given: $a = 48.1$; $\alpha = 4.8°$; $\beta = 82.0°$

$$\gamma = 180° - (\alpha + \beta) = 180° - (4.8° + 82.0°) = 93.2°$$

$$\frac{b}{\sin 82.0°} = \frac{c}{\sin 93.2°} = \frac{48.1}{\sin 4.8°}; \quad b = \frac{48.2 \sin 82.0°}{\sin 4.8°} \approx 569; \quad c = \frac{48.1 \sin 93.2°}{\sin 4.8°} \approx 574$$

35. Given: $c = 83.1$; $\beta = 81.5°$; $\gamma = 162°$

No solution; $\alpha + \beta + \gamma > 180°$

37. Given: $a = 8.1$; $\alpha = 28.5°$; $\gamma = 90.00°$; To two significant digits: $\alpha = 29°$; $\gamma = 90°$

$$\beta = 180° - (\alpha + \gamma) = 180° - (28.5° + 90.00°) = 61.5° \approx 62°$$

$$\frac{b}{\sin 61.5°} = \frac{c}{\sin 90.00°} = \frac{8.1}{\sin 28.5°}; \quad b = \frac{8.1 \sin 61.5°}{\sin 28.5°} \approx 15; \quad c = \frac{8.1 \sin 90.00°}{\sin 28.5°} \approx 17$$

39. Given: $b = 10.9$; $c = 4.45$; $\alpha = 16.2°$

$$a = \sqrt{4.45^2 + 10.9^2 - 2(4.45)(10.9)\cos 16.2°} = 6.74198855 \approx 6.74$$

$$\frac{\sin \gamma}{4.45} = \frac{\sin 16.2°}{6.74198855}; \quad \gamma = \sin^{-1}\left(\frac{4.45 \sin 16.2°}{6.74198855}\right) \approx 10.6°$$

$$\beta = 180° - (\alpha + \gamma) \approx 180° - (16.2° + 10.6°) = 153.2°$$

41. $\beta + 20° = 72°$, so $\beta = 52°$

$\alpha = 20°$ (alternate interior angles)

From the triangle: $\gamma = 180° - [52° + (90° + 20°)] = 18°$

$$\frac{h}{\sin 52°} = \frac{100}{\sin 18°}; \quad h = \frac{100 \sin 52°}{\sin 18°} \approx 260 \text{ ft}$$

179

43. $\beta + 8° = 35°$, so $\beta = 27°$

   $\alpha = 8°$ (alternate interior angles)

   From the triangle: $\gamma = 180° - [27° + (90° + 8°)] = 55°$

   $$\frac{h}{\sin 27°} = \frac{500}{\sin 55°}$$

   $$h = \frac{500 \sin 27°}{\sin 55°} \approx 280 \text{ ft}$$

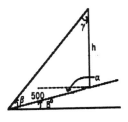

45. $$\frac{x}{\sin 1.90°} = \frac{y}{\sin 3.00°} = \frac{2,500}{\sin 175.10°}$$

   $$x = \frac{2,500 \sin 1.90°}{\sin 175.10°} \approx 970.4 \text{ m}$$

   $$y = \frac{2,500 \sin 3.00°}{\sin 175.10°} \approx 1,532 \text{ m}$$

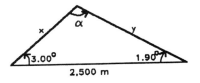

47. $\angle PBE = 11°$ (alternate interior angles)

   $\angle FBE = 11° - 5° = 6°$; $\angle FEB = 90° - 11° = 79°$

   $\angle BFE = 180° - (79° + 6°) = 95°$

   From $\triangle FBE$: $\dfrac{|EB|}{\sin 95°} = \dfrac{6}{\sin 6°}$;

   $|EB| = 57.18220664$

   From $\triangle EBC$: $\cos 11° = \dfrac{|EC|}{57.08220664}$

   $|EC| = 56.13160844$; $\sin 11° = \dfrac{|BC|}{57.18220664}$; $|BC| = 10.91087940$;

   From $\triangle ECT$: $\tan 20° = \dfrac{|CT|}{56.13160844}$; $|CT| = 20.43023468$;

   $\cos 20° = \dfrac{56.13160844}{|ET|}$; $|ET| \approx 60 \text{ ft}$; $|BT| = |BC| + |CT| \approx 31 \text{ ft}$

   The height of the building is 31 ft and it is 60 ft from Mr. T to the top.

49. See diagram in text.

   $\angle JFB = 180° - 35° = 145°$; $\angle JBF = 32°$ (alternate interior angles)

   From $\triangle JFB$: $\dfrac{|JB|}{\sin 145°} = \dfrac{50}{\sin 32°}$; $|JB| = 54.11922864$

   $\angle BJS = 32° - 28° = 4°$; $\angle BSJ = 28°$ (alternate interior angles)

   From $\triangle BJS$: $\dfrac{|SB|}{\sin 4°} = \dfrac{54.11922864}{\sin 28°}$; $|SB| \approx 8 \text{ ft}$

51. $\sin \alpha = \frac{h}{b},$ so $h = b \sin \alpha$

Also $\sin \beta = \frac{h}{a},$ so $h = a \sin \beta$

Thus $b \sin \alpha = a \sin \beta;$

so $\frac{1}{ab}(b \sin \alpha) = \frac{1}{ab}(a \sin \beta)$

Therefore, $\frac{\sin \alpha}{a} = \frac{\sin \beta}{b}$

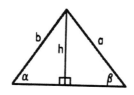

53. From Problem 51: $b \sin \alpha = a \sin \beta,$ so $\frac{1}{b \sin \beta}(b \sin \alpha) = \frac{1}{b \sin \beta}(a \sin \beta).$

Therefore, $\frac{\sin \alpha}{\sin \beta} = \frac{a}{b}$

55. $\frac{a-b}{a+b} = \frac{\sin \alpha - \sin \beta}{\sin \alpha + \sin \beta} = \frac{2 \cos \frac{1}{2}(\alpha + \beta) \sin \frac{1}{2}(\alpha - \beta)}{2 \sin \frac{1}{2}(\alpha + \beta) \cos \frac{1}{2}(\alpha - \beta)}$ (sum-product identities)

57. $\frac{b-c}{b+c} = \frac{2 \cos \frac{1}{2}(\beta + \gamma) \sin \frac{1}{2}(\beta - \gamma)}{2 \sin \frac{1}{2}(\beta + \gamma) \cos \frac{1}{2}(\beta - \gamma)}$ (from Problem 55)

$= \frac{\cos \frac{1}{2}(\beta + \gamma) \sin \frac{1}{2}(\beta - \gamma)}{\sin \frac{1}{2}(\beta + \gamma) \cos \frac{1}{2}(\beta - \gamma)}$

$= \cot \frac{1}{2}(\beta + \gamma) \cdot \tan \frac{1}{2}(\beta - \gamma)$

$= \frac{1}{\tan \frac{1}{2}(\beta + \gamma)} \cdot \tan \frac{1}{2}(\beta - \gamma)$

$= \frac{\tan \frac{1}{2}(\beta - \gamma)}{\tan \frac{1}{2}(\beta + \gamma)}$

59. From the law of sines: $\frac{a}{c} = \frac{\sin \alpha}{\sin \gamma}$ and $\frac{b}{c} = \frac{\sin \beta}{\sin \gamma}$. From the sum-product formula:

$\sin \alpha + \sin \beta = 2 \sin \frac{1}{2}(\alpha + \beta) \cos \frac{1}{2}(\alpha - \beta)$. From the double angle identity:

$\sin \gamma = \sin 2(\frac{1}{2}\gamma) = 2 \sin \frac{1}{2}\gamma \cos \frac{1}{2}\gamma$. Also since $\gamma = 180° - (\alpha + \beta)$,

$\frac{1}{2}\gamma = 90° - \frac{1}{2}(\alpha + \beta)$; so $\cos \frac{1}{2}\gamma = \cos[90° - \frac{1}{2}(\alpha + \beta)] = \sin \frac{1}{2}(\alpha + \beta)$, Finally,

$$\frac{a + b}{c} = \frac{a}{c} + \frac{b}{c}$$

$$= \frac{\sin \alpha}{\sin \gamma} + \frac{\sin \beta}{\sin \gamma}$$

$$= \frac{\sin \alpha + \sin \beta}{\sin \gamma}$$

$$= \frac{2 \sin \frac{1}{2}(\alpha + \beta) \cos \frac{1}{2}(\alpha - \beta)}{2 \sin \frac{1}{2}\gamma \cos \frac{1}{2}\gamma}$$

$$= \frac{\sin \frac{1}{2}(\alpha + \beta) \cos \frac{1}{2}(\alpha - \beta)}{\sin \frac{1}{2}\gamma \sin \frac{1}{2}(\alpha + \beta)}$$

$$= \frac{\cos \frac{1}{2}(\alpha - \beta)}{\sin \frac{1}{2}\gamma}$$

61. A complete discussion of at least five different solutions is found in the October 1977 issue of *Popular Science*. The problem is to find r where r is the rate of the car.

Step 1: $|AC| = $ rate $\cdot$ time $= r(\frac{1}{2})$

Step 2: $|AB| + |AC| = $ rate $\cdot$ time

$|AC| = $ rate $\cdot$ time $- |AB|$

$$= r\left(\frac{35}{60}\right) - 10$$

$$= \frac{7r}{12} - 10$$

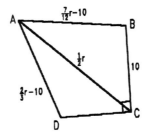

Step 3: $|AD| + |DC| = $ rate $\cdot$ time

$|AD| = $ rate $\cdot$ time $- |DC|$

$$= r\left(\frac{40}{60}\right) - 10$$

$$= \frac{2r}{3} - 10 \qquad \text{(continued on next page)}$$

182

Step 4:   Let $R = \frac{r}{12}$; then $|AB| = 7R - 10$; $|BC| = 10$; $|AC| = 6R$;

$|AD| = 8R - 10$, and $|DC| = 10$

Let $\angle ACD = \theta$; then $\angle ACB = 90° - \theta$ since $\angle DCB$ is a right angle.

Step 5: $\cos\theta = \dfrac{10^2 + (6R)^2 - (8R - 10)^2}{2(10)(6R)}$ from the Law of Cosines for $\triangle ADC$

$= \dfrac{160 - 28R}{120}$ Do not reduce this;

keep the common denominator for Step 7)

Step 6: $\cos(90° - \theta) = \dfrac{10^2 + (6R)^2 - (7R - 10)^2}{2(10)(6R)} = \dfrac{140 - 13R}{120}$

Step 7:   Since $\cos(90° - \theta) = \sin\theta$ and $\cos^2\theta + \sin^2\theta = 1$,

$\left(\dfrac{160 - 28R}{120}\right)^2 + \left(\dfrac{140 - 13R}{120}\right)^2 = 1$

Simplify to: $953R^2 - 12{,}600R + 30{,}800 = 0$

Solve for R using the quadratic formula:

$R = \dfrac{12{,}600 \pm \sqrt{12{,}600^2 - 4(953)(30{,}800)}}{2(953)}$

$= \dfrac{6{,}300 \pm 20\sqrt{25{,}844}}{953}$

$\approx 3.2369,\ 9.9845$

Thus, the rate of the car is 38.8428 mph or 199.8128 mph. Since the latter solution is highly unlikely, the published solution of 38.843 is correct to the nearest thousandth; a better approximation is 38.8430577.

PROBLEM SET 8.4

1.   Given: $a = 3.0$; $b = 4.0$; $\alpha = 125°$

$\alpha > 90°$ and OPP $\leq$ ADJ; no triangle formed

3.   Given: $a = 5.0$; $b = 7.0$; $\alpha = 75°$

$h = 7\sin 75° \approx 6.76$; $\alpha < 90°$ and OPP $< h <$ ADJ; no triangle formed

5. Given: $a = 5.0$; $b = 4.0$; $\alpha = 125°$

OPP $\geq$ ADJ so one solution; $\dfrac{\sin \beta}{4.0} = \dfrac{\sin 125°}{5.0}$; $\beta = \sin^{-1}\left(\dfrac{4.0 \sin 125°}{5.0}\right) \approx 41°$;

$\gamma = 180° - (\alpha + \beta) \approx 14°$; $\dfrac{c}{\sin \gamma} = \dfrac{5.0}{\sin 125°}$; $c = \dfrac{5.0 \sin \gamma}{\sin 125°} \approx 1.5$

7. Given: $a = 9.0$; $b = 7.0$; $\alpha = 75°$; OPP $\geq$ ADJ so one solution

$\dfrac{\sin \beta}{7.0} = \dfrac{\sin 75°}{9.0}$; $\beta = \sin^{-1}\left(\dfrac{7.0 \sin 75°}{9.0}\right) \approx 49°$

$\gamma = 180° - (\alpha + \beta) \approx 56°$; $\dfrac{c}{\sin \gamma} = \dfrac{9.0}{\sin 75°}$; $c = \dfrac{9.0 \sin \gamma}{\sin 75°} \approx 7.8$

9. Given: $a = 7.0$; $b = 9.0$; $\alpha = 52°$

$h = 9.0 \sin 52° \approx 7.09$; $\alpha < 90°$, OPP $< h <$ ADJ, no triangle formed

11. Given: $a = 8.629973679$; $b = 11.8$; $\alpha = 47.0°$; $a \approx 8.63$ (to two significant digits)

$h = 11.8 \sin 47.0° \approx 8.629973679 = a$; Right triangle; $\beta = 90.0°$

$\gamma = 90.0° - 47.0° = 43.0°$; $\cos 47° = \dfrac{c}{11.8}$; $c = 11.8 \cos 47.0° \approx 8.05$

13. Given: $a = 14.2$; $b = 16.3$; $\beta = 115.0°$

SSA; OPP $>$ ADJ so one solution; $\dfrac{\sin \alpha}{14.2} = \dfrac{\sin 115.0°}{16.3}$;

$\alpha = \sin^{-1}\left(\dfrac{14.2 \sin 115.0°}{16.3}\right) \approx 52.1°$; $\gamma = 180° - (\alpha + \beta) \approx 12.9°$;

$\dfrac{c}{\sin \gamma} = \dfrac{16.3}{\sin 115.0°}$; $c = \dfrac{16.3 \sin \gamma}{\sin 115.0°} \approx 4.00$

15. Given: $\beta = 15.0°$; $\gamma = 18.0°$; $b = 23.5$

AAS; $\alpha = 180° - (15.0° + 18.0°) = 147.0°$; $\dfrac{a}{\sin 147.0°} = \dfrac{c}{\sin 18.0°} = \dfrac{23.5}{\sin 15.0°}$;

$a = \dfrac{23.5 \sin 147.0°}{\sin 15.0°} \approx 49.5$; $c = \dfrac{23.5 \sin 18.0°}{\sin 15.0°} \approx 28.1$

17. Given: $b = 82.5$; $c = 52.2$; $\gamma = 32.1°$

SSA; $h = 82.5 \sin 32.1° \approx 43.8$; $h <$ OPP $<$ ADJ so two solutions.

Solution I: $\dfrac{\sin \beta}{82.5} = \dfrac{\sin 32.1°}{52.2}$; $\beta = \sin^{-1}\left(\dfrac{82.5 \sin 32.1°}{52.2}\right) \approx 57.1°$;

$\alpha = 180° - (\beta + \gamma) \approx 90.8°$; $\dfrac{a}{\sin \alpha} = \dfrac{52.2}{\sin 32.1°}$; $a = \dfrac{52.2 \sin \alpha}{\sin 32.1°} \approx 98.2$

Solution II: $\beta' = 180° - (\beta + \gamma) \approx 122.9°$; $\alpha' = 180° - (\beta' + \gamma) \approx 25.0°$;

$\dfrac{a'}{\sin \alpha'} = \dfrac{52.2}{\sin 32.1°}$; $a' = \dfrac{52.2 \sin \alpha'}{\sin 32.1°} \approx 41.6$

19. Given: $a = 68.2$; $\alpha = 145°$; $\beta = 52.4°$; $\alpha + \beta + \gamma > 180°$; No solution

21. Given: $a = 123$; $b = 225$; $c = 351$

    SSS; $a + b < c$; No solution, since the sum of the two smaller sides must be larger than the third.

23. Given: $c = 196$; $\alpha = 54.5°$; $\gamma = 63.0°$

    AAS; $\beta = 180° - (63.0° + 54.5°) = 62.5°$; $\dfrac{a}{\sin 54.5°} = \dfrac{b}{\sin 62.5°} = \dfrac{196}{\sin 63.0°}$;

    $a = \dfrac{196 \sin 54.5°}{\sin 63.0°} \approx 179$; $b = \dfrac{196 \sin 62.5°}{\sin 63.0°} \approx 195$

25. $K = \frac{1}{2}(15)(8)\sin 38° \approx 37$ square units

27. $K = \frac{1}{2}(14)(12)\sin 82° \approx 83$ square units

29. $\alpha = 180° - (100° + 50°) = 30°$; $K = \dfrac{30^2 \sin 50° \sin 100°}{2 \sin 30°} \approx 680$ square units

31. $\beta = 180° - (50° + 60°) = 70°$; $K = \dfrac{40^2 \sin 50° \sin 60°}{2 \sin 70°} \approx 560$ square units

33. $s = \frac{1}{2}(7.0 + 8.0 + 2.0) = 8.5$;

    $K = \sqrt{8.5(8.5 - 7.0)(8.5 - 8.0)(8.5 - 2.0)} \approx 6.4$ square units

35. $s = \frac{1}{2}(11 + 9.0 + 8.0) = 14$;

    $K = \sqrt{14(14 - 11)(14 - 9.0)(14 - 8.0)} \approx 35$ square units

37. Given: $a = 12.0$; $b = 9.00$; $\alpha = 52.0°$

    SSA; OPP > ADJ so one solution; $\dfrac{\sin \beta}{9.00} = \dfrac{\sin 52.0°}{12.0}$; $\beta = 36.22857624°$;

    $\gamma = 180° - (\alpha + \beta) = 91.77142376°$; $K = \frac{1}{2}(9.0)(12.0)\sin \gamma \approx 54.0$ square units

39. Given: $a = 10.2$; $b = 11.8$; $\alpha = 47.0°$

    SSA; $h = 11.8 \sin 47.0° \approx 8.63$; $h < \text{OPP} < \text{ADJ}$ so two solutions.

    Solution I: $\dfrac{\sin \beta}{11.8} = \dfrac{\sin 47.0°}{10.2}$; $\beta \approx 57.8°$; $\gamma \approx 180° - (47.0° + 57.8°) = 75.2°$;

    $K = \frac{1}{2}(10.2)(11.8)\sin \gamma \approx 58.2$ square units

    Solution: $\beta' = 180° - \beta \approx 122.2°$; $\gamma' = 180° - (47.0° + 122.2°) \approx 10.8°$;

    $K = \frac{1}{2}(10.2)(11.8)\sin \gamma' \approx 11.3$ square units

41. Given: $b = 82.5$; $c = 52.2$; $\gamma = 32.1°$

SSA: $h = 82.5 \sin 32.1° \approx 43.8$; $h < \text{OPP} < \text{ADJ}$ so two solutions.

Solution I: $\dfrac{\sin \beta}{82.5} = \dfrac{\sin 32.1°}{52.2}$; $\beta \approx 57.1°$; $\alpha \approx 180° - (57.1° + 32.1°) = 90.8°$;

$K = \frac{1}{2}(82.5)(52.2)\sin \alpha \approx 2{,}150$ square units

Solution: $\beta' = 180° - \beta \approx 122.9°$; $\alpha' \approx 180° - (122.9° + 32.1°) = 25.0°$;

$K = \frac{1}{2}(82.5)(52.2)\sin \alpha' \approx 911$ square units

43. Given: $\beta = 15.0°$; $\gamma = 18.0°$; $b = 23.5$

AAS; $\alpha = 180° - (15.0° + 18.0°) = 147.0°$;

$K = \dfrac{23.5^2 \sin 147.0° \sin 18.0°}{2 \sin 15.0°} \approx 180$ square units

45. Given: $a = 68.2$; $\alpha = 145°$; $\beta = 52.4°$

$\alpha + \beta + \gamma > 180°$; No triangle formed.

47. Given: $a = 124$; $b = 325$; $c = 351$

SSS; $s = \frac{1}{2}(124 + 325 + 351) = 400$;

$K = \sqrt{400(400 - 124)(400 - 325)(400 - 351)} \approx 20{,}100$ square units

49. $|TH| = \sqrt{4.30^2 + 5.20^2 - 2(4.30)(5.20)\cos 68.4°} \approx 5.39$ miles

51. $\angle U = 180° - (40.79° + 10.48°) = 128.73°$

$\dfrac{|UA|}{\sin 10.48°} = \dfrac{|UB|}{\sin 40.79°} = \dfrac{2.300}{\sin 128.73°}$

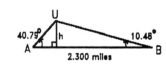

$|UA| = \dfrac{2.300 \sin 10.48°}{\sin 128.73°} \approx 0.5363$ miles

$|UB| = \dfrac{2.300 \sin 40.79°}{\sin 128.73°} \approx 1.926$ miles

$\sin 10.48° = \dfrac{h}{|UB|}$; $h = |UB| \sin 10.48° \approx .3503$ miles $\approx 1{,}850$ feet

The altitude is 1,850 feet and the distances are 1.926 miles (10,170 feet) and .5363 miles (2,832 feet).

53. $\theta = 90° + 5.45° = 95.45°$; $\beta = 180° - 95.45° = 84.55°$

   $\gamma = 180° - (20.24° + 84.55°) = 75.21°$

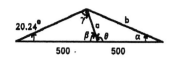

   $\dfrac{a}{\sin 20.24°} = \dfrac{500}{\sin 75.21°}$; $a \approx 178.9$

   $b = \sqrt{a^2 + 500^2 - 2(a)(500)\cos 95.45°} \approx 546.8$

   $\dfrac{\sin \alpha}{a} = \dfrac{\sin 95.45°}{b}$; $\alpha = \sin^{-1}\left(\dfrac{a \sin 95.45°}{b}\right) \approx 19.0°$

55. $54.4 = \tfrac{1}{2}(1.78)r^2$; $r = \sqrt{\dfrac{2(54.4)}{1.78}} \approx 7.82$ cm

57. $\theta = (20°) \cdot \left(\dfrac{\pi}{180°}\right) = \dfrac{\pi}{9}$; $A = \tfrac{1}{2}\left(\dfrac{\pi}{9}\right)(320^2) \approx 18{,}000$ m^2

59. $V = \dfrac{\pi}{3} \cdot (30^3) \cdot \tan 36° \approx 21{,}000$ ft^3 = 760 yd^3

61. $\alpha = 64.6° - 23.2° = 41.4°$; $\gamma = 64.6°$ (alternate interior angles);

   $\theta = 90° - 64.6° = 25.4°$;

   $\beta = 180° - (23.2° + 31.4°) = 125.4°$;

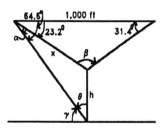

   $\dfrac{x}{\sin 31.4°} = \dfrac{1{,}000}{\sin 125.4°}$; $x \approx 639.2$;

   $\dfrac{h}{\sin 41.4°} = \dfrac{x}{\sin 25.4°}$; $h = \dfrac{x \sin 41.4°}{\sin 25.4°} \approx 985$ feet

63. a. If $d < 0$, then c is not a real number, so there is no triangle.

   b. If $d = 0$, then $a^2 = b^2 \sin \alpha$. Since $0° < \alpha < 90°$, $\sin \alpha > 0$. Also $a > 0$ and $b > 0$, so $a = b \sin \alpha = h$. Hence one solution.

   c. If $d > 0$, $a^2 > b^2 \sin \alpha$. Since a, b, and $\sin \alpha$ are positive, $a > b \sin \alpha$. Thus $a > h$ and there is at least one solution. If $b > a$ there will be two solutions, if $a > b$, one solution.

65. $K = \frac{1}{2}bc \sin \alpha$

$$= \frac{1}{2}bc \sin 2\left(\frac{1}{2}\alpha\right)$$

$$= bc \sin \tfrac{1}{2}\alpha \cos \tfrac{1}{2}\alpha$$

$$= bc\sqrt{\frac{1-\cos\alpha}{2}}\sqrt{\frac{1+\cos\alpha}{2}} \quad \text{(Positive since } 0° < \alpha < 180°)$$

$$= \frac{1}{2}bc\sqrt{1-\cos^2\alpha}$$

$$= \frac{1}{2}bc\sqrt{1-\left(\frac{b^2+c^2-a^2}{2bc}\right)^2} \quad \text{(law of cosines)}$$

$$= \frac{1}{2}bc\sqrt{\left(1+\frac{b^2+c^2-a^2}{2bc}\right)\left(1-\frac{b^2+c^2-a^2}{2bc}\right)}$$

$$= \frac{1}{2}bc\sqrt{\frac{(b^2+2bc+c^2-a^2)(a^2-b^2+2bc-c^2)}{4b^2c^2}}$$

$$= \frac{1}{4}\sqrt{\left[(b+c)^2-a^2\right]\cdot\left[a^2-(b-c)^2\right]}$$

$$= \frac{1}{4}\sqrt{(b+c+a)(b+c-a)(a+(b-c))(a-(b-c))}$$

$$= \sqrt{\frac{1}{16}(a+b+c)(b+c-a)(a+c-b)(a+b-c)}$$

$$= \sqrt{\tfrac{1}{2}(a+b+c)\tfrac{1}{2}(b+c-a)\tfrac{1}{2}(a+c-b)\tfrac{1}{2}(a+b-c)}$$

$$= \sqrt{s(s-a)(s-b)(s-c)}$$

CHAPTER 8 SUMMARY PROBLEMS

1.   If θ is an acute angle of a right triangle then:

$\cos\theta = \dfrac{\text{ADJ}}{\text{HYP}}$, $\sin\theta = \dfrac{\text{OPP}}{\text{HYP}}$, $\tan\theta = \dfrac{\text{OPP}}{\text{ADJ}}$

3.   Given: $b = 678$, $\beta = 55.0°$, and $\gamma = 90.0°$

$\alpha = 180° - (55.0° + 90.0°) = 35.0°$; $\sin 55.0° = \dfrac{678}{c}$; $c = \dfrac{678}{\sin 55.0°} \approx 828$;

$\tan 35° = \dfrac{a}{678}$; $a = 678 \tan 35° \approx 475$

5.   See text.   7.   See text.

9.   Procedure: $\cos\alpha = \dfrac{b^2+c^2-a^2}{2bc}$;   Solution: $\alpha = \cos^{-1}\left(\dfrac{b^2+c^2-a^2}{2bc}\right)$

11. Procedure: $\alpha + \beta + \gamma = 180°$; Solution: $\alpha = 180° - (\beta + \gamma)$

13. Given: $a = 24$; $c = 61$; $\beta = 58°$

SAS; $b = \sqrt{24^2 + 61^2 - 2(24)(61)\cos 58°} \approx 52$; $\dfrac{\sin \alpha}{24} = \dfrac{\sin 58°}{b}$;

$\alpha = \sin^{-1}\left(\dfrac{24 \sin 58°}{b}\right) \approx 23°$; $\gamma \approx 180° - (23° + 58°) = 99°$

15. Given: $b = 34$; $c = 21$; $\gamma = 16°$

SSA; $h = 34 \sin 16° \approx 9.4$; $h < $ OPP $<$ ADJ so two solutions.

Solution I: $\dfrac{\sin \beta}{34} = \dfrac{\sin 16°}{21}$; $\beta = \sin^{-1}\left(\dfrac{34 \sin 16°}{21}\right) \approx 27°$;

$\alpha \approx 180° - (27° + 16°) = 137°$; $\dfrac{a}{\sin \alpha} = \dfrac{21}{\sin 16°}$; $a = \dfrac{21 \sin \alpha}{\sin 16°} \approx 51$

Solution II: $\beta' = 180° - \beta \approx 153°$; $\alpha' \approx 180° - (153° + 16°) \approx 11°$;

$\dfrac{a'}{\sin \alpha'} = \dfrac{21}{\sin 16°}$; $a' = \dfrac{21 \sin \alpha'}{\sin 16°} \approx 14$

17. $K = \frac{1}{2}(25.5)(48.5)\sin 48° \approx 460$ ft^2

19. Given: $b = 17.2$; $c = 14.5$; $\gamma = 35.5°$

SSA; $h = 17.2 \sin 35.5° \approx 9.98$; $h < $ OPP $<$ ADJ so two solutions.

Solution I: $\dfrac{\sin \beta}{17.2} = \dfrac{\sin 35.5°}{14.5}$; $\beta \approx 43.5°$; $\alpha \approx 180° - (43.5° + 35.5°) = 101°$;

$K = \frac{1}{2}(17.2)(14.5)\sin \alpha \approx 122$ in^2

Solution II: $\beta' = 180° - \beta \approx 136.5°$; $\alpha' \approx 180° - (136.5° + 35.5°) = 8°$;

$K = \frac{1}{2}(17.2)(14.5)\sin \alpha' \approx 17.4$ in^2

21. distance from entrance to exit:

$x = \sqrt{382^2 + 485^2 - 2(382)(485)\cos 58.0°}$

$x = 429.8752328 \approx 430$ feet

angle of elevation:

$\dfrac{\sin \alpha}{382} = \dfrac{\sin 58.0°}{x}$;

$\alpha = \sin^{-1}\left(\dfrac{382 \sin 58.0°}{x}\right) \approx 48.9°$

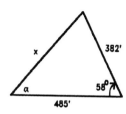

23. Using alternate interior angles:

$\angle PRQ = 67.5° - 38.4° = 29.1°$

$\tan 29.1° = \dfrac{|PQ|}{500};$

$|PQ| = 500 \tan 29.1° \approx 278$ feet

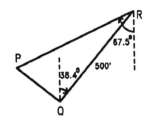

CUMULATIVE REVIEW II

1. Let θ be an angle in standard position with a point $P(x, y)$ on the terminal side a distance of r $(r \neq 0)$ from the origin. Then: $\cos \theta = \frac{x}{r}$; $\sin \theta = \frac{y}{r}$; $\tan \theta = \frac{y}{x}$ $(x \neq 0)$; $\sec \theta = \frac{r}{x}$ $(x \neq 0)$; $\csc \theta = \frac{r}{y}$ $(y \neq 0)$; $\cot \theta = \frac{x}{y}$ $(y \neq 0)$.

2. **a.** $\sin 300° = -\dfrac{\sqrt{3}}{2}$

b. $\cos \dfrac{5\pi}{6} = -\dfrac{\sqrt{3}}{2}$

c. $\tan 3\pi = 0$

d. $\mathrm{Sin}^{-1}\left(-\dfrac{\sqrt{3}}{2}\right) = -\dfrac{\pi}{3}$

e. $\mathrm{Arccos}\,\frac{1}{2} = \dfrac{\pi}{3}$

f. $\mathrm{Arccos}(-4.521)$ has no value

g. $\mathrm{Tan}^{-1}2.310 = 1.162253$

h. $\tan(\mathrm{Sec}^{-1}) = 2\sqrt{2}$

3. **a.**

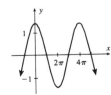

b.

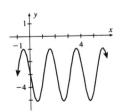

c.

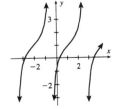

4. **a.** $\sec \theta = \dfrac{1}{\cos \theta}$ Proof: $\sec \theta = \dfrac{r}{x} = \dfrac{1}{x/r} = \dfrac{1}{\cos \theta}$

   $\csc \theta = \dfrac{1}{\sin \theta}$ Proof: $\csc \theta = \dfrac{r}{y} = \dfrac{1}{y/r} = \dfrac{1}{\sin \theta}$

   $\cot \theta = \dfrac{1}{\tan \theta}$ Proof: $\cot \theta = \dfrac{x}{y} = \dfrac{1}{y/x} = \dfrac{1}{\tan \theta}$

b. $\tan\theta = \frac{\sin\theta}{\cos\theta}$ Proof: $\tan\theta = \frac{y}{x} = \frac{y/r}{x/r} = \frac{\sin\theta}{\cos\theta}$

    $\cot\theta = \frac{\cos\theta}{\sin\theta}$ Proof: $\cot\theta = \frac{x}{y} = \frac{x/r}{y/r} = \frac{\cos\theta}{\sin\theta}$

c. $\sin^2\theta + \cos^2\theta = 1$ Proof: $\sin^2\theta + \cos^2\theta = \left(\frac{y}{r}\right)^2 + \left(\frac{x}{r}\right)^2$

$$= \frac{y^2 + x^2}{r^2}$$

$$= \frac{r^2}{r^2}$$

$$= 1$$

    $1 + \tan^2\theta = \sec^2\theta$ Proof: $1 + \tan^2\theta = 1 + \left(\frac{y}{x}\right)^2$

$$= \frac{x^2 + y^2}{x^2}$$

$$= \frac{r^2}{x^2}$$

$$= \sec^2\theta$$

    $\cot^2\theta + 1 = \csc^2\theta$ Proof: $\cot^2\theta + 1 = \left(\frac{x}{y}\right)^2 + 1$

$$= \frac{x^2 + y^2}{y^2}$$

$$= \frac{r^2}{y^2}$$

$$= \csc^2\theta$$

5.   a. $3\tan 3\theta = \sqrt{3}$; $\tan 3\theta = \frac{\sqrt{3}}{3}$; $3\theta' = \text{Tan}^{-1}\frac{\sqrt{3}}{3} = \frac{\pi}{6}$;   tangent is positive

in quadrants I and III.   so $3\theta = \frac{\pi}{6}, \frac{7\pi}{6}, \frac{13\pi}{6}, \frac{19\pi}{6}, \frac{25\pi}{6}, \frac{31\pi}{6}$.

Therefore, $\theta = \frac{\pi}{18}, \frac{7\pi}{18}, \frac{13\pi}{18}, \frac{19\pi}{18}, \frac{25\pi}{18}, \frac{31\pi}{18}$.

b. $4\sin^2\theta + 8\sin\theta - 1 = 0$; $\sin\theta = \dfrac{-8 \pm \sqrt{8^2 - 4(4)(-1)}}{2(4)}$;

$\sin\theta = \dfrac{-8 - \sqrt{80}}{8} \approx -2.1$ (no solution); $\sin\theta = \dfrac{-8 + \sqrt{80}}{8} \approx .1180$;

$\theta' \approx \sin^{-1} .1180 \approx .1183$; sine is positive in quadrants I and II.

$\theta \approx .1183,\ \pi - .1183 \approx .1183,\ 3.0238.$

6. a. $\dfrac{\sin 3\theta}{\sin\theta} - \dfrac{\cos 3\theta}{\cos\theta} = \dfrac{\sin 3\theta \cos\theta - \cos 3\theta \sin\theta}{\sin\theta \cos\theta}$

$= \dfrac{\sin(3\theta - \theta)}{\sin\theta \cos\theta}$

$= \dfrac{\sin 2\theta}{\sin\theta \cos\theta}$

$= \dfrac{2\sin\theta \cos\theta}{\sin\theta \cos\theta}$

$= 2$

b. $\dfrac{\sin\theta}{\csc\theta - \cot\theta} = \dfrac{\sin\theta}{\dfrac{1}{\sin\theta} - \dfrac{\cos\theta}{\sin\theta}}$

$= \dfrac{\sin^2\theta}{1 - \cos\theta}$

$= \dfrac{1 - \cos^2\theta}{1 - \cos\theta}$

$= \dfrac{(1 + \cos\theta)(1 - \cos\theta)}{1 - \cos\theta}$

$= 1 + \cos\theta$

c. $\cos 2\theta = \cos(\theta + \theta)$

$= \cos\theta \cos\theta - \sin\theta \sin\theta$

$= \cos^2\theta - \sin^2\theta$

$= (1 - \sin^2\theta) - \sin^2\theta$

$= 1 - 2\sin^2\theta$

7. a. $\dfrac{\sin\psi}{s} = \dfrac{\sin\theta}{t} = \dfrac{\sin\phi}{u}$

b. $u^2 = s^2 + t^2 - 2st\cos\phi$; $s^2 = t^2 + u^2 - 2tu\cos\psi$; $t^2 = s^2 + u^2 - 2su\cos\theta$

192

8. a.   Given: $a = 14$; $b = 27$; $c = 19$

$$\cos \beta = \frac{14^2 + 19^2 - 27^2}{2(14)(19)}; \quad \beta = \cos^{-1}\left(\frac{14^2 + 19^2 - 27^2}{2(14)(19)}\right) \approx 109°$$

$$\frac{\sin \alpha}{14} = \frac{\sin \beta}{27}; \quad \alpha = \sin^{-1}\left(\frac{14 \sin \beta}{27}\right) \approx 29°; \quad \gamma \approx 180° - (29° + 109°) = 42°$$

b.   Given: $b = 7.2$; $c = 15$; $\alpha = 113°$

$$a = \sqrt{7.2^2 + 15^2 - 2(7.2)(15)\cos 113°} \approx 19; \quad \frac{\sin \beta}{7.2} = \frac{\sin 113°}{a};$$

$$\beta = \sin^{-1}\left(\frac{7.2 \sin 113°}{a}\right) \approx 20°; \quad \gamma \approx 180° - (113° + 20°) = 47°$$

c.   Given: $a = 35$; $b = 45$; $\beta = 35°$

SSA; OPP > ADJ so one solution. $\dfrac{\sin \alpha}{35} = \dfrac{\sin 35°}{45}; \quad \alpha = \sin^{-1}\left(\dfrac{35 \sin 35°}{45}\right) \approx 26°;$

$$\gamma \approx 180° - (26° + 35°) = 119°; \quad \frac{c}{\sin \gamma} = \frac{45}{\sin 35°}; \quad c = \frac{45 \sin \gamma}{\sin 35°} \approx 69$$

d.   Given: $c = 85.7$; $\alpha = 50.8°$; $\beta = 83.5°$

$$\gamma = 180° - (50.8° + 83.5°) = 45.7°; \quad \frac{a}{\sin 50.8°} = \frac{85.7}{\sin 45.7°};$$

$$a = \frac{85.7 \sin 50.8°}{\sin 45.7°} \approx 92.8; \quad \frac{b}{\sin 83.5°} = \frac{85.7}{\sin 45.7°}; \quad b = \frac{85.7 \sin 83.5°}{\sin 45.7°} \approx 119$$

9. a. $K = \frac{1}{2}(16)(43) \sin 113° \approx 320 \text{ ft}^2$

b. $K = \dfrac{14.3^2 \sin 51.8° \sin 40.0°}{2 \sin 88.2°} \approx 51.7 \text{ in}^2$

c. $s = \frac{1}{2}(121 + 46 + 92) = 129.5;$

$$K = \sqrt{129.5(129.5 - 121)(129.5 - 46)(129.5 - 92)} \approx 1,900 \text{ cm}^2$$

10. a.   Given: $a = 63$; $c = 50$; $\gamma = 45°$

SSA: $h = 63 \sin 45° \approx 44.5$; $h < \text{OPP} < \text{ADJ}$ so two solutions

Solution I: $\dfrac{\sin \alpha}{63} = \dfrac{\sin 45°}{50};$

$$\alpha = \sin^{-1}\left(\frac{63 \sin 45°}{50}\right) \approx 63°;$$

$$\beta \approx 180° - (63° + 45°) = 72°$$

$$\frac{b}{\sin \beta} = \frac{50}{\sin 45°}; \quad b = \frac{50 \sin \beta}{\sin 45°} \approx 67 \text{ miles}$$

Solution II: $\alpha' = 180° - \alpha \approx 117°$; $\beta' \approx 180° - (117° + 45°) = 18°$

$$\frac{b'}{\sin \beta'} = \frac{50}{\sin 45°}; \quad b' = \frac{50 \sin \beta'}{\sin 45°}; \quad b' \approx 22 \text{ miles}$$

b. $\tan 88.76° = \frac{x}{100}$

$x = 100 \tan 88.76°$

$x \approx 4,620 \text{ feet}$

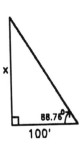

100'

EXTENDED APPLICATION: SOLAR POWER

1.
$$5 \begin{bmatrix} 3 \begin{bmatrix} 35° & 16:58 \\ 38° & \end{bmatrix} x \\ 40° & 16:44 \end{bmatrix} - 14$$

$$\frac{x}{-14} = \frac{3}{5}$$

$$x \approx -8$$

Sunset occurs at
$$16:58 + (-8) = 16:50 \text{ (4:50 P.M.)}$$

2.
$$10 \begin{bmatrix} 8 \begin{bmatrix} 20° & 5:23 \\ 28° & \end{bmatrix} x \\ 30° & 5:02 \end{bmatrix} - 21$$

$$\frac{x}{-21} = \frac{8}{10}$$

$$x \approx -17$$

Sunrise occurs at
$$5:23 + (-17) = 5:06 \text{ A.M.}$$

3.
$$10 \begin{bmatrix} 8 \begin{bmatrix} 20° & 18:43 \\ 28° & \end{bmatrix} x \\ 30° & 19:05 \end{bmatrix} 22$$

$$\frac{x}{22} = \frac{8}{10}$$

$$x \approx 18$$

Sunset occurs at
$$18:43 + (18) = 19:01 \text{ (7:01 P.M.)}$$

4.
$$15 \begin{bmatrix} 5 \begin{bmatrix} 45° & 17:21 \\ 50° & \end{bmatrix} x \\ 60° & 17:01 \end{bmatrix} - 20$$

$$\frac{x}{-20} = \frac{5}{15}$$

$$x \approx -7$$

Sunset occurs at
$$17:21 + (-7) = 17:14 \text{ (5:14 P.M.)}$$

5.

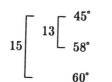

$$\frac{x}{20} = \frac{5}{15}$$

$$x \approx 7$$

Sunrise occurs at
$6:11 + (7) = 6:18$ A.M.

6.

$$\frac{x}{20} = \frac{13}{15}$$

$$x \approx 17$$

Sunrise occurs at
$6:39 + (17) = 6:56$ A.M.

7. See graph below

9. See graph below

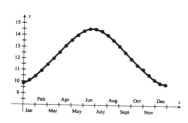

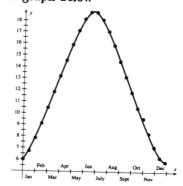

10. Answers vary   **11.** Answers vary   **12.** Answers vary

CHAPTER 9

PROBLEM SET 9.1

1.

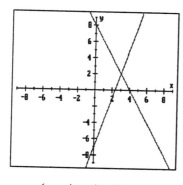

$(x, y) = (3, 2)$

3.

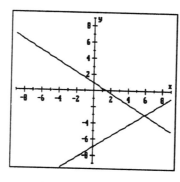

$(x, y) = (6, -3)$

5.

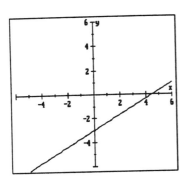

Dependent System

7.

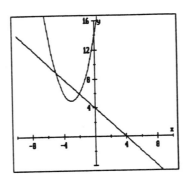

$(x, y) = (-2, 6), (-5, 9)$

9.

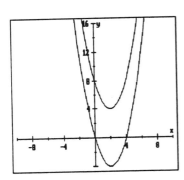

Inconsistent system

11.
$$\begin{cases} s+t = 1 \\ 3s+t = -5 \end{cases}$$

Solving the first equation for s, $s = 1-t$.
Substitute this into the second equation:

$$3(1-t)+t = -5$$

$$3-3t+t = -5$$

$$-2t = -8$$

$$t = 4 \qquad \text{Substitute this value into either equation and solve for } s.$$

$$s+4 = 1$$

$$s = -3 \qquad \text{So the solution is } (s, t) = (-3, 4).$$

By inspection $x_1 = s$, $x_2 = t$, $a_{11} = 1$, $a_{12} = 1$, $b_1 = 1$, $a_{21} = 3$, $a_{22} = 1$, and $b_2 = -5$.

13.

$$\begin{cases} v = \frac{3}{5}u + 2 \\ 3u - 5v = 10 \end{cases}$$

Substitute $v = \frac{3}{5}u + 2$ into the second equation:

$$3u - 5(\tfrac{3}{5}u + 2) = 10$$

$$3u - 3u - 10 = 10$$

$$0 = 20 \qquad \text{Inconsistent system}$$

Rewrite the first equation as $3u - 5v = -10$

$$\begin{cases} 3u - 5v = -10 \\ 3u - 5v = 10 \end{cases}$$

By inspection $x_1 = u$, $x_2 = v$, $a_{11} = 3$, $a_{12} = -5$, $b_1 = -10$, $a_{21} = 3$, $a_{22} = -5$, and $b_2 = 10$.

15.

$$\begin{cases} 3t_1 + 5t_2 = 1{,}541 \\ t_2 = 2t_1 + 160 \end{cases}$$

Substitute $t_2 = 2t_1 + 160$ into the first equation:

$$3t_1 + 5(2t_1 + 160) = 1{,}541$$
$$3t_1 + 10t_1 + 800 = 1{,}541$$
$$13t_1 = 741$$
$$t_1 = 57 \qquad \text{Substitute this value into either equation}$$

and solve for t_2.

$$t_2 = 2(57) + 160$$
$$t_2 = 274 \qquad \text{So the solution is } (t_1, \ t_2) = (57, \ 274)$$

Rewrite the second equation as $2t_1 - t_2 = -160$.

$$\begin{cases} 3t_1 + 5t_2 = 1{,}541 \\ 2t_1 - t_2 = -160 \end{cases}$$

By inspection: $x_1 = t_1$, $x_2 = t_2$, $a_{11} = 3$, $a_{12} = 5$, $b_1 = 1{,}541$, $a_{21} = 2$, $a_{22} = -1$, and $b_2 = -160$.

17.

$$\begin{cases} \gamma = 3\delta - 4 \\ 5\gamma - 4\delta = -9 \end{cases}$$

Substitute $\gamma = 3\delta - 4$ into the second equation.

$$5(3\delta - 4) - 4\delta = -9$$
$$15\delta - 20 - 4\delta = -9$$
$$11\delta = 11$$
$$\delta = 1 \qquad \text{Substitute this value into either equation and solve for } \gamma.$$
$$\gamma = 3(1) - 4$$
$$\gamma = -1 \qquad \text{So the solution is } (\gamma, \ \delta) = (-1, \ 1)$$

Rewrite the first equation as $\gamma - 3\delta = -4$.

$$\begin{cases} \gamma - 3\delta = -4 \\ 5\gamma - 4\delta = -9 \end{cases}$$

By inspection: $x_1 = \gamma$, $x_2 = \delta$, $a_{11} = 1$, $a_{12} = -3$, $b_1 = -4$, $a_{21} = 5$, $a_{22} = -4$, and $b_2 = -9$.

19. $+ \begin{cases} c + d = 2 \\ 2c - d = 1 \end{cases}$

$3c \quad = 3$

$c = 1$ Substitute this value into either equation and solve for d.

$1 + d = 2$

$d = 1$ So the solution is $(c, \ d) = (1, \ 1)$

By inspection: $x_1 = c$, $x_2 = d$, $a_{11} = 1$, $a_{12} = 1$, $b_1 = 2$, $a_{21} = 2$, $a_{22} = -1$, and $b_2 = 1$.

21. $\begin{matrix} 3 \\ 4 \end{matrix} \begin{cases} 3q_1 - 4q_2 = 3 \\ 5q_1 + 3q_2 = 5 \end{cases}$

$+ \begin{cases} 9q_1 - 12q_2 = 9 \\ 20q_1 + 12q_2 = 20 \end{cases}$

$29q_1 \quad = 29$

$q_1 = 1$ Substitute this value into either equation and solve for q_2.

$3(1) - 4q_2 = 3$

$-4q_2 = 0$

$q_2 = 0$ So the solution is $(q_1, \ q_2) = (1, \ 0)$

By inspection: $x_1 = q_1$, $x_2 = q_2$, $a_{11} = 3$, $a_{12} = -4$, $b_1 = 3$, $a_{21} = 5$, $a_{22} = 3$, and $b_2 = 5$.

23. $2 \begin{cases} 7x + y = 5 \\ 14x - 2y = -2 \end{cases}$

$+ \begin{cases} 14x + 2y = 10 \\ 14x - 2y = -2 \end{cases}$

$28x \quad = 8$

$x = \frac{2}{7}$ Substitute this value into either equation and solve for y.

$7(\frac{2}{7}) + y = 5$

$y = 3$ So the solution is $(x, \ y) = (\frac{2}{7}, \ 3)$

By inspection: $x_1 = x$, $x_2 = y$, $a_{11} = 7$, $a_{12} = 1$, $b_1 = 5$, $a_{21} = 14$, $a_{22} = -2$, and $b_2 = -2$.

25.
$$2 \begin{cases} \alpha + \beta = 12 \\ \alpha - 2\beta = -4 \end{cases}$$

$$+ \begin{cases} 2\alpha + 2\beta = 24 \\ \alpha - 2\beta = -4 \end{cases}$$

$$3\alpha \qquad = 20$$

$$\alpha = \tfrac{20}{3} \qquad \text{Substitute this value into either equation and solve for } \beta.$$

$$\tfrac{20}{3} + \beta = 12$$

$$\beta = \tfrac{16}{3} \qquad \text{So the solution is } (\alpha, \ \beta) = \left(\tfrac{20}{3}, \ \tfrac{16}{3} \right)$$

By inspection: $x_1 = \alpha$, $x_2 = \beta$, $a_{11} = 1$, $a_{12} = 1$, $b_1 = 12$, $a_{21} = 1$, $a_{22} = -2$, and $b_2 = -4$.

27.
$$\begin{array}{c} 3 \\ -2 \end{array} \begin{cases} 2\theta + 5\phi = 7 \\ 3\theta + 4\phi = 0 \end{cases}$$

$$+ \begin{cases} 6\theta + 15\phi = 21 \\ -6\theta - 8\phi = 0 \end{cases}$$

$$7\phi = 21$$

$$\phi = 3 \qquad \text{Substitute this value into either equation and solve for } \theta.$$

$$3\theta + 4(3) = 0$$

$$3\theta = -12$$

$$\theta = -4 \qquad \text{So the solution is } (\theta, \ \phi) = (-4, \ 3)$$

By inspection: $x_1 = \theta$, $x_2 = \phi$, $a_{11} = 2$, $a_{12} = 5$, $b_1 = 7$, $a_{21} = 3$, $a_{22} = 4$, and $b_2 = 0$.

29.
$$-2 \begin{cases} 3x + 2y = 1 \\ 6x + 4y = 2 \end{cases}$$

$$+ \begin{cases} -6x - 4y = -2 \\ 6x + 4y = 2 \end{cases}$$

$$0 = 0 \qquad \text{Dependent system}$$

31.
$$\begin{cases} 12x - 5y = -39 \\ y = 2x + 9 \end{cases} \qquad \text{Substitute the second equation into the first equation.}$$

$$12x - 5(2x + 9) = -39$$

$$12x - 10x - 45 = -39$$

$$2x = 6$$

$$x = 3 \qquad \text{Substitute this value into either equation and solve for } y.$$

$$y = 2(3) + 9$$

$$y = 15 \qquad \text{Solution: } (3, 15)$$

33.
$$\begin{cases} y = \frac{2}{3}x - 5 \\ y = -\frac{4}{3}x + 7 \end{cases}$$ Substitute $y = \frac{2}{3}x - 5$ into the second equation

$$\frac{2}{3}x - 5 = -\frac{4}{3}x + 7$$

$$2x = 12$$

$$x = 6$$ Substitute this value into either equation and solve for y.

$$y = \frac{2}{3}(6) - 5$$

$$y = -1$$ Solution: $(6, -1)$

35. $-1 \begin{cases} x - y = \gamma \\ x - 2y = \delta \end{cases}$

$+ \begin{cases} -x + y = -\gamma \\ x - 2y = \delta \end{cases}$

$$-y = -\gamma + \delta$$

$$y = \gamma - \delta$$ Substitute this value into either equation and solve for x.

$$x - (\gamma - \delta) = \gamma$$

$$x = 2\gamma - \delta$$ Solution: $(2\gamma - \delta, \ \gamma - \delta)$

37. $+ \begin{cases} x + y = a \\ x - y = b \end{cases}$

$$2x = a + b$$

$$x = \frac{a+b}{2}$$ Substitute this value into either equation and solve for y.

$$\frac{a+b}{2} + y = a$$

$$y = \frac{2a}{2} - \frac{a+b}{2}$$

$$y = \frac{a-b}{2}$$ Solution: $\left(\frac{a+b}{2}, \ \frac{a-b}{2}\right)$

39.
$$+\begin{cases} 2x - y = c \\ 3x + y = d \end{cases}$$

$$5x \quad = c + d$$

$$x = \frac{c+d}{5}$$ Substitute this value into either equation and solve for y.

$$3\left(\frac{c+d}{5}\right) + y = d$$

$$y = \frac{5d}{5} - \frac{3c+3d}{5}$$

$$y = \frac{2d-3c}{5}$$ Solution: $\left(\frac{c+d}{5}, \frac{2d-3c}{5}\right)$

41.
$$\begin{cases} y = 2x + 6 \\ 2x^2 + 8x = -2 - 3y \end{cases}$$ Substitute $y = 3x + 6$ into the second equation.

$$2x^2 + 8x = -2 - 3(2x+6)$$

$$2x^2 + 8x = -2 - 6x - 18$$

$$2x^2 + 14x + 20 = 0$$

$$2(x+5)(x+2) = 0$$

$$x = -5 \quad \text{or} \quad x = -2$$

When $x = -5$, $y = 2(-5) + 6 = -4$. When $x = -2$, $y = 2(-2) + 6 = 2$.

Solutions: $(-2, 2), (-5, -4)$

43.
$$-3\begin{cases} 3x^2 + 4y^2 = 19 \\ x^2 + y^2 = 5 \end{cases}$$

$$+\begin{cases} 3x^2 + 4y^2 = 19 \\ -3x^2 - 3y^2 = -15 \end{cases}$$

$$y^2 = 4$$

$$y = \pm 2$$

When $y = 2$, $x^2 + 2^2 = 5$, $x^2 = 1$, so $x = \pm 1$.

When $y = -2$, $x^2 + (-2)^2 = 5$, $x^2 = 1$, so $x = \pm 1$.

Solutions: $(1, 2), (1, -2), (-1, 2), (-1, -2)$

45. $\begin{cases} x^2 - 12y - 1 = 0 \\ x^2 - 4y^2 - 9 = 0 \end{cases}$ Substitute the second equation into the first.

$$x^2 - 12y - 1 = x^2 - 4y^2 - 9$$

$$4y^2 - 12y + 8 = 0$$

$$4(y - 1)(y - 2) = 0$$

$$y = 1 \quad \text{or} \quad y = 2$$

When $y = 1$, $x^2 - 12(1) - 1 = 0$, $x^2 = 13$, so $x = \pm\sqrt{13}$.

When $y = 2$, $x^2 - 12(2) - 1 + 0$, $x^2 = 25$, so $x = \pm 5$.

Solution: $(\sqrt{13}, \ 1), \ (-\sqrt{13}, \ 1), \ (5, \ 2), \ (-5, \ 2)$

47. $+ \begin{cases} \sin(x + y) = \sin x \cos y + \cos x \sin y \\ \sin(x - y) = \sin x \cos y - \cos x \sin y \end{cases}$

$$\sin(x + y) + \sin(x - y) = 2 \sin x \cos y$$

$$2 \sin x \cos y = \sin(x + y) + \sin(x - y)$$

49. $-1 \begin{cases} \sin(x + y) = \sin x \cos y + \cos x \sin y \\ \sin(x - y) = \sin x \cos y - \cos x \sin y \end{cases}$

$+ \begin{cases} \sin(x + y) = \sin x \cos y + \cos x \sin y \\ -\sin(x - y) = -\sin x \cos y + \cos x \sin y \end{cases}$

$$\sin(x + y) - \sin(x - y) = 2 \cos x \sin y$$

$$2 \cos x \sin y = \sin(x + y) - \sin(x - y)$$

51. Let x and y represent the numbers. The system of equations is:

$\begin{cases} x - y = 4 \\ xy = -3 \end{cases}$ Solve the first equation for x, $x = y + 4$, and substitute into the second equation.

$$(y + 4)y = -3$$

$$y^2 + 4y + 3 = 0$$

$$(y + 1)(y + 3) = 0$$

$$y = -1 \quad \text{or} \quad y = -3$$

When $y = -1$, $x = (-1) + 4 = 3$. When $y = -3$, $x = (-3) + 4 = 1$.

Solutions: 1 and -3 or 3 and -1.

53. Let l = length and w = width. The system of equations is:

$$\begin{cases} lw = 60 \\ l^2 + w^2 = 13^2 \end{cases} \quad \begin{pmatrix} A = lw \\ \text{Pythagorean Theorem} \end{pmatrix}$$

From the first equation $l = \frac{60}{w}$. Substitute this value into the second equation.

$$\left(\frac{60}{w}\right)^2 + w^2 = 13^2 \quad \text{Multiply both sides by } w^2.$$

$$3,600 + w^4 = 169w^2$$

$$w^4 - 169w^2 + 3,600 = 0$$

$$(w^2 - 25)(w^2 - 144) = 0$$

$$w^2 = 25 \quad \text{or} \quad w^2 = 144$$

$w = \pm 5$ or $w = \pm 12$. Since both l and w must be positive, $w = 5$ or 12.

When $w = 5$, $l = \frac{60}{5} = 12$. When $w = 12$, $l = \frac{60}{12} = 5$.

So the two sides have length 5 ft and 12 ft.

55. $$+\begin{cases} \cos x - \sin y = 1 \\ \cos x + \sin y = 0 \end{cases}$$

$$2 \cos x \qquad = 1$$

$$\cos x = \frac{1}{2}$$

$$x = \frac{\pi}{3} \text{ or } x = \frac{5\pi}{3}$$

From the second equation, $\sin y = -\cos x$.

When $x = \frac{\pi}{3}$, $\sin y = -\frac{1}{2}$. So $y = \frac{7\pi}{6}$ or $\frac{11\pi}{6}$.

When $x = \frac{5\pi}{3}$, $\sin y = -\frac{1}{2}$. So $y = \frac{7\pi}{6}$ or $\frac{11\pi}{6}$.

Solutions: $\left(\frac{\pi}{3}, \frac{7\pi}{6}\right), \left(\frac{\pi}{3}, \frac{11\pi}{6}\right), \left(\frac{5\pi}{3}, \frac{7\pi}{6}\right), \left(\frac{5\pi}{3}, \frac{11\pi}{6}\right)$

57. $$+\begin{cases} x^2 - xy + y^2 = 21 \\ xy - y^2 = 15 \end{cases}$$

$$x^2 \qquad\qquad = 36$$

$$x = \pm 6$$

When $x = 6$, $\qquad (6)^2 - 6y + y^2 = 21$

$$y^2 - 6y + 15 = 0, \quad y = \frac{6 \pm \sqrt{36 - 60}}{2}, \text{ not a real number}$$

When $x = -6$, $\quad (-6)^2 + 6y + y^2 = 21$

$$y^2 + 6y + 15 = 0, \quad y = \frac{-6 \pm \sqrt{36 - 60}}{2}, \text{ not a real number}$$

Since there are no real solutions, the system is inconsistent.

59.
$$\begin{cases} x^2 + 2xy = 8 \\ x^2 - 4y^2 = 8 \end{cases}$$

Solve the first equation for y: $2xy = 8 - x^2$, so $y = \dfrac{8 - x^2}{2x}$, $x \neq 0$. Substitute into the second equation.

$$x^2 - 4\left(\frac{8 - x^2}{2x}\right)^2 = 8$$

$$x^2 - 4\left(\frac{64 - 16x^2 + x^4}{4x^2}\right) = 8 \qquad \text{Multiply both sides by } x^2$$

$$x^4 - 64 + 16x^2 - x^4 = 8x^2$$

$$8x^2 = 64$$

$$x^2 = 8$$

$$x = \pm 2\sqrt{2}$$

When $x = 2\sqrt{2}$, $y = \dfrac{8 - 8}{4\sqrt{2}} = 0$. When $x = -2\sqrt{2}$, $y = \dfrac{8 - 8}{-4\sqrt{2}} = 0$.

Solutions: $(2\sqrt{2}, 0)$, $(-2\sqrt{2}, 0)$

61.
$$\begin{cases} \dfrac{2}{x-1} - \dfrac{5}{y+2} = 12 \\ \dfrac{4}{x-1} - \dfrac{2}{y+2} = -12 \end{cases}$$

Let $p = \dfrac{1}{x-1}$ and $q = \dfrac{1}{y+2}$. The system of equations becomes:

$$\begin{cases} 2p - 5q = 12 \\ 4p - 2q = -12 \end{cases}$$

$$-2 \begin{cases} 2p - 5q = 12 \\ 4p - 2q = -12 \end{cases}$$

$$+ \begin{cases} -4p + 10q = -24 \\ 4p - 2q = -12 \end{cases}$$

$$8q = -36$$

$$q = -\frac{9}{2}$$

From the second equation, $4p - 2\left(-\dfrac{9}{2}\right) = -12$, $4p = -21$, $p = -\dfrac{21}{4}$, So

$$\frac{-21}{4} = \frac{1}{x-1} \qquad \text{and} \qquad -\frac{9}{2} = \frac{1}{y+2}$$

$$-21x + 21 = 4 \qquad\qquad\qquad -9y - 18 = 2$$

$$x = \frac{17}{21} \qquad \text{and} \qquad y = -\frac{20}{9}$$

Solution: $\left(\dfrac{17}{21}, -\dfrac{20}{9}\right)$

63.

$$4 \begin{cases} x^2 - 3xy = 2 \\ -3 \begin{cases} x^2 - 4xy = 3x \end{cases} \end{cases}$$

$$\begin{cases} 4x^2 - 12xy = 8 \\ -3x^2 + 12xy = -9x \end{cases}$$

$$x^2 \qquad = 8 - 9x$$

$$x^2 + 9x - 8 = 0$$

$$x = \frac{-9 \pm \sqrt{81 + 32}}{2} = \frac{-9 \pm \sqrt{113}}{2}$$

From the first equation: $3xy = x^2 - 2$, $y = \dfrac{x^2 - 2}{3x}$ $(x \neq 0)$.

When $x = \dfrac{-9 + \sqrt{113}}{2}$,

$$y = \frac{\left(\dfrac{-9 + \sqrt{113}}{2}\right)^2 - 2}{3\left(\dfrac{-9 + \sqrt{113}}{2}\right)}$$

$$y = \frac{\dfrac{81 - 18\sqrt{113} + 113}{4} - 2}{\dfrac{-27 + 3\sqrt{113}}{2}}$$

$$y = \frac{\dfrac{186 - 18\sqrt{113}}{4}}{\dfrac{-27 + 3\sqrt{113}}{2}}$$

$$y = \frac{93 - 9\sqrt{113}}{2} \cdot \frac{2}{-27 + 3\sqrt{113}}$$

$$y = \frac{31 - 3\sqrt{113}}{-9 + \sqrt{113}} \cdot \frac{-9 - \sqrt{113}}{-9 - \sqrt{113}}$$

$$y = \frac{60 - 4\sqrt{113}}{-32}$$

$$y = \frac{-15 + \sqrt{113}}{8}$$

Similarly, when $x = \dfrac{-9 - \sqrt{113}}{2}$, $y = \dfrac{-15 - \sqrt{113}}{8}$

Solutions: $\left(\dfrac{-9 + \sqrt{113}}{2}, \dfrac{-15 + \sqrt{113}}{8}\right), \left(\dfrac{-9 - \sqrt{113}}{2}, \dfrac{-15 - \sqrt{113}}{8}\right)$

65. For the parabola $-3(x - h)^2 = y - k$,

from the point $(2, 5)$ we have $-3(2 - h)^2 = 5 - k$, and

from the point $(-1, -4)$ we have $-3(-1 - h)^2 = -4 - k$. Thus the system is:

$$-1 \begin{cases} -3(2 - h)^2 = 5 - k \\ -3(-1 - h)^2 = -4 - k \end{cases}$$

$$+ \begin{cases} 3(2 - h)^2 = -5 + k \\ -3(-1 - h)^2 = -4 - k \end{cases}$$

$$3(2 - h)^2 - 3(-1 - h)^2 = -9$$
$$(4 - 4h + h^2) - (1 + 2h + h^2) = -3$$
$$-6h = -6$$
$$h = 1 \qquad \text{Substituting into the first equation}$$
$$-3(2 - 1)^2 = 5 - k$$
$$k = 8 \qquad \text{Solution: } (h, k) = (1, 8)$$

67. $\begin{cases} xy = 1 \\ x + y = a \end{cases}$ From the second equation, $y = a - x$. Substituting into the first equation:

$$x(a - x) = 1$$
$$ax - x^2 = 1$$
$$x^2 - ax + 1 = 0$$
$$x = \frac{a \pm \sqrt{a^2 - 4}}{2}$$
$$y = a - x = \frac{2a}{2} - \frac{a \pm \sqrt{a^2 - 4}}{2} = \frac{a \mp \sqrt{a^2 - 4}}{2}$$

Solutions: $\left(\frac{a + \sqrt{a^2 - 4}}{2}, \frac{a - \sqrt{a^2 - 4}}{2} \right), \left(\frac{a - \sqrt{a^2 - 4}}{2}, \frac{a + \sqrt{a^2 - 4}}{2} \right)$

PROBLEM SET 9.2

1. a. $\begin{bmatrix} 4 & 5 & \vdots & -16 \\ 3 & 2 & \vdots & 5 \end{bmatrix}$ b. $\begin{bmatrix} 1 & 1 & 1 & \vdots & 4 \\ 3 & 2 & 1 & \vdots & 7 \\ 1 & -3 & 2 & \vdots & 0 \end{bmatrix}$

c. $\begin{bmatrix} 1 & 3 & 1 & 1 & \vdots & 3 \\ 1 & 0 & -2 & 2 & \vdots & 0 \\ 0 & 0 & 1 & 5 & \vdots & -14 \\ 0 & 1 & -3 & -1 & \vdots & 2 \end{bmatrix}$

3.
$$\begin{bmatrix} 3 & 1 & 2 & \vdots & 1 \\ 0 & 2 & 4 & \vdots & 5 \\ 1 & 3 & -4 & \vdots & 9 \end{bmatrix} \xrightarrow{\text{R1} \leftrightarrow \text{R3}} \begin{bmatrix} 1 & 3 & -4 & \vdots & 9 \\ 0 & 2 & 4 & \vdots & 5 \\ 3 & 1 & 2 & \vdots & 1 \end{bmatrix}$$

5.
$$\begin{bmatrix} 2 & 4 & 10 & \vdots & -12 \\ 6 & 3 & 4 & \vdots & 6 \\ 10 & -1 & 0 & \vdots & 1 \end{bmatrix} \xrightarrow{\frac{1}{2}R_1} \begin{bmatrix} 1 & 2 & 5 & \vdots & -6 \\ 6 & 3 & 4 & \vdots & 6 \\ 10 & -1 & 0 & \vdots & 1 \end{bmatrix}$$

7.
$$\begin{bmatrix} 5 & 6 & -3 & \vdots & 4 \\ 4 & 1 & 9 & \vdots & 2 \\ 7 & 6 & 1 & \vdots & 3 \end{bmatrix} \xrightarrow{-\text{R2}+\text{R1}} \begin{bmatrix} 1 & 5 & -12 & \vdots & 2 \\ 4 & 1 & 9 & \vdots & 2 \\ 7 & 6 & 1 & \vdots & 3 \end{bmatrix}$$

9.
$$\begin{bmatrix} 1 & 2 & -3 & \vdots & 0 \\ 0 & 3 & 1 & \vdots & 4 \\ 2 & 5 & 1 & \vdots & 6 \end{bmatrix} \xrightarrow{-2\text{R1}+\text{R3}} \begin{bmatrix} 1 & 2 & -3 & \vdots & 0 \\ 0 & 3 & 1 & \vdots & 4 \\ 0 & 1 & 7 & \vdots & 6 \end{bmatrix}$$

11.
$$\begin{bmatrix} 1 & 2 & 4 & \vdots & 1 \\ -2 & 5 & 0 & \vdots & 2 \\ -4 & 5 & 1 & \vdots & 3 \end{bmatrix} \xrightarrow[4\text{R1}+\text{R3}]{2\text{R1}+\text{R2}} \begin{bmatrix} 1 & 2 & 4 & \vdots & 1 \\ 0 & 9 & 8 & \vdots & 4 \\ 0 & 13 & 17 & \vdots & 7 \end{bmatrix}$$

13.
$$\begin{bmatrix} 1 & 4 & -1 & 3 & \vdots & 3 \\ -3 & 4 & 6 & 4 & \vdots & 0 \\ 5 & 1 & 9 & 1 & \vdots & -2 \\ 1 & 0 & 2 & 0 & \vdots & 0 \end{bmatrix} \xrightarrow[\substack{-5\text{R1}+\text{R3} \\ -\text{R1}+\text{R4}}]{3\text{R1}+\text{R2}} \begin{bmatrix} 1 & 4 & -1 & 3 & \vdots & 3 \\ 0 & 16 & 3 & 13 & \vdots & 9 \\ 0 & -19 & 14 & -14 & \vdots & -17 \\ 0 & -4 & 3 & -3 & \vdots & -3 \end{bmatrix}$$

15.
$$\begin{bmatrix} 1 & 3 & 5 & \vdots & 2 \\ 0 & 2 & 6 & \vdots & -8 \\ 0 & 3 & 4 & \vdots & 1 \end{bmatrix} \xrightarrow{\frac{1}{2}\text{R2}} \begin{bmatrix} 1 & 3 & 5 & \vdots & 2 \\ 0 & 1 & 3 & \vdots & -4 \\ 0 & 3 & 4 & \vdots & 1 \end{bmatrix}$$

17.
$$\begin{bmatrix} 1 & 3 & -2 & \vdots & 4 \\ 0 & 5 & 1 & \vdots & 3 \\ 0 & 7 & 9 & \vdots & 2 \end{bmatrix} \xrightarrow{\frac{1}{5}\text{R2}} \begin{bmatrix} 1 & 3 & -2 & \vdots & 4 \\ 0 & 1 & \frac{1}{5} & \vdots & \frac{3}{5} \\ 0 & 7 & 9 & \vdots & 2 \end{bmatrix}$$

19.
$$\begin{bmatrix} 1 & 3 & -2 & \vdots & 0 \\ 0 & 4 & 2 & \vdots & 9 \\ 0 & 3 & 6 & \vdots & 1 \end{bmatrix} \xrightarrow{-\text{R3}+\text{R2}} \begin{bmatrix} 1 & 3 & -2 & \vdots & 0 \\ 0 & 1 & -4 & \vdots & 8 \\ 0 & 3 & 6 & \vdots & 1 \end{bmatrix}$$

21. $\begin{bmatrix} 1 & 5 & -3 & \vdots & 2 \\ 0 & 1 & 4 & \vdots & 5 \\ 0 & 3 & 4 & \vdots & 2 \end{bmatrix} \xrightarrow{-3R2+R3} \begin{bmatrix} 1 & 5 & -3 & \vdots & 2 \\ 0 & 1 & 4 & \vdots & 5 \\ 0 & 0 & -8 & \vdots & -13 \end{bmatrix}$

23. $\begin{bmatrix} 1 & 7 & 6 & 6 & \vdots & 2 \\ 0 & 1 & 9 & 2 & \vdots & 1 \\ 0 & -4 & 6 & 1 & \vdots & 1 \\ 0 & 5 & 8 & 10 & \vdots & 3 \end{bmatrix} \xrightarrow[\substack{-5R2+R4}]{\substack{4R2+R3}} \begin{bmatrix} 1 & 7 & 6 & 6 & \vdots & 2 \\ 0 & 1 & 9 & 2 & \vdots & 1 \\ 0 & 0 & 42 & 9 & \vdots & 5 \\ 0 & 0 & -37 & 0 & \vdots & -2 \end{bmatrix}$

25. $\begin{bmatrix} 1 & -2 & 5 & \vdots & 6 \\ 0 & 1 & 9 & \vdots & 1 \\ 0 & 0 & -3 & \vdots & 5 \end{bmatrix} \xrightarrow{R3/(-3)} \begin{bmatrix} 1 & -2 & 5 & \vdots & 6 \\ 0 & 1 & 9 & \vdots & 1 \\ 0 & 0 & 1 & \vdots & -\frac{5}{3} \end{bmatrix}$

27. $\begin{bmatrix} 1 & 3 & -1 & \vdots & 5 \\ 0 & 1 & 2 & \vdots & 6 \\ 0 & 0 & 1 & \vdots & 4 \end{bmatrix} \xrightarrow[\substack{-2R3+R2}]{\substack{R3+R1}} \begin{bmatrix} 1 & 3 & 0 & \vdots & 9 \\ 0 & 1 & 0 & \vdots & -2 \\ 0 & 0 & 1 & \vdots & 4 \end{bmatrix}$

29. $\begin{bmatrix} 1 & 6 & 3 & 0 & \vdots & -2 \\ 0 & 1 & -2 & 0 & \vdots & 1 \\ 0 & 0 & 1 & 0 & \vdots & 0 \\ 0 & 0 & 0 & 1 & \vdots & -6 \end{bmatrix} \xrightarrow[\substack{2R3+R2}]{\substack{-3R3+R1}} \begin{bmatrix} 1 & 6 & 0 & 0 & \vdots & -2 \\ 0 & 1 & 0 & 0 & \vdots & 1 \\ 0 & 0 & 1 & 0 & \vdots & 0 \\ 0 & 0 & 0 & 1 & \vdots & -6 \end{bmatrix}$

31. $\begin{bmatrix} 1 & -4 & 0 & \vdots & 3 \\ 0 & 1 & 0 & \vdots & -\frac{8}{3} \\ 0 & 0 & 1 & \vdots & \frac{3}{5} \end{bmatrix} \xrightarrow{4R2+R1} \begin{bmatrix} 1 & 0 & 0 & \vdots & -\frac{23}{3} \\ 0 & 1 & 0 & \vdots & -\frac{8}{3} \\ 0 & 0 & 1 & \vdots & \frac{3}{5} \end{bmatrix}$

so the solution is $(x,\ y,\ z) = (-\frac{23}{3},\ -\frac{8}{3},\ \frac{3}{5})$

33. $\begin{bmatrix} 4 & -7 & \vdots & -2 \\ -1 & 2 & \vdots & 1 \end{bmatrix} \xrightarrow{R1\leftrightarrow R2} \begin{bmatrix} -1 & 2 & \vdots & 1 \\ 4 & -7 & \vdots & -2 \end{bmatrix}$

$\xrightarrow{-R1} \begin{bmatrix} 1 & -2 & \vdots & -1 \\ 4 & -7 & \vdots & -2 \end{bmatrix} \xrightarrow{-4R1+R2} \begin{bmatrix} 1 & -2 & \vdots & -1 \\ 0 & 1 & \vdots & 2 \end{bmatrix}$

$\xrightarrow{2R2+R1} \begin{bmatrix} 1 & 0 & \vdots & 3 \\ 0 & 1 & \vdots & 2 \end{bmatrix} \qquad (x,\ y) = (3,\ 2)$

35. $\begin{bmatrix} 8 & 6 & \vdots & 17 \\ -1 & 1 & \vdots & \frac{1}{2} \end{bmatrix} \xrightarrow{R1 \leftrightarrow R2} \begin{bmatrix} -1 & 1 & \vdots & \frac{1}{2} \\ 8 & 6 & \vdots & 17 \end{bmatrix}$

$\xrightarrow{-R1} \begin{bmatrix} 1 & -1 & \vdots & -\frac{1}{2} \\ 8 & 6 & \vdots & 17 \end{bmatrix} \xrightarrow{-8R1 + R2} \begin{bmatrix} 1 & -1 & \vdots & -\frac{1}{2} \\ 0 & 14 & \vdots & 21 \end{bmatrix}$

$\xrightarrow{R2 \div 14} \begin{bmatrix} 1 & -1 & \vdots & -\frac{1}{2} \\ 0 & 1 & \vdots & \frac{3}{2} \end{bmatrix} \xrightarrow{R1 + R2} \begin{bmatrix} 1 & 0 & \vdots & 1 \\ 0 & 1 & \vdots & \frac{3}{2} \end{bmatrix}$

$$(x, y) = (1, \tfrac{3}{2})$$

37. $\begin{bmatrix} \frac{2}{3} & 1 & \vdots & 3 \\ 1 & -6 & \vdots & -3 \end{bmatrix} \xrightarrow{R1 \leftrightarrow R2} \begin{bmatrix} 1 & -6 & \vdots & -3 \\ \frac{2}{3} & 1 & \vdots & 3 \end{bmatrix}$

$\xrightarrow{-\frac{2}{3}R1 + R2} \begin{bmatrix} 1 & -6 & \vdots & -3 \\ 0 & 5 & \vdots & 5 \end{bmatrix} \xrightarrow{R2 \div 5} \begin{bmatrix} 1 & -6 & \vdots & -3 \\ 0 & 1 & \vdots & 1 \end{bmatrix}$

$\xrightarrow{6R2 + R1} \begin{bmatrix} 1 & 0 & \vdots & 3 \\ 0 & 1 & \vdots & 1 \end{bmatrix} \qquad (x, y) = (3, 1)$

39. $\begin{bmatrix} 1 & 1 & 1 & \vdots & 6 \\ 2 & -1 & 1 & \vdots & 3 \\ 1 & -2 & -3 & \vdots & -12 \end{bmatrix} \xrightarrow[-R1 + R3]{-2R1 + R2} \begin{bmatrix} 1 & 1 & 1 & \vdots & 6 \\ 0 & -3 & -1 & \vdots & -9 \\ 0 & -3 & -4 & \vdots & -18 \end{bmatrix}$

$\xrightarrow{R2 \div (-3)} \begin{bmatrix} 1 & 1 & 1 & \vdots & 6 \\ 0 & 1 & \frac{1}{3} & \vdots & 3 \\ 0 & -3 & -4 & \vdots & -18 \end{bmatrix} \xrightarrow[3R2 + R3]{-R2 + R1} \begin{bmatrix} 1 & 0 & \frac{2}{3} & \vdots & 3 \\ 0 & 1 & \frac{1}{3} & \vdots & 3 \\ 0 & 0 & -3 & \vdots & -9 \end{bmatrix}$

$\xrightarrow{R3 \div (-3)} \begin{bmatrix} 1 & 0 & \frac{2}{3} & \vdots & 3 \\ 0 & 1 & \frac{1}{3} & \vdots & 3 \\ 0 & 0 & 1 & \vdots & 3 \end{bmatrix} \xrightarrow[-\frac{1}{3}R3 + R2]{-\frac{2}{3}R3 + R1} \begin{bmatrix} 1 & 0 & 0 & \vdots & 1 \\ 0 & 1 & 0 & \vdots & 2 \\ 0 & 0 & 1 & \vdots & 3 \end{bmatrix}$

$$(x, y, z) = (1, 2, 3)$$

41.

$$\begin{bmatrix} 1 & 1 & 1 & \vdots & 4 \\ 1 & 3 & 2 & \vdots & 4 \\ 1 & -2 & 1 & \vdots & 7 \end{bmatrix} \xrightarrow[\,-R1+R3\,]{-R1+R2} \begin{bmatrix} 1 & 1 & 1 & \vdots & 4 \\ 0 & 2 & 1 & \vdots & 0 \\ 0 & -3 & 0 & \vdots & 3 \end{bmatrix}$$

$$\xrightarrow[R3 \div (-3)]{} \begin{bmatrix} 1 & 1 & 1 & \vdots & 4 \\ 0 & 2 & 1 & \vdots & 0 \\ 0 & 1 & 0 & \vdots & -1 \end{bmatrix} \xrightarrow[R2 \leftrightarrow R3]{} \begin{bmatrix} 1 & 1 & 1 & \vdots & 4 \\ 0 & 1 & 0 & \vdots & -1 \\ 0 & 2 & 1 & \vdots & 0 \end{bmatrix}$$

$$\xrightarrow[\,-2R2+R3\,]{-R2+R1} \begin{bmatrix} 1 & 0 & 1 & \vdots & 5 \\ 0 & 1 & 0 & \vdots & -1 \\ 0 & 0 & 1 & \vdots & 2 \end{bmatrix} \xrightarrow[-R3+R1]{} \begin{bmatrix} 1 & 0 & 0 & \vdots & 3 \\ 0 & 1 & 0 & \vdots & -1 \\ 0 & 0 & 1 & \vdots & 2 \end{bmatrix}$$

$$(x, \, y, \, z) = (3, \, -1, \, 2)$$

43.

$$\begin{bmatrix} 1 & 0 & 2 & \vdots & 7 \\ 1 & 1 & 0 & \vdots & 11 \\ 0 & -2 & 9 & \vdots & -3 \end{bmatrix} \xrightarrow[\,]{-R1+R2} \begin{bmatrix} 1 & 0 & 2 & \vdots & 7 \\ 0 & 1 & -2 & \vdots & 4 \\ 0 & -2 & 9 & \vdots & -3 \end{bmatrix}$$

$$\xrightarrow[2R2+R3]{} \begin{bmatrix} 1 & 0 & 2 & \vdots & 7 \\ 0 & 1 & -2 & \vdots & 4 \\ 0 & 0 & 5 & \vdots & 5 \end{bmatrix} \xrightarrow[R3 \div 5]{} \begin{bmatrix} 1 & 0 & 2 & \vdots & 7 \\ 0 & 1 & -2 & \vdots & 4 \\ 0 & 0 & 1 & \vdots & 1 \end{bmatrix}$$

$$\xrightarrow[2R3+R2]{-2R3+R1} \begin{bmatrix} 1 & 0 & 0 & \vdots & 5 \\ 0 & 1 & 0 & \vdots & 6 \\ 0 & 0 & 1 & \vdots & 1 \end{bmatrix} \quad (x, \, y, \, z) = (5, \, 6, \, 1)$$

45.

$$\begin{bmatrix} 6 & 1 & 20 & \vdots & 27 \\ 1 & -1 & 0 & \vdots & 0 \\ 0 & 1 & 1 & \vdots & 2 \end{bmatrix} \xrightarrow[\,]{R1 \leftrightarrow R2} \begin{bmatrix} 1 & -1 & 0 & \vdots & 0 \\ 6 & 1 & 20 & \vdots & 27 \\ 0 & 1 & 1 & \vdots & 2 \end{bmatrix}$$

$$\xrightarrow[\,]{-6R1+R2} \begin{bmatrix} 1 & -1 & 0 & \vdots & 0 \\ 0 & 7 & 20 & \vdots & 27 \\ 0 & 1 & 1 & \vdots & 2 \end{bmatrix} \xrightarrow[R2 \leftrightarrow R3]{} \begin{bmatrix} 1 & -1 & 0 & \vdots & 0 \\ 0 & 1 & 1 & \vdots & 2 \\ 0 & 7 & 20 & \vdots & 27 \end{bmatrix}$$

$$\xrightarrow[\,-7R2+R3\,]{R2+R1} \begin{bmatrix} 1 & 0 & 1 & \vdots & 2 \\ 0 & 1 & 1 & \vdots & 2 \\ 0 & 0 & 13 & \vdots & 13 \end{bmatrix} \xrightarrow[R3 \div 13]{} \begin{bmatrix} 1 & 0 & 1 & \vdots & 2 \\ 0 & 1 & 1 & \vdots & 2 \\ 0 & 0 & 1 & \vdots & 1 \end{bmatrix}$$

$$\xrightarrow[\,-R3+R2\,]{-R3+R1} \begin{bmatrix} 1 & 0 & 0 & \vdots & 1 \\ 0 & 1 & 0 & \vdots & 1 \\ 0 & 0 & 1 & \vdots & 1 \end{bmatrix} \quad (x, \, y, \, z) = (1, \, 1, \, 1)$$

47.
$$\begin{bmatrix} 2 & 1 & 3 & \vdots & 1 \\ 2 & 0 & -1 & \vdots & -11 \\ 0 & 3 & 2 & \vdots & 6 \end{bmatrix} \xrightarrow{R1 \div 2} \begin{bmatrix} 1 & \frac{1}{2} & \frac{3}{2} & \vdots & \frac{1}{2} \\ 2 & 0 & -1 & \vdots & -11 \\ 0 & 3 & 2 & \vdots & 6 \end{bmatrix}$$

$$\xrightarrow{-2R1+R2} \begin{bmatrix} 1 & \frac{1}{2} & \frac{3}{2} & \vdots & \frac{1}{2} \\ 0 & -1 & -4 & \vdots & -12 \\ 0 & 3 & 2 & \vdots & 6 \end{bmatrix} \xrightarrow{-R2} \begin{bmatrix} 1 & \frac{1}{2} & \frac{3}{2} & \vdots & \frac{1}{2} \\ 0 & 1 & 4 & \vdots & 12 \\ 0 & 3 & 2 & \vdots & 6 \end{bmatrix}$$

$$\xrightarrow[-3R2+R3]{-\frac{1}{2}R2+R1} \begin{bmatrix} 1 & 0 & -\frac{1}{2} & \vdots & -\frac{11}{2} \\ 0 & 1 & 4 & \vdots & 12 \\ 0 & 0 & -10 & \vdots & -30 \end{bmatrix} \xrightarrow{R3 \div (-10)} \begin{bmatrix} 1 & 0 & -\frac{1}{2} & \vdots & -\frac{11}{2} \\ 0 & 1 & 4 & \vdots & 12 \\ 0 & 0 & 1 & \vdots & 3 \end{bmatrix}$$

$$\xrightarrow[-4R3+R2]{\frac{1}{2}R3+R1} \begin{bmatrix} 1 & 0 & 0 & \vdots & -4 \\ 0 & 1 & 0 & \vdots & 0 \\ 0 & 0 & 1 & \vdots & 3 \end{bmatrix} \qquad (x, \, y, \, z) = (-4, \, 0, \, 3)$$

49.
$$\begin{bmatrix} 2 & -1 & 1 & \vdots & 4 \\ 3 & -2 & 2 & \vdots & 3 \\ 1 & -1 & 3 & \vdots & 2 \end{bmatrix} \xrightarrow{R1 \leftrightarrow R3} \begin{bmatrix} 1 & -1 & 3 & \vdots & 2 \\ 3 & -2 & 2 & \vdots & 3 \\ 2 & -1 & 1 & \vdots & 4 \end{bmatrix}$$

$$\xrightarrow[-2R1+R3]{-3R1+R2} \begin{bmatrix} 1 & -1 & 3 & \vdots & 2 \\ 0 & 1 & -7 & \vdots & -3 \\ 0 & 1 & -5 & \vdots & 0 \end{bmatrix} \xrightarrow[-R2+R3]{R2+R1} \begin{bmatrix} 1 & 0 & -4 & \vdots & -1 \\ 0 & 1 & -7 & \vdots & -3 \\ 0 & 0 & 2 & \vdots & 3 \end{bmatrix}$$

$$\xrightarrow{R3 \div 2} \begin{bmatrix} 1 & 0 & -4 & \vdots & -1 \\ 0 & 1 & -7 & \vdots & -3 \\ 0 & 0 & 1 & \vdots & \frac{3}{2} \end{bmatrix} \xrightarrow[7R3+R2]{4R3+R1} \begin{bmatrix} 1 & 0 & 0 & \vdots & 5 \\ 0 & 1 & 0 & \vdots & \frac{15}{2} \\ 0 & 0 & 1 & \vdots & \frac{3}{2} \end{bmatrix}$$

$$(x, \, y, \, z) = (5, \tfrac{15}{2}, \tfrac{3}{2})$$

51.

$$\begin{bmatrix} 1 & 5 & -3 & \vdots & 2 \\ 2 & 2 & 1 & \vdots & -1 \\ 3 & -1 & 5 & \vdots & 3 \end{bmatrix} \xrightarrow[-3R1+R3]{-2R1+R2} \begin{bmatrix} 1 & 5 & -3 & \vdots & 2 \\ 0 & -8 & 7 & \vdots & -5 \\ 0 & -16 & 14 & \vdots & -3 \end{bmatrix}$$

$$\xrightarrow{R2 \div (-8)} \begin{bmatrix} 1 & 5 & -3 & \vdots & 2 \\ 0 & 1 & -\frac{7}{8} & \vdots & \frac{5}{8} \\ 0 & -16 & 14 & \vdots & -3 \end{bmatrix} \xrightarrow[16R2+R3]{-5R2+R1} \begin{bmatrix} 1 & 0 & \frac{11}{8} & \vdots & -\frac{9}{8} \\ 0 & 1 & -\frac{7}{8} & \vdots & \frac{5}{8} \\ 0 & 0 & 0 & \vdots & 7 \end{bmatrix}$$

$$0 \neq 7$$

Inconsistent system

53.

$$\begin{bmatrix} 0 & 1 & 2 & 0 & \vdots & 5 \\ 0 & 0 & 0 & 1 & \vdots & 3 \\ 1 & 1 & 3 & 0 & \vdots & 6 \\ 2 & 4 & 0 & 0 & \vdots & 2 \end{bmatrix} \xrightarrow[R2 \leftrightarrow R4]{R1 \leftrightarrow R3} \begin{bmatrix} 1 & 1 & 3 & 0 & \vdots & 6 \\ 2 & 4 & 0 & 0 & \vdots & 2 \\ 0 & 1 & 2 & 0 & \vdots & 5 \\ 0 & 0 & 0 & 1 & \vdots & 3 \end{bmatrix}$$

$$\xrightarrow{-2R1+R2} \begin{bmatrix} 1 & 1 & 3 & 0 & \vdots & 6 \\ 0 & 2 & -6 & 0 & \vdots & -10 \\ 0 & 1 & 2 & 0 & \vdots & 5 \\ 0 & 0 & 0 & 1 & \vdots & 3 \end{bmatrix} \xrightarrow{R2 \div 2} \begin{bmatrix} 1 & 1 & 3 & 0 & \vdots & 6 \\ 0 & 1 & -3 & 0 & \vdots & -5 \\ 0 & 1 & 2 & 0 & \vdots & 5 \\ 0 & 0 & 0 & 1 & \vdots & 3 \end{bmatrix}$$

$$\xrightarrow[-R2+R3]{-R2+R1} \begin{bmatrix} 1 & 0 & 6 & 0 & \vdots & 11 \\ 0 & 1 & -3 & 0 & \vdots & -5 \\ 0 & 0 & 5 & 0 & \vdots & 10 \\ 0 & 0 & 0 & 1 & \vdots & 3 \end{bmatrix} \xrightarrow{R3 \div 5} \begin{bmatrix} 1 & 0 & 6 & 0 & \vdots & 11 \\ 0 & 1 & -3 & 0 & \vdots & -5 \\ 0 & 0 & 1 & 0 & \vdots & 2 \\ 0 & 0 & 0 & 1 & \vdots & 3 \end{bmatrix}$$

$$\xrightarrow[3R3+R2]{-6R3+R1} \begin{bmatrix} 1 & 0 & 0 & 0 & \vdots & -1 \\ 0 & 1 & 0 & 0 & \vdots & 1 \\ 0 & 0 & 1 & 0 & \vdots & 2 \\ 0 & 0 & 0 & 1 & \vdots & 3 \end{bmatrix}$$

$$(w,\ x,\ y,\ z) = (-1,\ 1,\ 2,\ 3)$$

55. $y = ax^2 + bx + c$

$(0, 5)$: $\quad a(0)^2 + b(0) + c = 5$

$(-1, 2)$: $\quad a(-1)^2 + b(-1) + c = 2 \quad\longrightarrow$

$(3, 26)$: $\quad a(3)^2 + b(3) + c = 26$

$$\begin{cases} \quad\quad\quad c = 5 \\ a - b + c = 2 \\ 9a + 3b + c = 26 \end{cases}$$

$$\begin{bmatrix} 0 & 0 & 1 & \vdots & 5 \\ 1 & -1 & 1 & \vdots & 2 \\ 9 & 3 & 1 & \vdots & 26 \end{bmatrix} \xrightarrow{\text{R1}\leftrightarrow\text{R2}} \begin{bmatrix} 1 & -1 & 1 & \vdots & 2 \\ 0 & 0 & 1 & \vdots & 5 \\ 9 & 3 & 1 & \vdots & 26 \end{bmatrix}$$

$$\xrightarrow{-9\text{R1}+\text{R3}} \begin{bmatrix} 1 & -1 & 1 & \vdots & 2 \\ 0 & 0 & 1 & \vdots & 5 \\ 0 & 12 & -8 & \vdots & 8 \end{bmatrix} \xrightarrow{\text{R2}\leftrightarrow\text{R3}} \begin{bmatrix} 1 & -1 & 1 & \vdots & 2 \\ 0 & 12 & -8 & \vdots & 8 \\ 0 & 0 & 1 & \vdots & 5 \end{bmatrix}$$

$$\xrightarrow{\text{R2}\div 12} \begin{bmatrix} 1 & -1 & 1 & \vdots & 2 \\ 0 & 1 & -\frac{2}{3} & \vdots & \frac{2}{3} \\ 0 & 0 & 1 & \vdots & 5 \end{bmatrix} \xrightarrow{\text{R2}+\text{R1}} \begin{bmatrix} 1 & 0 & \frac{1}{3} & \vdots & \frac{8}{3} \\ 0 & 1 & -\frac{2}{3} & \vdots & \frac{2}{3} \\ 0 & 0 & 1 & \vdots & 5 \end{bmatrix}$$

$$\xrightarrow[\frac{2}{3}\text{R3}+\text{R2}]{-\frac{1}{3}\text{R3}+\text{R1}} \begin{bmatrix} 1 & 0 & 0 & \vdots & 1 \\ 0 & 1 & 0 & \vdots & 4 \\ 0 & 0 & 1 & \vdots & 5 \end{bmatrix} \qquad \begin{aligned} (a, b, c) &= (1, 4, 5) \\ y &= x^2 + 4x + 5 \end{aligned}$$

57. $y = ax^2 + bx + c$

$(4, -4)$: $\quad a(4)^2 + b(4) + c = -4$

$(5, -5)$: $\quad a(5)^2 + b(5) + c = -5 \quad\longrightarrow$

$(7, -1)$ $\quad a(7)^2 + b(7) + c = -1$

$$\begin{cases} 16a + 4b + c = -4 \\ 25a + 5b + c = -5 \\ 49a + 7b + c = -1 \end{cases}$$

$$\begin{bmatrix} 16 & 4 & 1 & \vdots & -4 \\ 25 & 5 & 1 & \vdots & -5 \\ 49 & 7 & 1 & \vdots & -1 \end{bmatrix} \xrightarrow{-3\text{R1}+\text{R3}} \begin{bmatrix} 16 & 4 & 1 & \vdots & -4 \\ 25 & 5 & 1 & \vdots & -5 \\ 1 & -5 & -2 & \vdots & 11 \end{bmatrix}$$

$$\xrightarrow{\text{R1}\leftrightarrow\text{R3}} \begin{bmatrix} 1 & -5 & -2 & \vdots & 11 \\ 25 & 5 & 1 & \vdots & -5 \\ 16 & 4 & 1 & \vdots & -4 \end{bmatrix} \xrightarrow[-16\text{R1}+\text{R3}]{-25\text{R1}+\text{R2}} \begin{bmatrix} 1 & -5 & -2 & \vdots & 11 \\ 0 & 130 & 51 & \vdots & -280 \\ 0 & 84 & 33 & \vdots & -180 \end{bmatrix}$$

$$\xrightarrow{\text{R2}\div 130} \begin{bmatrix} 1 & -5 & -2 & \vdots & 11 \\ 0 & 1 & \frac{51}{130} & \vdots & -\frac{28}{13} \\ 0 & 84 & 33 & \vdots & -180 \end{bmatrix} \xrightarrow[-84\text{R2}+\text{R3}]{5\text{R2}+\text{R1}} \begin{bmatrix} 1 & 0 & -\frac{1}{26} & \vdots & \frac{3}{13} \\ 0 & 1 & \frac{51}{130} & \vdots & -\frac{28}{13} \\ 0 & 0 & \frac{3}{65} & \vdots & \frac{12}{13} \end{bmatrix}$$

(Continued on next page)

$$\xrightarrow[-\frac{65}{3}R3]{} \begin{bmatrix} 1 & 0 & -\frac{1}{26} & \vdots & \frac{3}{13} \\ 0 & 1 & \frac{51}{130} & \vdots & -\frac{28}{13} \\ 0 & 0 & 1 & \vdots & 20 \end{bmatrix} \xrightarrow[-\frac{51}{130}R3 + R2]{\frac{1}{26}R3 + R1} \begin{bmatrix} 1 & 0 & 0 & \vdots & 1 \\ 0 & 1 & 0 & \vdots & -10 \\ 0 & 0 & 1 & \vdots & 20 \end{bmatrix}$$

$$(a,\ b,\ c) = (1,\ -10,\ 20)$$
$$y = x^2 - 10x + 20$$

59. Let x be the number of bars used of the alloy which is 3 kg metal A and 5 kg metal B.

Let y be the number of bars used of the alloy which is 4 kg metal A and 7 kg metal B. Then,

$\quad 3x + 4y = 33 \qquad$ Total amount of metal A needed, and

$\quad 5x + 7y = 56 \qquad$ Total amount of metal B needed.

$$\begin{bmatrix} 3 & 4 & \vdots & 33 \\ 5 & 7 & \vdots & 56 \end{bmatrix} \xrightarrow[-2R1 + R2]{} \begin{bmatrix} 3 & 4 & \vdots & 33 \\ -1 & -1 & \vdots & -10 \end{bmatrix}$$

$$\xrightarrow[R1 \leftrightarrow R2]{} \begin{bmatrix} -1 & -1 & \vdots & -10 \\ 3 & 4 & \vdots & 33 \end{bmatrix} \xrightarrow[-R1]{} \begin{bmatrix} 1 & 1 & \vdots & 10 \\ 3 & 4 & \vdots & 33 \end{bmatrix}$$

$$\xrightarrow[-3R1 + R2]{} \begin{bmatrix} 1 & 1 & \vdots & 10 \\ 0 & 1 & \vdots & 3 \end{bmatrix} \xrightarrow[-R2 + R1]{} \begin{bmatrix} 1 & 0 & \vdots & 7 \\ 0 & 1 & \vdots & 3 \end{bmatrix}$$

She needs 7 bars of alloy I and 3 bars of alloy II

61. Let x, y, and z be the amount of Type I, II, and III candy produced respectively. Then,

$\qquad 7x + 3y + 4z = 62 \qquad$ Total amount of chocolate used

$\qquad 5x + 2y + 3z = 44 \qquad$ Total amount of milk used

$\qquad x + 2y + 3z = 32 \qquad$ Total amount of almonds used

$$\begin{bmatrix} 7 & 3 & 4 & \vdots & 62 \\ 5 & 2 & 3 & \vdots & 44 \\ 1 & 2 & 3 & \vdots & 32 \end{bmatrix} \xrightarrow[R1 \leftrightarrow R3]{} \begin{bmatrix} 1 & 2 & 3 & \vdots & 32 \\ 5 & 2 & 3 & \vdots & 44 \\ 7 & 3 & 4 & \vdots & 62 \end{bmatrix}$$

$$\xrightarrow[-7R1 + R3]{-5R1 + R2} \begin{bmatrix} 1 & 2 & 3 & \vdots & 32 \\ 0 & -8 & -12 & \vdots & -116 \\ 0 & -11 & -17 & \vdots & -162 \end{bmatrix} \xrightarrow[-3R3 + 4R2]{} \begin{bmatrix} 1 & 2 & 3 & \vdots & 32 \\ 0 & 1 & 3 & \vdots & 22 \\ 0 & -11 & -17 & \vdots & -162 \end{bmatrix}$$

(Continued on next page)

$$\xrightarrow[\begin{array}{c} -2R2 + R1 \\ \hline 11R2 + R3 \end{array}]{} \left[\begin{array}{ccc:c} 1 & 0 & -3 & -12 \\ 0 & 1 & 3 & 22 \\ 0 & 0 & 16 & 80 \end{array}\right] \xrightarrow[R3 \div 16]{} \left[\begin{array}{ccc:c} 1 & 0 & -3 & -12 \\ 0 & 1 & 3 & 22 \\ 0 & 0 & 1 & 5 \end{array}\right]$$

$$\xrightarrow[\begin{array}{c} 3R3 + R1 \\ \hline -3R3 + R2 \end{array}]{} \left[\begin{array}{ccc:c} 1 & 0 & 0 & 3 \\ 0 & 1 & 0 & 7 \\ 0 & 0 & 1 & 5 \end{array}\right]$$

The candy maker can produce 3 units of I, 7 units of II, and 5 units of III.

PROBLEM SET 9.3

1. $E + F = \begin{bmatrix} 1 & 0 & 2 \\ 3 & -1 & 2 \\ 4 & 1 & 0 \end{bmatrix} + \begin{bmatrix} 1 & 4 & 0 \\ 3 & -1 & 2 \\ -2 & 1 & 5 \end{bmatrix} = \begin{bmatrix} 1+1 & 0+4 & 2+0 \\ 3+3 & -1+(-1) & 2+2 \\ 4+(-2) & 1+1 & 0+5 \end{bmatrix}$

$$= \begin{bmatrix} 2 & 4 & 2 \\ 6 & -2 & 4 \\ 2 & 2 & 5 \end{bmatrix}$$

3. $EG = \begin{bmatrix} 1 & 0 & 2 \\ 3 & -1 & 2 \\ 4 & 1 & 0 \end{bmatrix} \cdot \begin{bmatrix} 8 & 1 & 6 \\ 3 & 5 & 7 \\ 4 & 9 & 2 \end{bmatrix}$

$$= \begin{bmatrix} 8+0+8 & 1+0+18 & 6+0+4 \\ 24-3+8 & 3-5+18 & 18-7+4 \\ 32+3+0 & 4+5+0 & 24+7+0 \end{bmatrix} = \begin{bmatrix} 16 & 19 & 10 \\ 29 & 16 & 15 \\ 35 & 9 & 31 \end{bmatrix}$$

5. $E(F + G) = \begin{bmatrix} 1 & 0 & 2 \\ 3 & -1 & 2 \\ 4 & 1 & 0 \end{bmatrix} \left(\begin{bmatrix} 1 & 4 & 0 \\ 3 & -1 & 2 \\ -2 & 1 & 5 \end{bmatrix} + \begin{bmatrix} 8 & 1 & 6 \\ 3 & 5 & 7 \\ 4 & 9 & 2 \end{bmatrix} \right)$

$$= \begin{bmatrix} 1 & 0 & 2 \\ 3 & -1 & 2 \\ 4 & 1 & 0 \end{bmatrix} \cdot \begin{bmatrix} 1+8 & 4+1 & 0+6 \\ 3+3 & -1+5 & 2+7 \\ -2+4 & 1+9 & 5+2 \end{bmatrix}$$

$$= \begin{bmatrix} 1 & 0 & 2 \\ 3 & -1 & 2 \\ 4 & 1 & 0 \end{bmatrix} \cdot \begin{bmatrix} 9 & 5 & 6 \\ 6 & 4 & 9 \\ 2 & 10 & 7 \end{bmatrix}$$

$$= \begin{bmatrix} 9+0+4 & 5+0+20 & 6+0+14 \\ 27-6+4 & 15-4+20 & 18-9+14 \\ 36+6+0 & 20+4+0 & 24+9+0 \end{bmatrix}$$

$$= \begin{bmatrix} 13 & 25 & 20 \\ 25 & 31 & 23 \\ 42 & 24 & 33 \end{bmatrix}$$

7. $FG = \begin{bmatrix} 1 & 4 & 0 \\ 3 & -1 & 2 \\ -2 & 1 & 5 \end{bmatrix} \cdot \begin{bmatrix} 8 & 1 & 6 \\ 3 & 5 & 7 \\ 4 & 9 & 2 \end{bmatrix}$

$$= \begin{bmatrix} 8+12+0 & 1+20+0 & 6+28+0 \\ 24-3+8 & 3-5+18 & 18-7+4 \\ -16+3+20 & -2+5+45 & -12+7+10 \end{bmatrix} = \begin{bmatrix} 20 & 21 & 34 \\ 29 & 16 & 15 \\ 7 & 48 & 5 \end{bmatrix}$$

9. $(EF)G = \left(\begin{bmatrix} 1 & 0 & 2 \\ 3 & -1 & 2 \\ 4 & 1 & 0 \end{bmatrix} \cdot \begin{bmatrix} 1 & 4 & 0 \\ 3 & -1 & 2 \\ -2 & 1 & 5 \end{bmatrix} \right) \begin{bmatrix} 8 & 1 & 6 \\ 3 & 5 & 7 \\ 4 & 9 & 2 \end{bmatrix}$

$= \begin{bmatrix} 1+0-4 & 4+0+2 & 0+0+10 \\ 3-3-4 & 12+1+2 & 0-2+10 \\ 4+3+0 & 16-1+0 & 0+2+0 \end{bmatrix} \cdot \begin{bmatrix} 8 & 1 & 6 \\ 3 & 5 & 7 \\ 4 & 9 & 2 \end{bmatrix}$

$= \begin{bmatrix} -3 & 6 & 10 \\ -4 & 15 & 8 \\ 7 & 15 & 2 \end{bmatrix} \cdot \begin{bmatrix} 8 & 1 & 6 \\ 3 & 5 & 7 \\ 4 & 9 & 2 \end{bmatrix}$

$= \begin{bmatrix} -24+18+40 & -3+30+90 & -18+42+20 \\ -32+45+32 & -4+75+72 & -24+105+16 \\ 56+45+8 & 7+75+18 & 42+105+4 \end{bmatrix} = \begin{bmatrix} 34 & 117 & 44 \\ 45 & 143 & 97 \\ 109 & 100 & 151 \end{bmatrix}$

11. $B^2 = \begin{bmatrix} 4 & 2 \\ -1 & 3 \end{bmatrix} \cdot \begin{bmatrix} 4 & 2 \\ -1 & 3 \end{bmatrix} = \begin{bmatrix} 16-2 & 8+6 \\ -4-3 & -2+9 \end{bmatrix} = \begin{bmatrix} 14 & 14 \\ -7 & 7 \end{bmatrix}$

13. $\begin{bmatrix} 1 & 2 \\ 2 & 3 \end{bmatrix} \cdot \begin{bmatrix} -3 & 2 \\ 2 & -1 \end{bmatrix} = \begin{bmatrix} -3+4 & 2-2 \\ -6+6 & 4-3 \end{bmatrix} = \begin{bmatrix} 1 & 0 \\ 0 & 1 \end{bmatrix}$

The matrices are inverses.

15. $\begin{bmatrix} 4 & 3 \\ 2 & 2 \end{bmatrix} \cdot \begin{bmatrix} 1 & -\frac{3}{2} \\ -1 & 2 \end{bmatrix} = \begin{bmatrix} 4-3 & -6+6 \\ 2-2 & -3+4 \end{bmatrix} = \begin{bmatrix} 1 & 0 \\ 0 & 1 \end{bmatrix}$

The matrices are inverses.

217

17.
$$\begin{bmatrix} 3 & -2 & 4 \\ 2 & 1 & 2 \\ 5 & 3 & 5 \end{bmatrix} \cdot \begin{bmatrix} -1 & 22 & -8 \\ 0 & -5 & 2 \\ 1 & -19 & 7 \end{bmatrix}$$

$$= \begin{bmatrix} -3+0+4 & 66+10-76 & -24-4+28 \\ -2+0+2 & 44-5-38 & -16+2+14 \\ -5+0+5 & 110-15-95 & -40+6+35 \end{bmatrix} = \begin{bmatrix} 1 & 0 & 0 \\ 0 & 1 & 0 \\ 0 & 0 & 1 \end{bmatrix}$$

The matrices are inverses.

19.
$$\begin{bmatrix} 4 & -7 & | & 1 & 0 \\ -1 & 2 & | & 0 & 1 \end{bmatrix} \xrightarrow{R1 \leftrightarrow R2} \begin{bmatrix} -1 & 2 & | & 0 & 1 \\ 4 & -7 & | & 1 & 0 \end{bmatrix}$$

$$\xrightarrow{-R1} \begin{bmatrix} 1 & -2 & | & 0 & -1 \\ 4 & -7 & | & 1 & 0 \end{bmatrix} \xrightarrow{-4R1+R2} \begin{bmatrix} 1 & -2 & | & 0 & -1 \\ 0 & 1 & | & 1 & 4 \end{bmatrix}$$

$$\xrightarrow{2R2+R1} \begin{bmatrix} 1 & 0 & | & 2 & 7 \\ 0 & 1 & | & 1 & 4 \end{bmatrix} \quad \text{The inverse is} \begin{bmatrix} 2 & 7 \\ 1 & 4 \end{bmatrix}.$$

21.
$$\begin{bmatrix} 1 & 3 & | & 1 & 0 \\ 2 & 0 & | & 0 & 1 \end{bmatrix} \xrightarrow{-2R1+R2} \begin{bmatrix} 1 & 3 & | & 1 & 0 \\ 0 & -6 & | & -2 & 1 \end{bmatrix}$$

$$\xrightarrow{R2 \div (-6)} \begin{bmatrix} 1 & 3 & | & 1 & 0 \\ 0 & 1 & | & \frac{1}{3} & -\frac{1}{6} \end{bmatrix} \xrightarrow{-3R2+R1} \begin{bmatrix} 1 & 0 & | & 0 & \frac{1}{2} \\ 0 & 1 & | & \frac{1}{3} & -\frac{1}{6} \end{bmatrix}$$

The inverse is $\begin{bmatrix} 0 & \frac{1}{2} \\ \frac{1}{3} & -\frac{1}{6} \end{bmatrix}$

23. $\begin{bmatrix} 6 & 1 & 20 & | & 1 & 0 & 0 \\ 1 & -1 & 0 & | & 0 & 1 & 0 \\ 0 & 1 & 3 & | & 0 & 0 & 1 \end{bmatrix} \xrightarrow{R1 \leftrightarrow R2} \begin{bmatrix} 1 & -1 & 0 & | & 0 & 1 & 0 \\ 6 & 1 & 20 & | & 1 & 0 & 0 \\ 0 & 1 & 3 & | & 0 & 0 & 1 \end{bmatrix}$

$\xrightarrow{-6R1 + R2} \begin{bmatrix} 1 & -1 & 0 & | & 0 & 1 & 0 \\ 0 & 7 & 20 & | & 1 & -6 & 0 \\ 0 & 1 & 3 & | & 0 & 0 & 1 \end{bmatrix}$

$\xrightarrow{R2 \leftrightarrow R3} \begin{bmatrix} 1 & -1 & 0 & | & 0 & 1 & 0 \\ 0 & 1 & 3 & | & 0 & 0 & 1 \\ 0 & 7 & 20 & | & 1 & -6 & 0 \end{bmatrix}$

$\xrightarrow[-7R2 + R3]{R2 + R1} \begin{bmatrix} 1 & 0 & 3 & | & 0 & 1 & 1 \\ 0 & 1 & 3 & | & 0 & 0 & 1 \\ 0 & 0 & -1 & | & 1 & -6 & -7 \end{bmatrix}$

$\xrightarrow{-R3} \begin{bmatrix} 1 & 0 & 3 & | & 0 & 1 & 1 \\ 0 & 1 & 3 & | & 0 & 0 & 1 \\ 0 & 0 & 1 & | & -1 & 6 & 7 \end{bmatrix}$

$\xrightarrow[-3R3 + R1]{-3R3 + R1} \begin{bmatrix} 1 & 0 & 0 & | & 3 & -17 & -20 \\ 0 & 1 & 0 & | & 3 & -18 & -20 \\ 0 & 0 & 1 & | & -1 & 6 & 7 \end{bmatrix}$

The inverse is $\begin{bmatrix} 3 & -17 & -20 \\ 3 & -18 & -20 \\ -1 & 6 & 7 \end{bmatrix}$

25.

$$\left[\begin{array}{cccc|cccc} 1 & 0 & 0 & 1 & 1 & 0 & 0 & 0 \\ 0 & 2 & 0 & 0 & 0 & 1 & 0 & 0 \\ 0 & 0 & 0 & 1 & 0 & 0 & 1 & 0 \\ 2 & 0 & 1 & 0 & 0 & 0 & 0 & 1 \end{array}\right]$$

$$\xrightarrow{-2R1 + R4} \left[\begin{array}{cccc|cccc} 1 & 0 & 0 & 1 & 1 & 0 & 0 & 0 \\ 0 & 2 & 0 & 0 & 0 & 1 & 0 & 0 \\ 0 & 0 & 0 & 1 & 0 & 0 & 1 & 0 \\ 0 & 0 & 1 & -2 & -2 & 0 & 0 & 1 \end{array}\right]$$

$$\xrightarrow[\substack{R2 \div 2 \\ R3 \leftrightarrow R4}]{} \left[\begin{array}{cccc|cccc} 1 & 0 & 0 & 1 & 1 & 0 & 0 & 0 \\ 0 & 1 & 0 & 0 & 0 & \frac{1}{2} & 0 & 0 \\ 0 & 0 & 1 & -2 & -2 & 0 & 0 & 1 \\ 0 & 0 & 0 & 1 & 0 & 0 & 1 & 0 \end{array}\right]$$

$$\xrightarrow[\substack{-R4 + R1 \\ 2R4 + R3}]{} \left[\begin{array}{cccc|cccc} 1 & 0 & 0 & 0 & 1 & 0 & -1 & 0 \\ 0 & 1 & 0 & 0 & 0 & \frac{1}{2} & 0 & 0 \\ 0 & 0 & 1 & 0 & -2 & 0 & 2 & 1 \\ 0 & 0 & 0 & 1 & 0 & 0 & 1 & 0 \end{array}\right]$$

The inverse is $\left[\begin{array}{cccc} 1 & 0 & -1 & 0 \\ 0 & \frac{1}{2} & 0 & 0 \\ -2 & 0 & 2 & 1 \\ 0 & 0 & 1 & 0 \end{array}\right]$

$$27. \begin{bmatrix} 1 & 2 & 0 & 0 & \vdots & 1 & 0 & 0 & 0 \\ 0 & 0 & 1 & 0 & \vdots & 0 & 1 & 0 & 0 \\ 1 & 3 & 0 & 1 & \vdots & 0 & 0 & 1 & 0 \\ 4 & 0 & 0 & 2 & \vdots & 0 & 0 & 0 & 1 \end{bmatrix}$$

$$\xrightarrow[\substack{-R1+R3 \\ -4R1+R4}]{} \begin{bmatrix} 1 & 2 & 0 & 0 & \vdots & 1 & 0 & 0 & 0 \\ 0 & 0 & 1 & 0 & \vdots & 0 & 1 & 0 & 0 \\ 0 & 1 & 0 & 1 & \vdots & -1 & 0 & 1 & 0 \\ 0 & -8 & 0 & 2 & \vdots & -4 & 0 & 0 & 1 \end{bmatrix}$$

$$\xrightarrow[\substack{R2 \leftrightarrow R3}]{} \begin{bmatrix} 1 & 2 & 0 & 0 & \vdots & 1 & 0 & 0 & 0 \\ 0 & 1 & 0 & 1 & \vdots & -1 & 0 & 1 & 0 \\ 0 & 0 & 1 & 0 & \vdots & 0 & 1 & 0 & 0 \\ 0 & -8 & 0 & 2 & \vdots & -4 & 0 & 0 & 1 \end{bmatrix}$$

$$\xrightarrow[\substack{-2R2+R1 \\ 8R2+R4}]{} \begin{bmatrix} 1 & 0 & 0 & -2 & \vdots & 3 & 0 & -2 & 0 \\ 0 & 1 & 0 & 1 & \vdots & -1 & 0 & 1 & 0 \\ 0 & 0 & 1 & 0 & \vdots & 0 & 1 & 0 & 0 \\ 0 & 0 & 0 & 10 & \vdots & -12 & 0 & 8 & 1 \end{bmatrix}$$

$$\xrightarrow[\substack{R4 \div 10}]{} \begin{bmatrix} 1 & 0 & 0 & -2 & \vdots & 3 & 0 & -2 & 0 \\ 0 & 1 & 0 & 1 & \vdots & -1 & 0 & 1 & 0 \\ 0 & 0 & 1 & 0 & \vdots & 0 & 1 & 0 & 0 \\ 0 & 0 & 0 & 1 & \vdots & -1.2 & 0 & .8 & .1 \end{bmatrix}$$

$$\xrightarrow[\substack{2R4+R1 \\ -R4+R2}]{} \begin{bmatrix} 1 & 0 & 0 & 0 & \vdots & .6 & 0 & -.4 & .2 \\ 0 & 1 & 0 & 0 & \vdots & .2 & 0 & .2 & -.1 \\ 0 & 0 & 1 & 0 & \vdots & 0 & 1 & 0 & 0 \\ 0 & 0 & 0 & 1 & \vdots & -1.2 & 0 & .8 & .1 \end{bmatrix}$$

The inverse is $\begin{bmatrix} .6 & 0 & -.4 & .2 \\ .2 & 0 & .2 & -.1 \\ 0 & 1 & 0 & 0 \\ -1.2 & 0 & .8 & .1 \end{bmatrix}$

29. $X = A^{-1}B$

$$\begin{bmatrix} x \\ y \end{bmatrix} = \begin{bmatrix} 2 & 7 \\ 1 & 4 \end{bmatrix} \cdot \begin{bmatrix} -65 \\ 18 \end{bmatrix} = \begin{bmatrix} 2(-65)+7(18) \\ 1(-65)+4(18) \end{bmatrix} = \begin{bmatrix} -4 \\ 7 \end{bmatrix}$$

$$(x,\ y) = (-4,\ 7)$$

31. $X = A^{-1}B$

$$\begin{bmatrix} x \\ y \end{bmatrix} = \begin{bmatrix} 2 & 7 \\ 1 & 4 \end{bmatrix} \cdot \begin{bmatrix} 2 \\ 3 \end{bmatrix} = \begin{bmatrix} 2(2)+7(3) \\ 1(2)+4(3) \end{bmatrix} = \begin{bmatrix} 25 \\ 14 \end{bmatrix}$$

$$(x,\ y) = (25,\ 14)$$

33. $X = A^{-1}B$

$$\begin{bmatrix} x \\ y \end{bmatrix} = \begin{bmatrix} 2 & 7 \\ 1 & 4 \end{bmatrix} \cdot \begin{bmatrix} -3 \\ 8 \end{bmatrix} = \begin{bmatrix} 2(-3)+7(8) \\ 1(-3)+4(8) \end{bmatrix} = \begin{bmatrix} 50 \\ 29 \end{bmatrix}$$

$$(x,\ y) = (50,\ 29)$$

Problems 34–39 all use the same inverse:

$$\begin{bmatrix} 8 & 6 & | & 1 & 0 \\ -2 & 4 & | & 0 & 1 \end{bmatrix} \xrightarrow{R1 \leftrightarrow R2} \begin{bmatrix} -2 & 4 & | & 0 & 1 \\ 8 & 6 & | & 1 & 0 \end{bmatrix}$$

$$\xrightarrow{R1 \div (-2)} \begin{bmatrix} 1 & -2 & | & 0 & -\frac{1}{2} \\ 8 & 6 & | & 1 & 0 \end{bmatrix} \xrightarrow{-8R1 + R2} \begin{bmatrix} 1 & -2 & | & 0 & -\frac{1}{2} \\ 0 & 22 & | & 1 & 4 \end{bmatrix}$$

$$\xrightarrow{R2 \div 22} \begin{bmatrix} 1 & -2 & | & 0 & -\frac{1}{2} \\ 0 & 1 & | & \frac{1}{22} & \frac{2}{11} \end{bmatrix} \xrightarrow{2R2 + R1} \begin{bmatrix} 1 & 0 & | & \frac{1}{11} & -\frac{3}{22} \\ 0 & 1 & | & \frac{1}{22} & \frac{2}{11} \end{bmatrix}$$

The inverse is $\begin{bmatrix} \frac{1}{11} & -\frac{3}{22} \\ \frac{1}{22} & \frac{2}{11} \end{bmatrix}$

35. $X = A^{-1}B$

$$\begin{bmatrix} x \\ y \end{bmatrix} = \begin{bmatrix} \frac{1}{11} & -\frac{3}{22} \\ \frac{1}{22} & \frac{2}{11} \end{bmatrix} \cdot \begin{bmatrix} 16 \\ 18 \end{bmatrix} = \begin{bmatrix} \frac{1}{11}(16)+(-\frac{3}{22})(18) \\ \frac{1}{22}(16)+\frac{2}{11}(18) \end{bmatrix} = \begin{bmatrix} -1 \\ 4 \end{bmatrix}$$

$$(x,\ y) = (-1,\ 4)$$

37. $X = A^{-1}B$

$$\begin{bmatrix} x \\ y \end{bmatrix} = \begin{bmatrix} \frac{1}{11} & -\frac{3}{22} \\ \frac{1}{22} & \frac{2}{11} \end{bmatrix} \cdot \begin{bmatrix} -28 \\ 18 \end{bmatrix} = \begin{bmatrix} \frac{1}{11}(-28) + (-\frac{3}{22})(18) \\ \frac{1}{22}(-28) + \frac{2}{11}(18) \end{bmatrix} = \begin{bmatrix} -5 \\ 2 \end{bmatrix}$$

$$(x,\ y) = (-5,\ 2)$$

39. $X = A^{-1}B$

$$\begin{bmatrix} x \\ y \end{bmatrix} = \begin{bmatrix} \frac{1}{11} & -\frac{3}{22} \\ \frac{1}{22} & \frac{2}{11} \end{bmatrix} \cdot \begin{bmatrix} -36 \\ -2 \end{bmatrix} = \begin{bmatrix} \frac{1}{11}(-36) + (-\frac{3}{22})(-2) \\ \frac{1}{22}(-36) + \frac{2}{11}(-2) \end{bmatrix} = \begin{bmatrix} -3 \\ -2 \end{bmatrix}$$

$$(x,\ y) = (-3,\ -2)$$

Problems 40–45 all use the same inverse:

$$\left[\begin{array}{cc|cc} 2 & 3 & 1 & 0 \\ 1 & -6 & 0 & 1 \end{array}\right] \xrightarrow{R1 \leftrightarrow R2} \left[\begin{array}{cc|cc} 1 & -6 & 0 & 1 \\ 2 & 3 & 1 & 0 \end{array}\right]$$

$$\xrightarrow{-2R1 + R2} \left[\begin{array}{cc|cc} 1 & -6 & 0 & 1 \\ 0 & 15 & 1 & -2 \end{array}\right] \xrightarrow{R2 \div 15} \left[\begin{array}{cc|cc} 1 & -6 & 0 & -\frac{1}{2} \\ 0 & 1 & \frac{1}{15} & -\frac{2}{15} \end{array}\right]$$

$$\xrightarrow{-6R2 + R1} \left[\begin{array}{cc|cc} 1 & 0 & \frac{2}{5} & \frac{1}{5} \\ 0 & 1 & \frac{1}{15} & -\frac{2}{15} \end{array}\right]$$

The inverse is $\begin{bmatrix} \frac{2}{5} & \frac{1}{5} \\ \frac{1}{15} & -\frac{2}{15} \end{bmatrix}$

41. $X = A^{-1}B$

$$\begin{bmatrix} x \\ y \end{bmatrix} = \begin{bmatrix} \frac{2}{5} & \frac{1}{5} \\ \frac{1}{15} & -\frac{2}{15} \end{bmatrix} \cdot \begin{bmatrix} 2 \\ 16 \end{bmatrix} = \begin{bmatrix} \frac{2}{5}(2) + \frac{1}{5}(16) \\ \frac{1}{15}(2) + (-\frac{2}{15})(16) \end{bmatrix} = \begin{bmatrix} 4 \\ -2 \end{bmatrix}$$

$$(x,\ y) = (4,\ -2)$$

43. $X = A^{-1}B$

$$\begin{bmatrix} x \\ y \end{bmatrix} = \begin{bmatrix} \frac{2}{5} & \frac{1}{5} \\ \frac{1}{15} & -\frac{2}{15} \end{bmatrix} \cdot \begin{bmatrix} 9 \\ 42 \end{bmatrix} = \begin{bmatrix} \frac{2}{5}(9) + \frac{1}{5}(42) \\ \frac{1}{15}(9) + (-\frac{2}{15})(42) \end{bmatrix} = \begin{bmatrix} 12 \\ -5 \end{bmatrix}$$

$$(x,\ y) = (12,\ -5)$$

45. $X = A^{-1}B$

$$\begin{bmatrix} x \\ y \end{bmatrix} = \begin{bmatrix} \frac{2}{5} & \frac{1}{5} \\ \frac{1}{15} & -\frac{2}{15} \end{bmatrix} \cdot \begin{bmatrix} 12 \\ -24 \end{bmatrix} = \begin{bmatrix} \frac{2}{5}(12) + \frac{1}{5}(-24) \\ \frac{1}{15}(12) + (-\frac{2}{15})(-24) \end{bmatrix} = \begin{bmatrix} 0 \\ 4 \end{bmatrix}$$

$$(x,\ y) = (0,\ 4)$$

Problems 46–51 all use the inverse from problem 22.

$$\begin{bmatrix} 1 & 0 & 2 & | & 1 & 0 & 0 \\ 2 & 1 & 0 & | & 0 & 1 & 0 \\ 0 & -2 & 9 & | & 0 & 0 & 1 \end{bmatrix} \xrightarrow{-2R1+R2} \begin{bmatrix} 1 & 0 & 2 & | & 1 & 0 & 0 \\ 0 & 1 & -4 & | & -2 & 1 & 0 \\ 0 & -2 & 9 & | & 0 & 0 & 1 \end{bmatrix}$$

$$\xrightarrow[2R2+R3]{} \begin{bmatrix} 1 & 0 & 2 & | & 1 & 0 & 0 \\ 0 & 1 & -4 & | & -2 & 1 & 0 \\ 0 & 0 & 1 & | & -4 & 2 & 1 \end{bmatrix} \xrightarrow[4R3+R2]{-2R3+R1} \begin{bmatrix} 1 & 0 & 0 & | & 9 & -4 & -2 \\ 0 & 1 & 0 & | & -18 & 9 & 4 \\ 0 & 0 & 1 & | & -4 & 2 & 1 \end{bmatrix}$$

The inverse is $\begin{bmatrix} 9 & -4 & -2 \\ -18 & 9 & 4 \\ -4 & 2 & 1 \end{bmatrix}$

47. $X = A^{-1}B$

$$\begin{bmatrix} x \\ y \\ z \end{bmatrix} = \begin{bmatrix} 9 & -4 & -2 \\ -18 & 9 & 4 \\ -4 & 2 & 1 \end{bmatrix} \cdot \begin{bmatrix} 4 \\ 0 \\ 19 \end{bmatrix} = \begin{bmatrix} 9(4) + (-4)(0) + (-2)(19) \\ (-18)(4) + 9(0) + 4(19) \\ (-4)(4) + 2(0) + 1(19) \end{bmatrix} = \begin{bmatrix} -2 \\ 4 \\ 3 \end{bmatrix}$$

$$(x,\ y,\ z) = (-2,\ 4,\ 3)$$

49. $X = A^{-1}B$

$$\begin{bmatrix} x \\ y \\ z \end{bmatrix} = \begin{bmatrix} 9 & -4 & -2 \\ -18 & 9 & 4 \\ -4 & 2 & 1 \end{bmatrix} \cdot \begin{bmatrix} 7 \\ 1 \\ 28 \end{bmatrix} = \begin{bmatrix} 9(7)+(-4)(1)+(-2)(28) \\ (-18)(7)+9(1)+4(28) \\ (-4)(7)+2(1)+1(28) \end{bmatrix} = \begin{bmatrix} 3 \\ -5 \\ 2 \end{bmatrix}$$

$$(x,\ y,\ z) = (3,\ -5,\ 2)$$

51. $X = A^{-1}B$

$$\begin{bmatrix} x \\ y \\ z \end{bmatrix} = \begin{bmatrix} 9 & -4 & -2 \\ -18 & 9 & 4 \\ -4 & 2 & 1 \end{bmatrix} \cdot \begin{bmatrix} 5 \\ 8 \\ 9 \end{bmatrix} = \begin{bmatrix} 9(5)+(-4)(8)+(-2)(9) \\ (-18)(5)+9(8)+4(9) \\ (-4)(5)+2(8)+1(9) \end{bmatrix} = \begin{bmatrix} -5 \\ 18 \\ 5 \end{bmatrix}$$

$$(x,\ y,\ z) = (3,\ -5,\ 2)$$

53. $X = A^{-1}B$ (Use the inverse from problem 23)

$$\begin{bmatrix} x \\ y \\ z \end{bmatrix} = \begin{bmatrix} 3 & -17 & -20 \\ 3 & -18 & -20 \\ -1 & 6 & 7 \end{bmatrix} \cdot \begin{bmatrix} 14 \\ 1 \\ 1 \end{bmatrix} = \begin{bmatrix} 3(14)+(-17)(1)+(-20)(1) \\ 3(14)+(-18)(1)+(-20)(1) \\ (-1)(14)+6(1)+7(1) \end{bmatrix} = \begin{bmatrix} 5 \\ 4 \\ -1 \end{bmatrix}$$

$$(x,\ y,\ z) = (5,\ 4,\ -1)$$

Problems 55-57 all use the inverse found in problem 24.

$$\begin{bmatrix} 4 & 1 & 0 & | & 1 & 0 & 0 \\ 2 & -1 & 4 & | & 0 & 1 & 0 \\ -3 & 2 & 1 & | & 0 & 0 & 1 \end{bmatrix} \xrightarrow{R3+R1} \begin{bmatrix} 1 & 3 & 1 & | & 1 & 0 & 1 \\ 2 & -1 & 4 & | & 0 & 1 & 0 \\ -3 & 2 & 1 & | & 0 & 0 & 1 \end{bmatrix}$$

$$\xrightarrow[3R1+R3]{-2R1+R2} \begin{bmatrix} 1 & 3 & 1 & | & 1 & 0 & 1 \\ 0 & -7 & 2 & | & -2 & 1 & -2 \\ 0 & 11 & 4 & | & 3 & 0 & 4 \end{bmatrix} \xrightarrow{2R3+3R2} \begin{bmatrix} 1 & 3 & 1 & | & 1 & 0 & 1 \\ 0 & 1 & 14 & | & 0 & 3 & 2 \\ 0 & 11 & 4 & | & 3 & 0 & 4 \end{bmatrix}$$

$$\xrightarrow[-11R2+R3]{-3R2+R1} \begin{bmatrix} 1 & 0 & -41 & | & 1 & -9 & -5 \\ 0 & 1 & 14 & | & 0 & 3 & 2 \\ 0 & 0 & -150 & | & 3 & -33 & -18 \end{bmatrix} \xrightarrow{R3 \div (-150)} \begin{bmatrix} 1 & 0 & -41 & | & 1 & -9 & -5 \\ 0 & 1 & 14 & | & 0 & 3 & 2 \\ 0 & 0 & 1 & | & -\frac{1}{50} & \frac{11}{50} & \frac{3}{25} \end{bmatrix}$$

$$\xrightarrow[-14R3+R2]{41R3+R1} \begin{bmatrix} 1 & 0 & 0 & | & \frac{9}{50} & \frac{1}{50} & -\frac{2}{25} \\ 0 & 1 & 0 & | & \frac{7}{25} & -\frac{2}{25} & \frac{8}{25} \\ 0 & 0 & 1 & | & -\frac{1}{50} & \frac{11}{50} & \frac{3}{25} \end{bmatrix} \qquad \text{The inverse is} \begin{bmatrix} \frac{9}{50} & \frac{1}{50} & -\frac{2}{25} \\ \frac{7}{25} & -\frac{2}{25} & \frac{8}{25} \\ -\frac{1}{50} & \frac{11}{50} & \frac{3}{25} \end{bmatrix}$$

55. $X = A^{-1}B$

$$\begin{bmatrix} x \\ y \\ z \end{bmatrix} = \begin{bmatrix} \frac{9}{50} & \frac{1}{50} & -\frac{2}{25} \\ \frac{7}{25} & -\frac{2}{25} & \frac{8}{25} \\ -\frac{1}{50} & \frac{11}{50} & \frac{3}{25} \end{bmatrix} \cdot \begin{bmatrix} 6 \\ 12 \\ 4 \end{bmatrix} = \begin{bmatrix} \frac{54}{50} + \frac{12}{50} + \left(-\frac{8}{25}\right) \\ \frac{42}{25} + \left(-\frac{24}{25}\right) + \frac{32}{25} \\ -\frac{6}{50} + \frac{132}{50} + \frac{12}{25} \end{bmatrix} = \begin{bmatrix} 1 \\ 2 \\ 3 \end{bmatrix}$$

$$(x, \ y, \ z) = (1, \ 2, \ 3)$$

57. $X = A^{-1}B$

$$\begin{bmatrix} x \\ y \\ z \end{bmatrix} = \begin{bmatrix} \frac{9}{50} & \frac{1}{50} & -\frac{2}{25} \\ \frac{7}{25} & -\frac{2}{25} & \frac{8}{25} \\ -\frac{1}{50} & \frac{11}{50} & \frac{3}{25} \end{bmatrix} \cdot \begin{bmatrix} -10 \\ 20 \\ 20 \end{bmatrix} = \begin{bmatrix} -\frac{90}{50} + \frac{20}{50} + \left(-\frac{40}{25}\right) \\ -\frac{70}{25} + \left(-\frac{40}{25}\right) + \frac{160}{25} \\ \frac{10}{50} + \frac{220}{50} + \frac{60}{25} \end{bmatrix} = \begin{bmatrix} -3 \\ 2 \\ 7 \end{bmatrix}$$

$$(x, \ y, \ z) = (-3, \ 2, \ 7)$$

59. Uses the inverse from problem 26.

$$\left[\begin{array}{cccc|cccc} 0 & 1 & 2 & 0 & 1 & 0 & 0 & 0 \\ 0 & 0 & 0 & 1 & 0 & 1 & 0 & 0 \\ 1 & 1 & 3 & 0 & 0 & 0 & 1 & 0 \\ 2 & 4 & 0 & 0 & 0 & 0 & 0 & 1 \end{array}\right]$$

$$\xrightarrow[\text{R2}\leftrightarrow\text{R4}]{\text{R1}\leftrightarrow\text{R3}} \left[\begin{array}{cccc|cccc} 1 & 1 & 3 & 0 & 0 & 0 & 1 & 0 \\ 2 & 4 & 0 & 0 & 0 & 0 & 0 & 1 \\ 0 & 1 & 2 & 0 & 1 & 0 & 0 & 0 \\ 0 & 0 & 0 & 1 & 0 & 1 & 0 & 0 \end{array}\right]$$

$$\xrightarrow{-2\text{R1}+\text{R2}} \left[\begin{array}{cccc|cccc} 1 & 1 & 3 & 0 & 0 & 0 & 1 & 0 \\ 0 & 2 & -6 & 0 & 0 & 0 & -2 & 1 \\ 0 & 1 & 2 & 0 & 1 & 0 & 0 & 0 \\ 0 & 0 & 0 & 1 & 0 & 1 & 0 & 0 \end{array}\right]$$

$$\xrightarrow{\text{R2}\leftrightarrow\text{R3}} \left[\begin{array}{cccc|cccc} 1 & 1 & 3 & 0 & 0 & 0 & 1 & 0 \\ 0 & 1 & 2 & 0 & 1 & 0 & 0 & 0 \\ 0 & 2 & -6 & 0 & 0 & 0 & -2 & 1 \\ 0 & 0 & 0 & 1 & 0 & 1 & 0 & 0 \end{array}\right]$$

(Continued on next page)

$$\xrightarrow[\;-2R2+R3\;]{-R2+R1}\quad \left[\begin{array}{cccc|cccc} 1 & 0 & 1 & 0 & -1 & 0 & 1 & 0 \\ 0 & 1 & 2 & 0 & 1 & 0 & 0 & 0 \\ 0 & 0 & -10 & 0 & -2 & 0 & -2 & 1 \\ 0 & 0 & 0 & 1 & 0 & 1 & 0 & 0 \end{array}\right]$$

$$\xrightarrow[\;R3\div(-10)\;]{}\quad \left[\begin{array}{cccc|cccc} 1 & 0 & 1 & 0 & -1 & 0 & 1 & 0 \\ 0 & 1 & 2 & 0 & 1 & 0 & 0 & 0 \\ 0 & 0 & 1 & 0 & .2 & 0 & .2 & -.1 \\ 0 & 0 & 0 & 1 & 0 & 1 & 0 & 0 \end{array}\right]$$

$$\xrightarrow[\;-2R3+R2\;]{-R3+R1}\quad \left[\begin{array}{cccc|cccc} 1 & 0 & 0 & 0 & -1.2 & 0 & .8 & .1 \\ 0 & 1 & 0 & 0 & .6 & 0 & -.4 & .2 \\ 0 & 0 & 1 & 0 & .2 & 0 & .2 & -.1 \\ 0 & 0 & 0 & 1 & 0 & 1 & 0 & 0 \end{array}\right]$$

The inverse is $\left[\begin{array}{cccc} -1.2 & 0 & .8 & .1 \\ .6 & 0 & -.4 & .2 \\ .2 & 0 & .2 & -.1 \\ 0 & 1 & 0 & 0 \end{array}\right]$

$X = A^{-1}B$

$$\begin{bmatrix} w \\ x \\ y \\ z \end{bmatrix} = \begin{bmatrix} -1.2 & 0 & .8 & .1 \\ .6 & 0 & -.4 & .2 \\ .2 & 0 & .2 & -.1 \\ 0 & 1 & 0 & 0 \end{bmatrix} \cdot \begin{bmatrix} 0 \\ -4 \\ 4 \\ -2 \end{bmatrix} = \begin{bmatrix} 0+0+3.2-.2 \\ 0+0-1.6-.4 \\ 0+0+.8+.2 \\ 0-4+0+0 \end{bmatrix} = \begin{bmatrix} 3 \\ -2 \\ 1 \\ -4 \end{bmatrix}$$

$$(w,\ x,\ y,\ z) = (3,\ -2,\ 1,\ -4)$$

61.
$$(MN)(MN)^{-1} = I$$
$$M^{-1}(MN)(MN)^{-1} = M^{-1}I$$
$$(M^{-1}M)N(MN)^{-1} = M^{-1}I$$
$$IN(MN)^{-1} = M^{-1}$$
$$N(MN)^{-1} = M^{-1}$$
$$N^{-1}N(MN)^{-1} = N^{-1}M^{-1}$$
$$I(MN)^{-1} = N^{-1}M^{-1}$$
$$(MN)^{-1} = N^{-1}M^{-1}$$

63.
$$MN = M$$
$$M^{-1}(MN) = M^{-1}M$$
$$(M^{-1}M)N = M^{-1}M$$
$$I_3N = I_3$$
$$N = I_3$$

PROBLEM SET 9.4

1.
$$\begin{vmatrix} 3 & 1 \\ 2 & 4 \end{vmatrix} = (3)(4) - (1)(2) = 12 - 2 = 10$$

3.
$$\begin{vmatrix} -2 & 4 \\ 1 & 3 \end{vmatrix} = (-2)(3) - (4)(1) = -6 - 4 = -10$$

5.
$$\begin{vmatrix} 6 & -3 \\ 2 & 5 \end{vmatrix} = (6)(5) - (-3)(2) = 30 + 6 = 36$$

7.
$$\begin{vmatrix} 6 & 8 \\ -2 & -1 \end{vmatrix} = (6)(-1) - (8)(-2) = -6 + 16 = 10$$

9.
$$\begin{vmatrix} 8 & -3 \\ 4 & -5 \end{vmatrix} = (8)(-5) - (-3)(4) = -40 + 12 = -28$$

11.
$$\begin{vmatrix} \sqrt{2} & 3 \\ 1 & \sqrt{2} \end{vmatrix} = (\sqrt{2})(\sqrt{2}) - (3)(1) = 2 - 3 = -1$$

13.
$$\begin{vmatrix} \sin\theta & -\cos\theta \\ \cos\theta & \sin\theta \end{vmatrix} = (\sin\theta)(\sin\theta) - (-\cos\theta)(\cos\theta) = \sin^2\theta + \cos^2\theta = 1$$

15. $\begin{vmatrix} \csc\theta & 1 \\ 1 & \csc\theta \end{vmatrix} = (\csc\theta)(\csc\theta) - (1)(1) = \csc^2\theta - 1 = \cot^2\theta$

17. $x = \dfrac{\begin{vmatrix} 1 & 1 \\ -5 & 1 \end{vmatrix}}{\begin{vmatrix} 1 & 1 \\ 3 & 1 \end{vmatrix}} = \dfrac{1+5}{1-3} = \dfrac{6}{-2} = -3$

$y = \dfrac{\begin{vmatrix} 1 & 1 \\ 3 & -5 \end{vmatrix}}{-2} = \dfrac{-5-3}{-2} = \dfrac{-8}{-2} = 4$

$(x,\ y) = (-3,\ 4)$

19. $x = \dfrac{\begin{vmatrix} 2 & 1 \\ 1 & -1 \end{vmatrix}}{\begin{vmatrix} 1 & 1 \\ 2 & -1 \end{vmatrix}} = \dfrac{-2-1}{-1-2} = \dfrac{-3}{-3} = 1$

$y = \dfrac{\begin{vmatrix} 1 & 2 \\ 2 & 1 \end{vmatrix}}{-3} = \dfrac{1-4}{-3} = \dfrac{-3}{-3} = 1$

$(c,\ d) = (1,\ 1)$

21. $s_1 = \dfrac{\begin{vmatrix} 3 & 3 \\ 1 & -2 \end{vmatrix}}{\begin{vmatrix} 2 & 3 \\ 5 & -2 \end{vmatrix}} = \dfrac{-6-3}{-4-15} = \dfrac{-9}{-19} = \dfrac{9}{19}$

$s_2 = \dfrac{\begin{vmatrix} 2 & 3 \\ 5 & 1 \end{vmatrix}}{-19} = \dfrac{2-15}{-19} = \dfrac{-13}{-19} = \dfrac{13}{19}$

$(s_1,\ s_2) = \left(\dfrac{9}{19},\ \dfrac{13}{19}\right)$

23. $x = \dfrac{\begin{vmatrix} 5 & 3 \\ 2 & 2 \end{vmatrix}}{\begin{vmatrix} 4 & 3 \\ 3 & 2 \end{vmatrix}} = \dfrac{10-6}{8-9} = \dfrac{4}{-1} = -4$

$y = \dfrac{\begin{vmatrix} 4 & 5 \\ 3 & 2 \end{vmatrix}}{-1} = \dfrac{8-15}{-1} = \dfrac{-7}{-1} = 7$

$(x,\ y) = (-4,\ 7)$

25. Rewrite the first equation as $2x - 3y = 21$.

$$x = \frac{\begin{vmatrix} 21 & -3 \\ 3 & 3 \end{vmatrix}}{\begin{vmatrix} 2 & -3 \\ 2 & 3 \end{vmatrix}} = \frac{63 + 9}{6 + 6} = \frac{72}{12} = 6$$

$$y = \frac{\begin{vmatrix} 2 & 21 \\ 2 & 3 \end{vmatrix}}{12} = \frac{6 - 42}{12} = \frac{-36}{12} = -3$$

$$(x, \ y) = (6, \ -3)$$

27. Rewrite the second equation as $2x - 3y = 15$.

$$x = \frac{\begin{vmatrix} 9 & -3 \\ 15 & -3 \end{vmatrix}}{\begin{vmatrix} 2 & -3 \\ 2 & -3 \end{vmatrix}} = \frac{-27 + 45}{-6 + 6} = \frac{18}{0}$$

Inconsistent system

29. Rewrite the first equation as $2x - 3y = -15$

$$x = \frac{\begin{vmatrix} -15 & -3 \\ 1 & 7 \end{vmatrix}}{\begin{vmatrix} 2 & -3 \\ 1 & 7 \end{vmatrix}} = \frac{-105 + 3}{14 + 3} = \frac{-102}{17} = -6,$$

$$y = \frac{\begin{vmatrix} 2 & -15 \\ 1 & 1 \end{vmatrix}}{17} = \frac{2 + 15}{17} = \frac{17}{17} = 1$$

$$(x, \ y) = (-6, \ 1)$$

31. Rewrite the second equation as $3x - 2y = 16$

$$x = \frac{\begin{vmatrix} 4 & -3 \\ 16 & -2 \end{vmatrix}}{\begin{vmatrix} 2 & -3 \\ 3 & -2 \end{vmatrix}} = \frac{-8 + 48}{-4 + 9} = \frac{40}{5} = 8$$

$$y = \frac{\begin{vmatrix} 2 & 4 \\ 3 & 16 \end{vmatrix}}{5} = \frac{32 - 12}{5} = \frac{20}{5} = 4$$

$$(x, \ y) = (8, \ 4)$$

33. Rewrite the system as: $\begin{cases} 2x - 3y = 5 \\ 3x - 5y = -2 \end{cases}$

$$x = \frac{\begin{vmatrix} 5 & -3 \\ -2 & -5 \end{vmatrix}}{\begin{vmatrix} 2 & -3 \\ 3 & -5 \end{vmatrix}} = \frac{-25 - 6}{-10 + 9} = \frac{-31}{-1} = 31 \qquad y = \frac{\begin{vmatrix} 2 & 5 \\ 3 & -2 \end{vmatrix}}{-1} = \frac{-4 - 15}{-1} = \frac{-19}{-1} = 19$$

$$(x, \ y) = (31, \ 19)$$

35. Rewrite the system as: $\begin{cases} x - 2y = 8 \\ 2x - 3y = -15 \end{cases}$

$$x = \frac{\begin{vmatrix} 8 & -2 \\ -15 & -3 \end{vmatrix}}{\begin{vmatrix} 1 & -2 \\ 2 & -3 \end{vmatrix}} = \frac{-24 - 30}{-3 + 4} = \frac{-54}{1} = -54 \qquad y = \frac{\begin{vmatrix} 1 & 8 \\ 2 & -15 \end{vmatrix}}{1} = \frac{-15 - 16}{1} = -31$$

$$(x, \ y) = (-54, \ -31)$$

37. $x = \dfrac{\begin{vmatrix} a & 3 \\ b & -5 \end{vmatrix}}{\begin{vmatrix} 2 & 3 \\ 1 & -5 \end{vmatrix}} = \dfrac{-5a - 3b}{-10 - 3} = \dfrac{-5a - 3b}{-13} = \dfrac{5a + 3b}{13}$

$$y = \frac{\begin{vmatrix} 2 & a \\ 1 & b \end{vmatrix}}{-13} = \frac{2b - a}{-13} = \frac{a - 2b}{13} \qquad (x, \ y) = \left(\frac{5a + 3b}{13}, \ \frac{a - 2b}{13} \right)$$

39. $x = \dfrac{\begin{vmatrix} 1 & b \\ 0 & a \end{vmatrix}}{\begin{vmatrix} a & b \\ b & a \end{vmatrix}} = \dfrac{a - 0}{a^2 - b^2} = \dfrac{a}{a^2 - b^2} \qquad y = \dfrac{\begin{vmatrix} a & 1 \\ b & 0 \end{vmatrix}}{a^2 - b^2} = \dfrac{0 - b}{a^2 - b^2} = \dfrac{-b}{a^2 - b^2}$

$$= (x, \ y) = \left(\frac{a}{a^2 - b^2}, \ \frac{-b}{a^2 - b^2} \right)$$

41. $x = \dfrac{\begin{vmatrix} s & d \\ t & f \\ \hline c & d \\ e & f \end{vmatrix}} = \dfrac{sf - td}{cf - ed}$

$y = \dfrac{\begin{vmatrix} c & s \\ e & t \end{vmatrix}}{cf - ed} = \dfrac{ct - es}{cf - ed}$

$(x, \ y) = \left(\dfrac{sf - td}{cf - ed}, \ \dfrac{ct - es}{cf - ed} \right)$

43. $x = \dfrac{\begin{vmatrix} \alpha & b \\ \beta & d \\ \hline a & b \\ c & d \end{vmatrix}} = \dfrac{\alpha d - \beta b}{ad - bc}$

$y = \dfrac{\begin{vmatrix} a & \alpha \\ c & \beta \end{vmatrix}}{ad - bc} = \dfrac{a\beta - c\alpha}{ad - bc}$

$(x, \ y) = \left(\dfrac{\alpha d - \beta b}{ad - bc}, \ \dfrac{a\beta - c\alpha}{ad - bc} \right)$

45. $x = \dfrac{\begin{vmatrix} 108 & .06 \\ 1000 & 1 \\ \hline .12 & .06 \\ 1 & 1 \end{vmatrix}} = \dfrac{108 - 60}{.12 - .06} = \dfrac{48}{.06} = 800$

$y = \dfrac{\begin{vmatrix} .12 & 108 \\ 1 & 1000 \end{vmatrix}}{.06} = \dfrac{120 - 108}{.06} = \dfrac{12}{.06} = 200$

$(x, \ y) = (800, \ 200)$

47. $x = \dfrac{\begin{vmatrix} 228 & .06 \\ 2000 & 1 \\ \hline .12 & .06 \\ 1 & 1 \end{vmatrix}} = \dfrac{228 - 120}{.12 - .06} = \dfrac{108}{.06} = 1800$

$y = \dfrac{\begin{vmatrix} .12 & 228 \\ 1 & 2000 \end{vmatrix}}{.06} = \dfrac{240 - 228}{.06} = \dfrac{12}{.06} = 200$

$(x, \ y) = (1800, \ 200)$

49. $x = \dfrac{\begin{vmatrix} 147 & 1 \\ 24.15 & .1 \\ \hline 1 & 1 \\ .25 & .1 \end{vmatrix}} = \dfrac{14.7 - 24.15}{.1 - .25} = \dfrac{-9.45}{-.15} = 63$

$y = \dfrac{\begin{vmatrix} 1 & 147 \\ .25 & 24.15 \end{vmatrix}}{-.15} = \dfrac{24.15 - 36.75}{-.15} = 84$

$(x, \ y) = (63, \ 84)$

51. $x = \dfrac{\begin{vmatrix} \cos^2 45° & 3 \\ -\sin^2 45° & -1 \end{vmatrix}}{\begin{vmatrix} 2 & 3 \\ 1 & -1 \end{vmatrix}} = \dfrac{\begin{vmatrix} \frac{1}{2} & 3 \\ -\frac{1}{2} & -1 \end{vmatrix}}{-2-3} = \dfrac{-\frac{1}{2}+\frac{3}{2}}{-5} = \dfrac{1}{-5} = -\dfrac{1}{5}$

$y = \dfrac{\begin{vmatrix} 2 & \cos^2 45° \\ 1 & -\sin^2 45° \end{vmatrix}}{-5} = \dfrac{\begin{vmatrix} 2 & \frac{1}{2} \\ 1 & -\frac{1}{2} \end{vmatrix}}{-5} = \dfrac{-1-\frac{1}{2}}{-5} = \dfrac{-\frac{3}{2}}{-5} = \dfrac{3}{10}$

$(x,\ y) = \left(-\dfrac{1}{5},\ \dfrac{3}{10}\right)$

53. $x = \dfrac{\begin{vmatrix} \csc^2 60° & 3 \\ \cot^2 60° & -2 \end{vmatrix}}{\begin{vmatrix} 1 & 3 \\ 1 & -2 \end{vmatrix}} = \dfrac{\begin{vmatrix} \frac{4}{3} & 3 \\ \frac{1}{3} & -2 \end{vmatrix}}{-2-3} = \dfrac{-\frac{8}{3}-\frac{3}{3}}{-5} = \dfrac{-\frac{11}{3}}{-5} = \dfrac{11}{15}$

$y = \dfrac{\begin{vmatrix} 1 & \csc^2 60° \\ 1 & \cot^2 60° \end{vmatrix}}{-5} = \dfrac{\begin{vmatrix} 1 & \frac{4}{3} \\ 1 & \frac{1}{3} \end{vmatrix}}{-5} = \dfrac{\frac{1}{3}-\frac{4}{3}}{-5} = \dfrac{-1}{-5} = \dfrac{1}{5}$

$(x,\ y) = \left(\dfrac{11}{15},\ \dfrac{1}{5}\right)$

55. $|A| = \begin{vmatrix} a & b \\ c & d \end{vmatrix} = ad - bc$

$\begin{vmatrix} c & d \\ a & b \end{vmatrix} = cb - ad = bc - ad = -(ad - bc) = -|A|$

57. $|A| = \begin{vmatrix} a & b \\ c & d \end{vmatrix} = ad - bc$

$\begin{vmatrix} a & a+b \\ c & c+d \end{vmatrix} = a(c+d) - c(a+b) = ac + ad - ac - bc = ad - bc = |A|$

59. $|A| = \begin{vmatrix} a & b \\ c & d \end{vmatrix} = ad - bc$

$\begin{vmatrix} a & b \\ c+ka & d+kb \end{vmatrix} = a(d+kb) - b(c+ka) = ad + akb - bc - akb = ad - bc = |A|$

$\begin{vmatrix} a+kc & b+kd \\ c & d \end{vmatrix} = d(a+kc) - c(b+kd) = ad + kcd - bc - kcd = ad - bc = |A|$

Therefore, $\begin{vmatrix} a & b \\ c+ka & d+kb \end{vmatrix} = \begin{vmatrix} a+kc & b+kd \\ c & d \end{vmatrix} = |A|$

61. $x^2 = \dfrac{\begin{vmatrix} 169 & 12 \\ 12 & 1 \end{vmatrix}}{\begin{vmatrix} 13 & 12 \\ 1 & 1 \end{vmatrix}} = \dfrac{169 - 144}{13 - 12} = \dfrac{25}{1} = 25, \quad \text{so } x = \pm 5$

$y^2 = \dfrac{\begin{vmatrix} 13 & 169 \\ 1 & 12 \end{vmatrix}}{1} = \dfrac{156 - 169}{1} = \dfrac{-13}{1} = -13, \quad \text{since } y^2 = -13, \ y \text{ has no real values.}$

The system is inconsistent.

63. Let $u = \dfrac{1}{x-1}$ and $v = \dfrac{1}{y+2}$, then the system becomes: $\begin{cases} 2u - 5v = 12 \\ 4u - 2v = -12 \end{cases}$

$$u = \frac{\begin{vmatrix} 12 & -5 \\ -12 & -2 \end{vmatrix}}{\begin{vmatrix} 2 & -5 \\ 4 & -2 \end{vmatrix}} = \frac{-24 - 60}{-4 + 20} = \frac{-84}{16} = -\frac{21}{4} \qquad v = \frac{\begin{vmatrix} 2 & 12 \\ 4 & -12 \end{vmatrix}}{16} = \frac{-24 - 48}{16} = \frac{-72}{16} = -\frac{9}{2}$$

$$u = \frac{1}{x-1} \qquad\qquad\qquad v = \frac{1}{y+2}$$

$$-\frac{21}{4} = \frac{1}{x-1} \qquad\qquad -\frac{9}{2} = \frac{1}{y+2}$$

$$-21x + 21 = 4 \qquad\qquad -9y - 18 = 2$$

$$-21x = -17 \qquad\qquad\quad -9y = 20$$

$$x = \frac{17}{21} \qquad\qquad\qquad y = -\frac{20}{9}$$

$$(x,\ y) = \left(\frac{17}{21},\ -\frac{20}{9} \right)$$

PROBLEM SET 9.5

1. $\begin{vmatrix} 3 & 0 & 0 \\ 2 & 1 & 4 \\ 3 & 6 & -1 \end{vmatrix} = (3) \begin{vmatrix} 1 & 4 \\ 6 & -1 \end{vmatrix} - (0) \begin{vmatrix} 2 & 4 \\ 3 & -1 \end{vmatrix} + (0) \begin{vmatrix} 2 & 1 \\ 3 & 6 \end{vmatrix}$ (Using row 1)

$$= (3)(-1 - 24) - (0)(-2 - 12) + (0)(12 - 3)$$
$$= (3)(-25) - (0)(-14) + (0)(9)$$
$$= -75$$

3. $\begin{vmatrix} 4 & -2 & 6 \\ 3 & 1 & 4 \\ 0 & 0 & -2 \end{vmatrix} = (0) \begin{vmatrix} -2 & 6 \\ 1 & 4 \end{vmatrix} - (0) \begin{vmatrix} 4 & 6 \\ 3 & 4 \end{vmatrix} + (-2) \begin{vmatrix} 4 & -2 \\ 3 & 1 \end{vmatrix}$ (Using row 3)

$$= (0)(-8 - 6) - (0)(16 - 18) + (-2)(4 + 6)$$
$$= (0)(-14) - (0)(-2) + (-2)(10)$$
$$= -20$$

$$5. \quad \begin{vmatrix} 1 & -2 & 3 \\ -2 & 0 & 4 \\ 3 & 0 & 5 \end{vmatrix} = -(-2)\begin{vmatrix} -2 & 4 \\ 3 & 5 \end{vmatrix} + (0)\begin{vmatrix} 1 & 3 \\ 3 & 5 \end{vmatrix} - (0)\begin{vmatrix} 1 & 3 \\ -2 & 4 \end{vmatrix} \quad \text{(Using col 2)}$$

$$= -(-2)(-10-12) + (0)(5-9) - (0)(4+6)$$

$$= (2)(-22) + (0)(-4) - (0)(10)$$

$$= -44$$

$$7. \quad \begin{vmatrix} 1 & 1 & 1 \\ 1 & 3 & 2 \\ 1 & -2 & 1 \end{vmatrix} = (1)\begin{vmatrix} 3 & 2 \\ -2 & 1 \end{vmatrix} - (1)\begin{vmatrix} 1 & 2 \\ 1 & 1 \end{vmatrix} + (1)\begin{vmatrix} 1 & 3 \\ 1 & -2 \end{vmatrix} \quad \text{(Using row 1)}$$

$$= (1)(3+4) - (1)(1-2) + (1)(-2-3)$$

$$= 7 + 1 - 5$$

$$= 3$$

$$9. \quad \begin{vmatrix} 1 & 4 & 1 \\ 1 & 4 & 2 \\ 1 & 7 & 1 \end{vmatrix} = (1)\begin{vmatrix} 4 & 2 \\ 7 & 1 \end{vmatrix} - (4)\begin{vmatrix} 1 & 2 \\ 1 & 1 \end{vmatrix} + (1)\begin{vmatrix} 1 & 4 \\ 1 & 7 \end{vmatrix} \quad \text{(Using row 1)}$$

$$= (1)(4-14) - (4)(1-2) + (1)(7-4)$$

$$= -10 + 4 + 3$$

$$= -3$$

$$11. \quad \begin{vmatrix} 2 & -3 & 1 \\ 1 & 14 & -3 \\ 3 & 12 & -1 \end{vmatrix} = (2)\begin{vmatrix} 14 & -3 \\ 12 & -1 \end{vmatrix} - (-3)\begin{vmatrix} 1 & -3 \\ 3 & -1 \end{vmatrix} + (1)\begin{vmatrix} 1 & 14 \\ 3 & 12 \end{vmatrix} \quad \text{(Using row 1)}$$

$$= (2)(-14+36) + (3)(-1+9) + (1)(12-42)$$

$$= (2)(22) + (3)(8) + (1)(-30)$$

$$= 38$$

$$13. \quad \begin{vmatrix} 2 & 4 & 3 \\ -2 & 3 & -2 \\ 4 & 3 & 5 \end{vmatrix} = (2)\begin{vmatrix} 3 & -2 \\ 3 & 5 \end{vmatrix} - (4)\begin{vmatrix} -2 & -2 \\ 4 & 5 \end{vmatrix} + (3)\begin{vmatrix} -2 & 3 \\ 4 & 3 \end{vmatrix} \quad \text{(Using row 1)}$$

$$= (2)(15+6) - (4)(-10+8) + (3)(-6-12)$$

$$= (2)(21) - (4)(-2) + (3)(-18)$$

$$= 42 + 8 - 54$$

$$= -4$$

15.
$$\begin{vmatrix} 4 & 8 & 5 \\ 3 & 2 & 3 \\ 5 & 5 & 4 \end{vmatrix} = (4)\begin{vmatrix} 2 & 3 \\ 5 & 4 \end{vmatrix} - (8)\begin{vmatrix} 3 & 3 \\ 5 & 4 \end{vmatrix} + (5)\begin{vmatrix} 3 & 2 \\ 5 & 5 \end{vmatrix}$$

$$= (4)(8 - 15) - (8)(12 - 15) + (5)(15 - 10)$$
$$= (4)(-7) - (8)(-3) + (5)(5)$$
$$= -28 + 24 + 25$$
$$= 21$$

17.
$$\begin{vmatrix} 5 & 2 & 6 & -11 \\ -3 & 0 & 3 & 1 \\ 4 & 0 & 0 & 6 \\ 5 & 0 & 0 & -1 \end{vmatrix} = -(2)\begin{vmatrix} -3 & 3 & 1 \\ 4 & 0 & 6 \\ 5 & 0 & -1 \end{vmatrix}$$ (Using column 2)

$$= -(2)(-(3))\begin{vmatrix} 4 & 6 \\ 5 & -1 \end{vmatrix}$$ (Using column 2)

$$= (6)(-4 - 30)$$
$$= -204$$

19.
$$\begin{vmatrix} 7 & 2 & -5 & 3 \\ 1 & 0 & -3 & 2 \\ 4 & 0 & 1 & -5 \\ 0 & 0 & 3 & 0 \end{vmatrix} = -(2)\begin{vmatrix} 1 & -3 & 2 \\ 4 & 1 & -5 \\ 0 & 3 & 0 \end{vmatrix}$$ (Using column 2)

$$= -(2)(-(3))\begin{vmatrix} 1 & 2 \\ 4 & -5 \end{vmatrix}$$ (Using row 3)

$$= (6)(-5 - 8)$$
$$= -78$$

237

21.

$$\begin{vmatrix} 6 & 3 & 0 & -2 \\ 4 & 3 & 4 & -1 \\ 1 & 2 & 0 & 5 \\ 3 & -2 & 0 & 5 \end{vmatrix} = -(4) \begin{vmatrix} 6 & 3 & -2 \\ 1 & 2 & 5 \\ 3 & -2 & 5 \end{vmatrix} \begin{matrix} \leftarrow \\ \times (-6) \\ \rightarrow \\ \times (-3) \end{matrix} = -(4) \begin{vmatrix} 0 & -9 & -32 \\ 1 & 2 & 5 \\ 0 & -8 & -10 \end{vmatrix}$$

$$= -(4)(-(1)) \begin{vmatrix} -9 & -32 \\ -8 & -10 \end{vmatrix}$$

$$= (4)(90 - 256)$$

$$= -664$$

23.

$$\begin{vmatrix} 2 & 1 & 2 & 4 \\ 3 & -1 & 2 & 5 \\ -3 & 2 & 3 & -4 \\ -3 & 2 & 8 & -4 \end{vmatrix} \begin{matrix} \rightarrow \\ \leftarrow \times (-1) \end{matrix} = \begin{vmatrix} 2 & 1 & 2 & 4 \\ 3 & -1 & 2 & 5 \\ -3 & 2 & 3 & -4 \\ 0 & 0 & 5 & 0 \end{vmatrix} = -(5) \begin{vmatrix} 2 & 1 & 4 \\ 3 & -1 & 5 \\ -3 & 2 & -4 \end{vmatrix}$$

$$\begin{matrix} \uparrow \quad \downarrow\downarrow \quad \uparrow \\ \times (-2) \quad \times (-4) \end{matrix}$$

$$= -(5) \begin{vmatrix} 0 & 1 & 0 \\ 5 & -1 & 9 \\ -7 & 2 & -12 \end{vmatrix} = -(5)(-(1)) \begin{vmatrix} 5 & 9 \\ -7 & -12 \end{vmatrix}$$

$$= 5(-60 + 63)$$

$$= 15$$

25.

$$\begin{vmatrix} 2 & 1 & 3 & 1 \\ 6 & 3 & -3 & 2 \\ 2 & 0 & 5 & 1 \\ 3 & 5 & -2 & -1 \end{vmatrix} \begin{matrix} \uparrow \quad \uparrow\_\downarrow \\ \times (-5)\downarrow \\ \times (-2) \end{matrix} = \begin{vmatrix} 0 & 1 & -2 & 1 \\ 2 & 3 & -13 & 2 \\ 0 & 0 & 0 & 1 \\ 5 & 5 & 3 & -1 \end{vmatrix} = -(1) \begin{vmatrix} 0 & 1 & -2 \\ 2 & 3 & -13 \\ 5 & 5 & 3 \end{vmatrix}$$

$$\begin{matrix} \downarrow\_\uparrow \\ \times 2 \end{matrix}$$

$$= -(1) \begin{vmatrix} 0 & 1 & 0 \\ 2 & 3 & -7 \\ 5 & 5 & 13 \end{vmatrix} = -(1)(-(1)) \begin{vmatrix} 2 & -7 \\ 5 & 13 \end{vmatrix}$$

$$= (1)(26 + 35)$$

$$= 61$$

27.
$$\begin{vmatrix} 0 & -3 & 5 & 6 & -1 \\ 0 & 0 & 1 & 1 & 3 \\ 5 & -3 & 8 & -5 & 1 \\ 0 & 0 & 2 & 0 & 0 \\ 0 & 0 & 4 & -1 & 2 \end{vmatrix} = (5)\begin{vmatrix} -3 & 5 & 6 & -1 \\ 0 & 1 & 1 & 3 \\ 0 & 2 & 0 & 0 \\ 0 & 4 & -1 & 2 \end{vmatrix} = (5)(-3)\begin{vmatrix} 1 & 1 & 3 \\ 2 & 0 & 0 \\ 4 & -1 & 2 \end{vmatrix}$$

$$= (-15)(-(2))\begin{vmatrix} 1 & 3 \\ -1 & 2 \end{vmatrix} = (30)(2+3) = 150$$

29.
$$\begin{vmatrix} -3 & 4 & 5 & 8 & -9 \\ 0 & 0 & 0 & 0 & 6 \\ 3 & 0 & 0 & -1 & 4 \\ 0 & 0 & 4 & 0 & 9 \\ -3 & 0 & 0 & 1 & 8 \end{vmatrix} = -(4)\begin{vmatrix} 0 & 0 & 0 & 6 \\ 3 & 0 & -1 & 4 \\ 0 & 4 & 0 & 9 \\ -3 & 0 & 1 & 8 \end{vmatrix} = -(4)(-(6))\begin{vmatrix} 3 & 0 & -1 \\ 0 & 4 & 0 \\ -3 & 0 & 1 \end{vmatrix}$$

$$= (24)(4)\begin{vmatrix} 3 & -1 \\ -3 & 1 \end{vmatrix} = 96(3-3) = 0$$

31.
$$\begin{vmatrix} 2 & -1 & 0 & 1 & -1 \\ 3 & 0 & 0 & 2 & 0 \\ 2 & 1 & 3 & 1 & 3 \\ 0 & 0 & 0 & -2 & 0 \\ 1 & 3 & -1 & 2 & 1 \end{vmatrix} = (-2)\begin{vmatrix} 2 & -1 & 0 & -1 \\ 3 & 0 & 0 & 0 \\ 2 & 1 & 3 & 3 \\ 1 & 3 & -1 & 1 \end{vmatrix} = (-2)(-(3))\begin{vmatrix} -1 & 0 & -1 \\ 1 & 3 & 3 \\ 3 & -1 & 1 \end{vmatrix}$$

$$\times\ (-1)$$

$$= (6)\begin{vmatrix} -1 & 0 & 0 \\ 1 & 3 & 2 \\ 3 & -1 & -2 \end{vmatrix} = (6)((-1))\begin{vmatrix} 3 & 2 \\ -1 & -2 \end{vmatrix}$$

$$= (-6)(-6+2) = 24$$

33. $D = \begin{vmatrix} 1 & 1 & 1 \\ 2 & -1 & 1 \\ 1 & -2 & -3 \end{vmatrix} \times 2 = \begin{vmatrix} 1 & 1 & 1 \\ 3 & 0 & 2 \\ 3 & 0 & -1 \end{vmatrix} = -(1) \begin{vmatrix} 3 & 2 \\ 3 & -1 \end{vmatrix}$

$$= -(1)(-3-6) = 9$$

$D_x = \begin{vmatrix} 6 & 1 & 1 \\ 3 & -1 & 1 \\ 6 & -2 & -3 \end{vmatrix} \times 2 = \begin{vmatrix} 6 & 1 & 1 \\ 9 & 0 & 2 \\ 18 & 0 & -1 \end{vmatrix} = -(1) \begin{vmatrix} 9 & 2 \\ 18 & -1 \end{vmatrix}$

$$= -(1)(-9-36) = 45$$

$D_y = \begin{vmatrix} 1 & 6 & 1 \\ 2 & 3 & 1 \\ 1 & 6 & -3 \end{vmatrix} \times(-1) = \begin{vmatrix} 1 & 6 & 1 \\ 2 & 3 & 1 \\ 0 & 0 & -4 \end{vmatrix} = (-4) \begin{vmatrix} 1 & 6 \\ 2 & 3 \end{vmatrix}$

$$= (-4)(3-12) = 36$$

$D_z = \begin{vmatrix} 1 & 1 & 6 \\ 2 & -1 & 3 \\ 1 & -2 & 6 \end{vmatrix} \times(-1) = \begin{vmatrix} 1 & 1 & 6 \\ 2 & -1 & 3 \\ 0 & -3 & 0 \end{vmatrix} = -(-3) \begin{vmatrix} 1 & 6 \\ 2 & 3 \end{vmatrix}$

$$= (3)(3-12) = -27$$

$$x = \frac{D_x}{D} = \frac{45}{9} = 5 \qquad y = \frac{D_y}{D} = \frac{36}{9} = 4 \qquad z = \frac{D_z}{D} = \frac{-27}{9} = -3$$

The solution is $(x,\ y,\ z) = (5,\ 4,\ -3)$.

35. $D = \begin{vmatrix} 1 & 1 & 1 \\ 1 & 3 & 2 \\ 1 & -2 & 1 \end{vmatrix} = \begin{vmatrix} 1 & 1 & 0 \\ 1 & 3 & 1 \\ 1 & -2 & 0 \end{vmatrix} = -(1) \begin{vmatrix} 1 & 1 \\ 1 & -2 \end{vmatrix} = -(1)(-2-1) = 3$

$\times(-1)$

$D_x = \begin{vmatrix} 4 & 1 & 1 \\ 4 & 3 & 2 \\ 7 & -2 & 1 \end{vmatrix} = \begin{vmatrix} 0 & 1 & 0 \\ -8 & 3 & -1 \\ 15 & -2 & 3 \end{vmatrix} = -(1) \begin{vmatrix} -8 & -1 \\ 15 & 3 \end{vmatrix} = -(1)(-24+15) = 9$

$\times(-4) \qquad \times(-1)$

(Continued on next page)

$$D_y = \begin{vmatrix} 1 & 4 & 1 \\ 1 & 4 & 2 \\ 1 & 7 & 1 \end{vmatrix} = \begin{vmatrix} 1 & 4 & 0 \\ 1 & 4 & 1 \\ 1 & 7 & 0 \end{vmatrix} = -(1)\begin{vmatrix} 1 & 4 \\ 1 & 7 \end{vmatrix} = -(1)(7-4) = -3$$

$\times(-1)$

$$D_z = \begin{vmatrix} 1 & 1 & 4 \\ 1 & 3 & 4 \\ 1 & -2 & 7 \end{vmatrix} = \begin{vmatrix} 1 & 1 & 0 \\ 1 & 3 & 0 \\ 1 & -2 & 3 \end{vmatrix} = (3)\begin{vmatrix} 1 & 1 \\ 1 & 3 \end{vmatrix} = (3)(3-1) = 6$$

$\times(-4)$

$$x = \frac{D_x}{D} = \frac{9}{3} = 3 \qquad y = \frac{D_y}{D} = \frac{-3}{3} = -1 \qquad z = \frac{D_z}{D} = \frac{6}{3} = 2$$

The solution is $(x, y, z) = (3, -1, 2)$.

37. $$D = \begin{vmatrix} 1 & 1 & 1 \\ 0 & 3 & -1 \\ 1 & 0 & 0 \end{vmatrix} = (1)\begin{vmatrix} 1 & 1 \\ 3 & -1 \end{vmatrix} = (1)(-1-3) = -4$$

$$D_x = \begin{vmatrix} 3 & 1 & 1 \\ -11 & 3 & -1 \\ 4 & 0 & 0 \end{vmatrix} = (4)\begin{vmatrix} 1 & 1 \\ 3 & -1 \end{vmatrix} = (4)(-1-3) = -16$$

$$D_y = \begin{vmatrix} 1 & 3 & 1 \\ 0 & -11 & -1 \\ 1 & 4 & 0 \end{vmatrix} \times(-1) = \begin{vmatrix} 1 & 3 & 1 \\ 0 & -11 & -1 \\ 0 & 1 & -1 \end{vmatrix} = (1)\begin{vmatrix} -11 & -1 \\ 1 & -1 \end{vmatrix} = (1)(11+1) = 12$$

$$D_z = \begin{vmatrix} 1 & 1 & 3 \\ 0 & 3 & -11 \\ 1 & 0 & 4 \end{vmatrix} \times(-1) = \begin{vmatrix} 1 & 1 & 3 \\ 0 & 3 & -11 \\ 0 & -1 & 1 \end{vmatrix} = (1)\begin{vmatrix} 3 & -11 \\ -1 & 1 \end{vmatrix} = (1)(3-11) = -8$$

$$x = \frac{D_x}{D} = \frac{-16}{-4} = 4 \qquad y = \frac{D_y}{D} = \frac{12}{-4} = -3 \qquad z = \frac{D_z}{D} = \frac{-8}{-4} = 2$$

The solution is $(x, y, z) = (4, -3, 2)$.

39. $D = \begin{vmatrix} 2 & 2 & 3 \\ 2 & 0 & -1 \\ 0 & 3 & 2 \end{vmatrix} = \begin{vmatrix} 8 & 2 & 3 \\ 0 & 0 & -1 \\ 4 & 3 & 2 \end{vmatrix} = -(-1)\begin{vmatrix} 8 & 2 \\ 4 & 3 \end{vmatrix} = (1)(24-8) = 16$

$\times 2$

$D_x = \begin{vmatrix} 1 & 2 & 3 \\ -11 & 0 & -1 \\ 6 & 3 & 2 \end{vmatrix} = \begin{vmatrix} -32 & 2 & 3 \\ 0 & 0 & -1 \\ -16 & 3 & 2 \end{vmatrix} = -(-1)\begin{vmatrix} -32 & 2 \\ -16 & 3 \end{vmatrix} = (1)(-96+32) = -64$

$\times (-11)$

$D_y = \begin{vmatrix} 2 & 1 & 3 \\ 2 & -11 & -1 \\ 0 & 6 & 2 \end{vmatrix} = \begin{vmatrix} 2 & -8 & 3 \\ 2 & -8 & -1 \\ 0 & 0 & 2 \end{vmatrix} = (2)\begin{vmatrix} 2 & -8 \\ 2 & -8 \end{vmatrix} = (2)(-16+16) = 0$

$\times (-3)$

$D_z = \begin{vmatrix} 2 & 2 & 1 \\ 2 & 0 & -11 \\ 0 & 3 & 6 \end{vmatrix} = \begin{vmatrix} 2 & 2 & -3 \\ 2 & 0 & -11 \\ 0 & 3 & 0 \end{vmatrix} = -(3)\begin{vmatrix} 2 & -3 \\ 2 & -11 \end{vmatrix} = -(3)(-22+6) = 48$

$\times (-2)$

$x = \dfrac{D_x}{D} = \dfrac{-64}{16} = -4 \qquad y = \dfrac{D_y}{D} = \dfrac{0}{16} = 0 \qquad z = \dfrac{D_z}{D} = \dfrac{48}{16} = 3$

The solution is $(x,\ y,\ z) = (-4,\ 0,\ 3)$.

41. $D = \begin{vmatrix} 2 & -1 & 1 \\ 3 & -2 & 2 \\ 1 & -1 & 3 \end{vmatrix} = \begin{vmatrix} 2 & 0 & 1 \\ 3 & 0 & 2 \\ 1 & 2 & 3 \end{vmatrix} = -(2)\begin{vmatrix} 2 & 1 \\ 3 & 2 \end{vmatrix} = -(2)(4-3) = -2$

$\times 1$

$D_x = \begin{vmatrix} 4 & -1 & 1 \\ 3 & -2 & 2 \\ 2 & -1 & 3 \end{vmatrix} = \begin{vmatrix} 4 & 0 & 1 \\ 3 & 0 & 2 \\ 2 & 2 & 3 \end{vmatrix} = -(2)\begin{vmatrix} 4 & 1 \\ 3 & 2 \end{vmatrix} = -(2)(8-3) = -10$

$\times 1$

(Continued on next page)

$$D_y = \begin{vmatrix} 2 & 4 & 1 \\ 3 & 3 & 2 \\ 1 & 2 & 3 \end{vmatrix} = \begin{vmatrix} 2 & 0 & 1 \\ 3 & -3 & 2 \\ 1 & 0 & 3 \end{vmatrix} = (-3) \begin{vmatrix} 2 & 1 \\ 1 & 3 \end{vmatrix} = (-3)(6-1) = -15$$

$\downarrow\underline{\qquad}\uparrow$
$\times\,(-2)$

$$D_z = \begin{vmatrix} 2 & -1 & 4 \\ 3 & -2 & 3 \\ 1 & -1 & 2 \end{vmatrix} = \begin{vmatrix} 2 & -1 & 0 \\ 3 & -2 & -3 \\ 1 & -1 & 0 \end{vmatrix} = -(-3) \begin{vmatrix} 2 & -1 \\ 1 & -1 \end{vmatrix} = (3)(-2+1) = -3$$

$\downarrow\underline{\qquad\qquad}\uparrow$
$\times\,(-2)$

$$x = \frac{D_x}{D} = \frac{-10}{-2} = 5 \qquad y = \frac{D_y}{D} = \frac{-15}{-2} = \frac{15}{2} \qquad z = \frac{D_z}{D} = \frac{-3}{-2} = \frac{3}{2}$$

The solution is $(x,\ y,\ z) = (-4,\ 0,\ 3)$.

43. $\begin{vmatrix} x & y & 1 \\ -3 & -2 & 1 \\ 1 & -3 & 1 \end{vmatrix} \quad \times(-1) \quad = 0$

$$\begin{vmatrix} x-1 & y+3 & 0 \\ -4 & 1 & 0 \\ -2 & -3 & 1 \end{vmatrix} = 0$$

$$(1) \begin{vmatrix} x-1 & y+3 \\ -4 & 1 \end{vmatrix} = 0$$

$$(x-1) - (-4)(y+3) = 0$$
$$x - 1 + 4y + 12 = 0$$
$$x + 4y + 11 = 0$$

45. $\begin{vmatrix} x & y & 1 \\ 1 & 5 & 1 \\ -2 & -3 & 1 \end{vmatrix} \quad \times(-1) \quad = 0$

$$\begin{vmatrix} x+2 & y+3 & 0 \\ 3 & 8 & 0 \\ -2 & -3 & 1 \end{vmatrix} = 0$$

$$(1) \begin{vmatrix} x+2 & y+3 \\ 3 & 8 \end{vmatrix} =$$

$$8(x+2) - 3(y+3) = 0$$
$$8x + 16 - 3y - 9 = 0$$
$$8x - 3y + 7 = 0$$

47. $A = \frac{1}{2} \begin{vmatrix} 1 & 1 & 1 \\ -2 & -3 & 1 \\ 11 & -3 & 1 \end{vmatrix} = \frac{1}{2} \begin{vmatrix} 0 & 0 & 1 \\ -3 & -4 & 1 \\ 10 & -4 & 1 \end{vmatrix} = \frac{1}{2}(1) \begin{vmatrix} -3 & -4 \\ 10 & -4 \end{vmatrix} = \frac{1}{2}(1)(12+40) = 26$

$\times\,(-1)$
$\times\,(-1)$

49. $A = \frac{1}{2}\begin{vmatrix} -8 & 0 & 1 \\ 12 & 10 & 1 \\ 4 & -5 & 1 \end{vmatrix} = \frac{1}{2}\begin{vmatrix} 0 & 0 & 1 \\ 20 & 10 & 1 \\ 12 & -5 & 1 \end{vmatrix} = \frac{1}{2}(1)\begin{vmatrix} 20 & 10 \\ 12 & -5 \end{vmatrix}$

$$= \frac{1}{2}(1)(-100 - 120) = -110, \quad |-110| = 110$$

51. $D = \begin{vmatrix} 1 & 1 & -1 & 1 \\ 2 & 0 & 1 & -3 \\ 1 & 2 & 0 & -1 \\ -1 & 0 & 1 & 1 \end{vmatrix} = \begin{vmatrix} 1 & 1 & -1 & 1 \\ 2 & 0 & 1 & -3 \\ -1 & 0 & 2 & -3 \\ -1 & 0 & 1 & 1 \end{vmatrix} = -(1)\begin{vmatrix} 2 & 1 & -3 \\ -1 & 2 & -3 \\ -1 & 1 & 1 \end{vmatrix}$

$$= (-1)\begin{vmatrix} 3 & 1 & -4 \\ 1 & 2 & -5 \\ 0 & 1 & 0 \end{vmatrix} = (-1)(-(1))\begin{vmatrix} 3 & -4 \\ 1 & -5 \end{vmatrix} = (1)(-15 + 4) = -11$$

$D_w = \begin{vmatrix} 7 & 1 & -1 & 1 \\ 1 & 0 & 1 & -3 \\ 4 & 2 & 0 & -1 \\ -4 & 0 & 1 & 1 \end{vmatrix} = \begin{vmatrix} 7 & 1 & -1 & 1 \\ 1 & 0 & 1 & -3 \\ -10 & 0 & 2 & -3 \\ -4 & 0 & 1 & 1 \end{vmatrix} = -(1)\begin{vmatrix} 1 & 1 & -3 \\ -10 & 2 & -3 \\ -4 & 1 & 1 \end{vmatrix}$

$$= \begin{vmatrix} 5 & 1 & -4 \\ -2 & 2 & -5 \\ 0 & 1 & 0 \end{vmatrix} = (-1)(-(1))\begin{vmatrix} 5 & -4 \\ -2 & -5 \end{vmatrix} = (1)(-25 - 8) = -33$$

$D_x = \begin{vmatrix} 1 & 7 & -1 & 1 \\ 2 & 1 & 1 & -3 \\ 1 & 4 & 0 & -1 \\ -1 & -4 & 1 & 1 \end{vmatrix} = \begin{vmatrix} 3 & 8 & 0 & -2 \\ 2 & 1 & 1 & -3 \\ 1 & 4 & 0 & -1 \\ -3 & -5 & 0 & 4 \end{vmatrix} = -(1)\begin{vmatrix} 3 & 8 & -2 \\ 1 & 4 & -1 \\ -3 & -5 & 4 \end{vmatrix}$

$$= -(1)\begin{vmatrix} 3 & -4 & 1 \\ 1 & 0 & 0 \\ -3 & 7 & 1 \end{vmatrix} = -(1)(-(1))\begin{vmatrix} -4 & 1 \\ 7 & 1 \end{vmatrix} = (1)(-4 - 7) = -11$$

(Continued on next page)

$$D_y = \begin{vmatrix} 1 & 1 & 7 & 1 \\ 2 & 0 & 1 & -3 \\ 1 & 2 & 4 & -1 \\ -1 & 0 & -4 & 1 \end{vmatrix} \;\;\times(-2) = \begin{vmatrix} 1 & 1 & 7 & 1 \\ 2 & 0 & 1 & -3 \\ -1 & 0 & -10 & -3 \\ -1 & 0 & -4 & 1 \end{vmatrix} = -(1)\begin{vmatrix} 2 & 1 & -3 \\ -1 & -10 & -3 \\ -1 & -4 & 1 \end{vmatrix}$$

$$\times(-2)\;\;\times 3$$

$$= -(1)\begin{vmatrix} 0 & 1 & 0 \\ 19 & -10 & -33 \\ 7 & -4 & -11 \end{vmatrix} = -(1)(-(1))\begin{vmatrix} 19 & -33 \\ 7 & -11 \end{vmatrix} = (1)(-209+231) = 22$$

$$D_z = \begin{vmatrix} 1 & 1 & -1 & 7 \\ 2 & 0 & 1 & 1 \\ 1 & 2 & 0 & 4 \\ -1 & 0 & 1 & -4 \end{vmatrix} \;\;\times(-2) = \begin{vmatrix} 1 & 1 & -1 & 7 \\ 2 & 0 & 1 & 1 \\ -1 & 0 & 2 & -10 \\ -1 & 0 & 1 & -4 \end{vmatrix} = -(1)\begin{vmatrix} 2 & 1 & 1 \\ -1 & 2 & -10 \\ -1 & 1 & -4 \end{vmatrix}$$

$$\times 1\;\;\times 4$$

$$= -(1)\begin{vmatrix} 3 & 1 & 5 \\ 1 & 2 & -2 \\ 0 & 1 & 0 \end{vmatrix} = -(1)(-(1))\begin{vmatrix} 3 & 5 \\ 1 & -2 \end{vmatrix} = (1)(-6-5) = -11$$

$$w = \frac{D_w}{D} = \frac{-33}{-11} = 3, \quad x = \frac{D_x}{D} = \frac{-11}{-11} = 1, \quad y = \frac{D_y}{D} = \frac{22}{-11} = -2, \quad z = \frac{D_z}{D} = \frac{-11}{-11} = 1$$

$$(w,\ x,\ y,\ z) = (3,\ 1,\ -2,\ 1)$$

53. $$D = \begin{vmatrix} 1 & 3 & -2 & 0 \\ 0 & -1 & 1 & 2 \\ 0 & -1 & -1 & 1 \\ 1 & 0 & 0 & -2 \end{vmatrix} \;\;\times(-1) = \begin{vmatrix} 1 & 3 & -2 & 0 \\ 0 & -1 & 1 & 2 \\ 0 & -1 & -1 & 1 \\ 0 & -3 & 2 & -2 \end{vmatrix} = (1)\begin{vmatrix} -1 & 1 & 2 \\ -1 & -1 & 1 \\ -3 & 2 & -2 \end{vmatrix}$$

$$\times 1$$

$$= (1)\begin{vmatrix} -1 & 1 & 3 \\ -1 & -1 & 0 \\ -3 & 2 & 0 \end{vmatrix} = (1)((3))\begin{vmatrix} -1 & -1 \\ -3 & 2 \end{vmatrix} = (3)(-2-3) = -15$$

$$D_s = \begin{vmatrix} 4 & 3 & -2 & 0 \\ -5 & -1 & 1 & 2 \\ 0 & -1 & -1 & 1 \\ -1 & 0 & 0 & -2 \end{vmatrix} = \begin{vmatrix} 4 & 3 & -2 & -8 \\ -5 & -1 & 1 & 12 \\ 0 & -1 & -1 & 1 \\ -1 & 0 & 0 & 0 \end{vmatrix} = -(-1)\begin{vmatrix} 3 & -2 & -8 \\ -1 & 1 & 12 \\ -1 & -1 & 1 \end{vmatrix}$$

$$\times(-2)$$

$$\times 2$$

$$\times 1$$

(Continued on next page)

$$= (1) \begin{vmatrix} 1 & 0 & 16 \\ -1 & 1 & 12 \\ -2 & 0 & 13 \end{vmatrix} = (1)(1) \begin{vmatrix} 1 & 16 \\ -2 & 13 \end{vmatrix} = (1)(13+32) = 45$$

$$D_t = \begin{vmatrix} 1 & 4 & -2 & 0 \\ 0 & -5 & 1 & 2 \\ 0 & 0 & -1 & 1 \\ 1 & -1 & 0 & -2 \end{vmatrix} \times (-1) = \begin{vmatrix} 1 & 4 & -2 & 0 \\ 0 & -5 & 1 & 2 \\ 0 & 0 & -1 & 1 \\ 0 & -5 & 2 & -2 \end{vmatrix} = (1) \begin{vmatrix} -5 & 1 & 2 \\ 0 & -1 & 1 \\ -5 & 2 & -2 \end{vmatrix}$$

$$\times 1$$

$$= (1) \begin{vmatrix} -5 & 1 & 3 \\ 0 & -1 & 0 \\ -5 & 2 & 0 \end{vmatrix} = (1)(3) \begin{vmatrix} 0 & -1 \\ -5 & 2 \end{vmatrix} = (3)(0-5) = -15$$

$$D_u = \begin{vmatrix} 1 & 3 & 4 & 0 \\ 0 & -1 & -5 & 2 \\ 0 & -1 & 0 & 1 \\ 1 & 0 & -1 & -2 \end{vmatrix} \times (-1) = \begin{vmatrix} 1 & 3 & 4 & 0 \\ 0 & -1 & -5 & 2 \\ 0 & -1 & 0 & 1 \\ 0 & -3 & -5 & -2 \end{vmatrix} = (1) \begin{vmatrix} -1 & -5 & 2 \\ -1 & 0 & 1 \\ -3 & -5 & -2 \end{vmatrix}$$

$$\times 1$$

$$= (1) \begin{vmatrix} 1 & -5 & 2 \\ 0 & 0 & 1 \\ -5 & -5 & -2 \end{vmatrix} = (1)(-(1)) \begin{vmatrix} 1 & -5 \\ -5 & -5 \end{vmatrix} = (-1)(-5-25) = 30$$

$$D_x = \begin{vmatrix} 1 & 3 & -2 & 4 \\ 0 & -1 & 1 & -5 \\ 0 & -1 & -1 & 0 \\ 1 & 0 & 0 & -1 \end{vmatrix} \times (-1) = \begin{vmatrix} 1 & 3 & -2 & 4 \\ 0 & -1 & 1 & -5 \\ 0 & -1 & -1 & 0 \\ 0 & -3 & 2 & -5 \end{vmatrix} = (1) \begin{vmatrix} -1 & 1 & -5 \\ -1 & -1 & 0 \\ -3 & 2 & -5 \end{vmatrix}$$

$$\times (-1)$$

$$= (1) \begin{vmatrix} -2 & 1 & -5 \\ 0 & -1 & 0 \\ -5 & 2 & -5 \end{vmatrix} = (1)(-1) \begin{vmatrix} -2 & -5 \\ -5 & -5 \end{vmatrix} = (-1)(10-25) = 15$$

$$s = \frac{D_s}{D} = \frac{45}{-15} = -3, \quad t = \frac{D_t}{D} = \frac{-15}{-15} = 1, \quad u = \frac{D_u}{D} = \frac{30}{-15} = -2, \quad x = \frac{D_x}{D} = \frac{15}{-15} = -1$$

$$(s, \; t, \; u, \; x) = (-3, \; 1, \; -2, \; -1)$$

246

55.

$$\begin{vmatrix} x & y & 1 \\ x_1 & y_1 & 1 \\ x_2 & y_2 & 1 \end{vmatrix} = x \begin{vmatrix} y_1 & 1 \\ y_2 & 1 \end{vmatrix} - y \begin{vmatrix} x_1 & 1 \\ x_2 & 1 \end{vmatrix} + (1) \begin{vmatrix} x_1 & y_1 \\ x_2 & y_2 \end{vmatrix}$$

$$= x(y_1 - y_2) - y(x_1 - x_2) + (x_1 y_2 - y_1 x_2) = 0$$

Solving for y results in: $y = \left(\dfrac{y_1 - y_2}{x_1 - x_2}\right)x + \dfrac{x_1 y_2 - y_1 x_2}{x_1 - x_2}$. Clearly this is a linear equation, so

we need only show that each point satisfies the equation.

$$y_1 \overset{?}{=} \left(\dfrac{y_1 - y_2}{x_1 - x_2}\right)x_1 + \dfrac{x_1 y_2 - y_1 x_2}{x_1 - x_2}$$

$$y_1(x_1 - x_2) \overset{?}{=} (y_1 - y_2)x_1 + (x_1 y_2 - y_1 x_2)$$

$$y_1 x_1 - y_1 x_2 \overset{?}{=} y_1 x_1 - y_2 x_1 + x_1 y_2 - y_1 x_2$$

$$y_1 x_1 - y_1 x_2 = y_1 x_1 - y_1 x_2 \quad \checkmark$$

$$y_2 \overset{?}{=} \left(\dfrac{y_1 - y_2}{x_1 - x_2}\right)x_2 + \dfrac{x_1 y_2 - y_1 x_2}{x_1 - x_2}$$

$$y_2(x_1 - x_2) \overset{?}{=} (y_1 - y_2)x_2 + (x_1 y_2 - y_1 x_2)$$

$$y_2 x_1 - y_2 x_2 \overset{?}{=} y_1 x_2 - y_2 x_2 + x_1 y_2 - y_1 x_2$$

$$y_2 x_1 - y_2 x_2 = x_1 y_2 - y_2 x_2 \quad \checkmark$$

57. The fourth roots of 1 are 1, -1, i, and $-i$.

$$\begin{vmatrix} 1 & -1 & i & -i \\ -1 & i & -i & 1 \\ i & -i & 1 & -1 \\ -i & 1 & -1 & i \end{vmatrix} = \begin{vmatrix} 1 & -1 & i & -i \\ 0 & i-1 & 0 & 1-i \\ i & -i & 1 & -1 \\ 0 & 1-i & 0 & i-1 \end{vmatrix}$$

$$= (1) \begin{vmatrix} i-1 & 0 & 1-i \\ -i & 1 & -1 \\ 1-i & 0 & i-1 \end{vmatrix} + (i) \begin{vmatrix} -1 & i & -i \\ i-1 & 0 & 1-i \\ 1-i & 0 & i-1 \end{vmatrix}$$

$$= (1)(1) \begin{vmatrix} i-1 & 1-i \\ 1-i & i-1 \end{vmatrix} + (i)(-i) \begin{vmatrix} i-1 & 1-i \\ 1-i & i-1 \end{vmatrix}$$

$$= (i-1)^2 - (1-i)^2 - i^2\left[(i-1)^2 - (1-i)^2\right]$$

$$= (i-1)^2 - (1-i)^2 + (i-1)^2 - (1-i)^2$$

$$= i^2 - 2i + 1 - 1 + 2i - i^2 + i^2 - 2i + 1 - 1 + 2i - i^2 = 0$$

247

59. $|A| = \begin{vmatrix} a_{11} & a_{12} & a_{13} \\ a_{21} & a_{22} & a_{23} \\ a_{31} & a_{32} & a_{33} \end{vmatrix}$

Expanding along the second row:

$$= -a_{21}\begin{vmatrix} a_{12} & a_{13} \\ a_{32} & a_{33} \end{vmatrix} + a_{22}\begin{vmatrix} a_{11} & a_{13} \\ a_{31} & a_{33} \end{vmatrix} - a_{23}\begin{vmatrix} a_{11} & a_{12} \\ a_{31} & a_{32} \end{vmatrix}$$

$$= -a_{21}(a_{12}a_{33} - a_{13}a_{32}) + a_{22}(a_{11}a_{33} - a_{13}a_{31}) - a_{23}(a_{11}a_{32} - a_{12}a_{31})$$

$$= -a_{21}a_{12}a_{33} + a_{21}a_{13}a_{32} + a_{22}a_{11}a_{33} - a_{22}a_{13}a_{31} - a_{23}a_{11}a_{32} + a_{23}a_{12}a_{31})$$

$$= -a_{12}(a_{21}a_{33} - a_{23}a_{31}) + a_{22}(a_{11}a_{33} - a_{13}a_{31}) - a_{32}(a_{11}a_{23} - a_{13}a_{21})$$

$$= -a_{12}\begin{vmatrix} a_{21} & a_{23} \\ a_{31} & a_{33} \end{vmatrix} + a_{22}\begin{vmatrix} a_{11} & a_{13} \\ a_{31} & a_{33} \end{vmatrix} - a_{32}\begin{vmatrix} a_{11} & a_{13} \\ a_{13} & a_{23} \end{vmatrix} \qquad \text{Expansion along the second column}$$

61. $|A| = \begin{vmatrix} a_{11} & a_{12} & a_{13} \\ a_{11} & a_{12} & a_{13} \\ a_{31} & a_{32} & a_{33} \end{vmatrix}$ Expand along the third row.

$$= a_{31}\begin{vmatrix} a_{12} & a_{13} \\ a_{12} & a_{13} \end{vmatrix} - a_{32}\begin{vmatrix} a_{11} & a_{13} \\ a_{11} & a_{13} \end{vmatrix} + a_{33}\begin{vmatrix} a_{11} & a_{12} \\ a_{11} & a_{12} \end{vmatrix}$$

$$= a_{31}(a_{12}a_{13} - a_{12}a_{13}) - a_{32}(a_{11}a_{13} - a_{11}a_{13}) + a_{33}(a_{11}a_{12} - a_{11}a_{12})$$

$$= a_{31}(0) - a_{32}(0) + a_{33}(0)$$

$$= 0$$

63. $|A| = \begin{vmatrix} a_{11} & a_{12} & a_{13} \\ a_{21} & a_{22} & a_{23} \\ a_{31} & a_{32} & a_{33} \end{vmatrix}$ Expanding along the first row

$$|A| = a_{11} \begin{vmatrix} a_{22} & a_{23} \\ a_{32} & a_{33} \end{vmatrix} - a_{12} \begin{vmatrix} a_{21} & a_{23} \\ a_{31} & a_{33} \end{vmatrix} + a_{13} \begin{vmatrix} a_{21} & a_{22} \\ a_{31} & a_{32} \end{vmatrix}$$

$$= a_{11}(a_{22}a_{33} - a_{23}a_{32}) - a_{12}(a_{21}a_{33} - a_{23}a_{31}) + a_{13}(a_{21}a_{32} - a_{22}a_{31})$$

$$= a_{11}a_{22}a_{33} - a_{11}a_{23}a_{32} - a_{12}a_{21}a_{33} + a_{12}a_{23}a_{31} + a_{13}a_{21}a_{32} - a_{13}a_{22}a_{31}$$

$\begin{vmatrix} a_{31} & a_{32} & a_{33} \\ a_{21} & a_{22} & a_{23} \\ a_{11} & a_{12} & a_{13} \end{vmatrix}$ Expand along the third row

$$= a_{11} \begin{vmatrix} a_{32} & a_{33} \\ a_{22} & a_{23} \end{vmatrix} - a_{12} \begin{vmatrix} a_{31} & a_{33} \\ a_{21} & a_{23} \end{vmatrix} + a_{13} \begin{vmatrix} a_{31} & a_{32} \\ a_{21} & a_{22} \end{vmatrix}$$

$$= a_{11}(a_{32}a_{23} - a_{33}a_{22}) - a_{12}(a_{31}a_{23} - a_{33}a_{21}) + a_{13}(a_{31}a_{22} - a_{32}a_{21})$$

$$= a_{11}a_{32}a_{23} - a_{11}a_{33}a_{22} - a_{12}a_{31}a_{23} + a_{12}a_{33}a_{21} + a_{13}a_{31}a_{22} - a_{13}a_{32}a_{21}$$

$$= -a_{11}a_{33}a_{22} + a_{11}a_{32}a_{23} + a_{12}a_{33}a_{21} - a_{12}a_{31}a_{23} - a_{13}a_{32}a_{21} + a_{13}a_{31}a_{22}$$

$$= -|A|$$

PROBLEM SET 9.6

1.

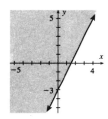

3.

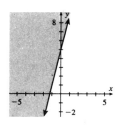

5.

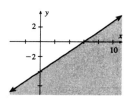

7.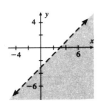

9.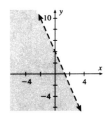

11.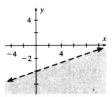

13.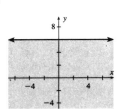

15.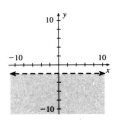

17.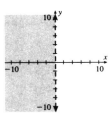

19.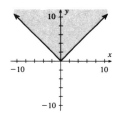

21.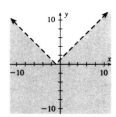

23.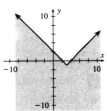

25.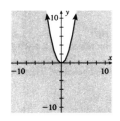

27.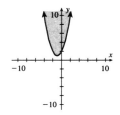

29.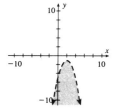

31.

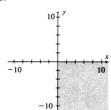

33.

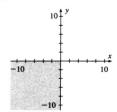

35.

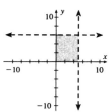

37.

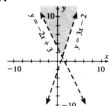

39.

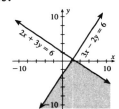

41.

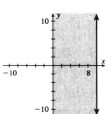

43.

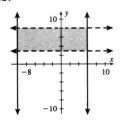

45.

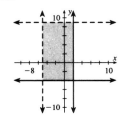

47.

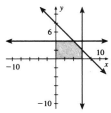

49.

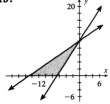

51.

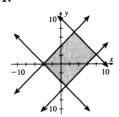

53.

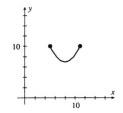

55.

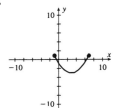

57.

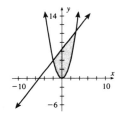

59.

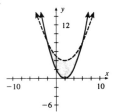

61.

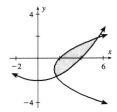

CHAPTER 9 SUMMARY

1.

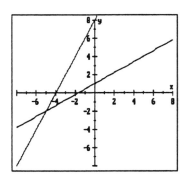

$$(x, y) = (-5, -2)$$

3.

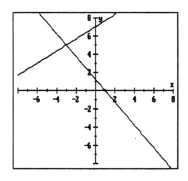

$$(x, y) = (-3, 5)$$

5.
$$\begin{cases} 4x + 3y = -18 \\ y = -\frac{2}{3}x - 2 \end{cases}$$

$$4x + 3\left(-\frac{2}{3}x - 2\right) = -18$$

$$4x - 2x - 6 = -18$$

$$2x = -12$$

$$x = -6$$

$$y = -\frac{2}{3}(-6) - 2 = 2$$

$$(x,\ y) = (-6,\ 2)$$

7.
$$\begin{cases} x + 2y = 26 \\ 5x - 2y = -122 \end{cases}$$

$$x = 26 - 2y$$

$$5(26 - 2y) - 2y = -122$$

$$130 - 10y - 2y = -122$$

$$-12y = -252$$

$$y = 21$$

$$x = 26 - 2(21) = -16$$

$$(x,\ y) = (-16,\ 21)$$

9.
$$\begin{array}{r} -2 \\ 3 \end{array} \begin{cases} 2x + 3y = 6 \\ 3x + 2y = -1 \end{cases}$$

$$+ \begin{cases} -4x - 6y = -12 \\ 9x + 6y = -3 \end{cases}$$

$$5x \quad = -15$$

$$x = -3$$

$$3(-3) + 2y = -1$$

$$2y = 8$$

$$y = 4$$

$$(x,\ y) = (-3,\ 4)$$

11.
$$\begin{array}{r} \\ -2 \end{array} \begin{cases} 4x + 3y = -7 \\ 2x - 5y = 55 \end{cases}$$

$$+ \begin{cases} 4x + 3y = -7 \\ -4x + 10y = -110 \end{cases}$$

$$13y = -117$$

$$y = -9$$

$$2x - 5(-9) = 55$$

$$2x = 10$$

$$x = 5$$

$$(x,\ y) = (5,\ -9)$$

13.
$$\left[\begin{array}{ccc:c} 2 & 3 & -1 & -2 \\ 1 & 1 & -5 & -3 \\ 5 & -7 & -10 & 60 \end{array}\right] \xrightarrow{R1 \leftrightarrow R2} \left[\begin{array}{ccc:c} 1 & 1 & -5 & -3 \\ 2 & 3 & -1 & -2 \\ 5 & -7 & -10 & 60 \end{array}\right]$$

$$\xrightarrow[-5R1+R3]{-2R1+R2} \left[\begin{array}{ccc:c} 1 & 1 & -5 & -3 \\ 0 & 1 & 9 & 4 \\ 0 & -12 & 15 & 75 \end{array}\right] \xrightarrow[12R2+R3]{-R2+R1} \left[\begin{array}{ccc:c} 1 & 0 & -14 & -7 \\ 0 & 1 & 9 & 4 \\ 0 & 0 & 123 & 123 \end{array}\right]$$

$$\xrightarrow{R3 \div 123} \left[\begin{array}{ccc:c} 1 & 0 & -14 & -7 \\ 0 & 1 & 9 & 4 \\ 0 & 0 & 1 & 1 \end{array}\right] \xrightarrow[-4R3+R2]{14R3+R1} \left[\begin{array}{ccc:c} 1 & 0 & 0 & 7 \\ 0 & 1 & 0 & -5 \\ 0 & 0 & 1 & 1 \end{array}\right]$$

$$(x,\ y,\ z) = (7,\ -5,\ 1)$$

15.
$$\begin{bmatrix} 2 & -1 & 1 & \vdots & -3 \\ 3 & 1 & -2 & \vdots & 11 \\ 5 & -2 & 3 & \vdots & -8 \end{bmatrix} \xrightarrow{-2R1+R3} \begin{bmatrix} 2 & -1 & 1 & \vdots & -3 \\ 3 & 1 & -2 & \vdots & 11 \\ 1 & 0 & 1 & \vdots & -2 \end{bmatrix}$$

$$\xrightarrow{R1 \leftrightarrow R3} \begin{bmatrix} 1 & 0 & 1 & \vdots & -2 \\ 3 & 1 & -2 & \vdots & 11 \\ 2 & -1 & 1 & \vdots & -3 \end{bmatrix} \xrightarrow[-2R1+R3]{-3R1+R2} \begin{bmatrix} 1 & 0 & 1 & \vdots & -2 \\ 0 & 1 & -5 & \vdots & 17 \\ 0 & -1 & -1 & \vdots & 1 \end{bmatrix}$$

$$\xrightarrow{R2+R3} \begin{bmatrix} 1 & 0 & 1 & \vdots & -2 \\ 0 & 1 & -5 & \vdots & 17 \\ 0 & 0 & -6 & \vdots & 18 \end{bmatrix} \xrightarrow{R3 \div (-6)} \begin{bmatrix} 1 & 0 & 1 & \vdots & -2 \\ 0 & 1 & -5 & \vdots & 17 \\ 0 & 0 & 1 & \vdots & -3 \end{bmatrix}$$

$$\xrightarrow[5R3+R2]{-R3+R1} \begin{bmatrix} 1 & 0 & 0 & \vdots & 1 \\ 0 & 1 & 0 & \vdots & 2 \\ 0 & 0 & 1 & \vdots & -3 \end{bmatrix} \qquad (x,\ y,\ z) = (1,\ 2,\ -3)$$

17.
$$\begin{bmatrix} 3 & -2 & 1 \\ -2 & 1 & 4 \\ 1 & -3 & 1 \end{bmatrix} + \begin{bmatrix} 2 & 1 & 0 \\ -1 & 7 & 3 \\ 2 & -3 & 5 \end{bmatrix} = \begin{bmatrix} 3+2 & -2+1 & 1+0 \\ -2+(-1) & 1+7 & 4+3 \\ 1+2 & -3+(-3) & 1+5 \end{bmatrix}$$

$$= \begin{bmatrix} 5 & -1 & 1 \\ -3 & 8 & 7 \\ 3 & -6 & 6 \end{bmatrix}$$

19.
$$\begin{bmatrix} 3 & -2 & 1 \\ -2 & 1 & 4 \\ 1 & -3 & 1 \end{bmatrix} \cdot \begin{bmatrix} 2 & 1 & 0 \\ -1 & 7 & 3 \\ 2 & -3 & 5 \end{bmatrix} = \begin{bmatrix} 6+2+2 & 3-14-3 & 0-6+5 \\ -4-1+8 & -2+7-12 & 0+3+20 \\ 2+3+2 & 1-21-3 & 0-9+5 \end{bmatrix}$$

$$= \begin{bmatrix} 10 & -14 & -1 \\ 3 & -7 & 23 \\ 7 & -23 & -4 \end{bmatrix}$$

21.
$$\begin{bmatrix} 1 & -2 & | & 1 & 0 \\ -3 & 7 & | & 0 & 1 \end{bmatrix} \xrightarrow{} \begin{bmatrix} 1 & -2 & | & 1 & 0 \\ 0 & 1 & | & 3 & 1 \end{bmatrix}$$
$$3R1 + R2$$

$$\xrightarrow{} \begin{bmatrix} 1 & 0 & | & 7 & 2 \\ 0 & 1 & | & 3 & 1 \end{bmatrix} \qquad \text{The inverse is } \begin{bmatrix} 7 & 2 \\ 3 & 1 \end{bmatrix}$$
$$2R2 + R1$$

23.
$$\begin{bmatrix} 1 & 1 & | & 1 & 0 \\ 1 & 1 & | & 0 & 1 \end{bmatrix} \xrightarrow{} \begin{bmatrix} 1 & 1 & | & 1 & 0 \\ 0 & 0 & | & -1 & 1 \end{bmatrix}$$
$$-R1 + R2$$

The inverse does not exist.

25. $X = A^{-1}B$
$$\begin{bmatrix} x \\ y \\ z \end{bmatrix} = \begin{bmatrix} 1 & 0 & 1 \\ 2 & 1 & 0 \\ 0 & 1 & -1 \end{bmatrix} \cdot \begin{bmatrix} 6 \\ -5 \\ -5 \end{bmatrix} = \begin{bmatrix} 6+0-5 \\ 12-5+0 \\ 0-5+5 \end{bmatrix} = \begin{bmatrix} 1 \\ 7 \\ 0 \end{bmatrix}$$

$$(x,\ y,\ z) = (1,\ 7,\ 0)$$

27. $X = A^{-1}B$
$$\begin{bmatrix} x \\ y \\ z \end{bmatrix} = \begin{bmatrix} -1 & 1 & -1 \\ 2 & -1 & 2 \\ 2 & -1 & 1 \end{bmatrix} \cdot \begin{bmatrix} 11 \\ 14 \\ -5 \end{bmatrix} = \begin{bmatrix} -11+14+5 \\ 22-14-10 \\ 22-14-5 \end{bmatrix} = \begin{bmatrix} 8 \\ -2 \\ 3 \end{bmatrix}$$

$$(x,\ y,\ z) = (8,\ -2,\ 3)$$

29. $\quad x = \dfrac{\begin{vmatrix} 2 & -1 \\ -3 & 5 \end{vmatrix}}{\begin{vmatrix} 3 & -1 \\ 1 & 5 \end{vmatrix}} = \dfrac{10-3}{15+1} = \dfrac{7}{16}$
$\qquad\qquad y = \dfrac{\begin{vmatrix} 3 & 2 \\ 1 & -3 \end{vmatrix}}{16} = \dfrac{-9-2}{16} = -\dfrac{11}{16}$

$$(x,\ y) = \left(\frac{7}{16},\ -\frac{11}{16}\right)$$

31. $x = \dfrac{\begin{vmatrix} 0 & 4 \\ 1 & 8 \end{vmatrix}}{\begin{vmatrix} 3 & 4 \\ 6 & 8 \end{vmatrix}} = \dfrac{0-4}{24-24} = \dfrac{-4}{0}$ Inconsistent System

33. $\begin{vmatrix} 1 & 3 & -2 \\ 4 & 5 & 1 \\ 3 & -2 & 4 \end{vmatrix} \begin{matrix} \leftarrow \\ \times\ 2 \\ \rightarrow \\ \times\ (-4) \end{matrix} = \begin{vmatrix} 9 & 13 & 0 \\ 4 & 5 & 1 \\ -13 & -22 & 0 \end{vmatrix} = (-1)\begin{vmatrix} 9 & 13 \\ -13 & -22 \end{vmatrix}$

$$= (-1)(-198 + 169) = 29$$

35. $\begin{vmatrix} 3 & -2 & 0 \\ 2 & 5 & 8 \\ 5 & 3 & 5 \end{vmatrix} = (0)\begin{vmatrix} 2 & 5 \\ 5 & 3 \end{vmatrix} - (8)\begin{vmatrix} 3 & -2 \\ 5 & 3 \end{vmatrix} + (5)\begin{vmatrix} 3 & -2 \\ 2 & 5 \end{vmatrix}$

$$= (0)(6-25) - (8)(9+10) + 5(15+4) = 0 - 152 + 95 = -57$$

37. $D = \begin{vmatrix} 3 & -2 & 1 \\ 2 & 5 & -3 \\ 1 & -3 & 2 \end{vmatrix} = \begin{vmatrix} 3 & 7 & -5 \\ 2 & 11 & -7 \\ 1 & 0 & 0 \end{vmatrix} = (1)\begin{vmatrix} 7 & -5 \\ 11 & -7 \end{vmatrix} = (1)(-49+55) = 6$

$D_x = \begin{vmatrix} 9 & -2 & 1 \\ 17 & 5 & -3 \\ -2 & -3 & 2 \end{vmatrix} \times (-2) = \begin{vmatrix} 9 & -2 & 1 \\ 44 & -1 & 0 \\ -20 & 1 & 0 \end{vmatrix}$

$$= (1)\begin{vmatrix} 44 & -1 \\ -20 & 1 \end{vmatrix} = (1)(44-20) = 24$$

$D_y = \begin{vmatrix} 3 & 9 & 1 \\ 2 & 17 & -3 \\ 1 & -2 & 2 \end{vmatrix} \times (-2) = \begin{vmatrix} 3 & 9 & 1 \\ 11 & 44 & 0 \\ -5 & -20 & 0 \end{vmatrix}$

$$= (1)\begin{vmatrix} 11 & 44 \\ -5 & -20 \end{vmatrix} = (1)(-220+220) = 0$$

(Continued on next page)

$$D_z = \begin{vmatrix} 3 & -2 & 9 \\ 2 & 5 & 17 \\ 1 & -3 & -2 \end{vmatrix} = \begin{bmatrix} 3 & 7 & 15 \\ 2 & 11 & 21 \\ 1 & 0 & 0 \end{bmatrix} = (1)\begin{vmatrix} 7 & 15 \\ 11 & 21 \end{vmatrix} = (1)(147 - 165) = -18$$

（×3, ×2 annotations under first determinant）

$$x = \frac{D_x}{D} = \frac{24}{6} = 4, \qquad y = \frac{D_y}{D} = \frac{0}{6} = 0, \qquad z = \frac{D_z}{D} = \frac{-18}{6} = -3$$

$$(x, y, z) = (4, 0, -3)$$

39. $$D = \begin{vmatrix} 2 & -1 & 1 \\ 3 & -2 & 1 \\ 1 & 1 & 3 \end{vmatrix} = \begin{vmatrix} 0 & 0 & 1 \\ 1 & -1 & 1 \\ -5 & 4 & 3 \end{vmatrix} = (1)\begin{vmatrix} 1 & -1 \\ -5 & 4 \end{vmatrix} = (1)(4 - 5) = -1$$

（×1, ×(−2) annotations）

$$D_x = \begin{vmatrix} 6 & -1 & 1 \\ 8 & -2 & 1 \\ 11 & 1 & 3 \end{vmatrix} = \begin{vmatrix} 0 & 0 & 1 \\ 2 & -1 & 1 \\ -7 & 4 & 3 \end{vmatrix} = (1)\begin{vmatrix} 2 & -1 \\ -7 & 4 \end{vmatrix} = (1)(8 - 7) = 1$$

（×1, ×(−6) annotations）

$$D_y = \begin{vmatrix} 2 & 6 & 1 \\ 3 & 8 & 1 \\ 1 & 11 & 3 \end{vmatrix} = \begin{vmatrix} 0 & 0 & 1 \\ 1 & 2 & 1 \\ -5 & -7 & 3 \end{vmatrix} = (1)\begin{vmatrix} 1 & 2 \\ -5 & -7 \end{vmatrix} = (1)(-7 + 10) = 3$$

（×(−6), ×(−2) annotations）

$$D_z = \begin{vmatrix} 2 & -1 & 6 \\ 3 & -2 & 8 \\ 1 & 1 & 11 \end{vmatrix} = \begin{vmatrix} 2 & -3 & -16 \\ 3 & -5 & -25 \\ 1 & 0 & 0 \end{vmatrix} = (1)\begin{vmatrix} -3 & -16 \\ -5 & -25 \end{vmatrix} = (1)(75 - 80) = -5$$

（×(−1), ×(−11) annotations）

$$x = \frac{D_x}{D} = \frac{1}{-1} = -1, \qquad y = \frac{D_y}{D} = \frac{3}{-1} = -3, \qquad z = \frac{D_z}{D} = \frac{-5}{-1} = 5$$

$$(x, y, z) = (-1, -3, 5)$$

41.

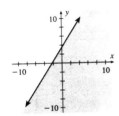

43.

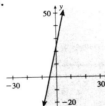

45.

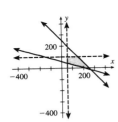

47.

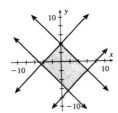

CHAPTER 10

PROBLEM SET 10.1

1. $P(1)$: $5 = 1[2(1) + 3]$; $5 = 5$; TRUE

3. $P(1)$: $2^2 = \dfrac{2 \cdot 1(1+1)(2 \cdot 1 + 1)}{3}$; $2^2 = \dfrac{2(2)(3)}{3}$; $4 = 4$; TRUE

5. $P(1)$: $\cos(\theta + \pi) = (-1)^1 \cos\theta$; $\cos\theta\cos\pi - \sin\theta\sin\pi = -\cos\theta$; $-\cos\theta = -\cos\theta$; TRUE

7. $P(1)$: $1^2 + 1 = 2$ is even; TRUE

9. $P(1)$: $\left(\dfrac{2}{3}\right)^{1+1} < \left(\dfrac{2}{3}\right)^1$; $\left(\dfrac{2}{3}\right)^2 < \left(\dfrac{2}{3}\right)^1$; $\dfrac{4}{9} < \dfrac{2}{3}$; TRUE

11. $P(k)$: $5 + 9 + 13 + \cdots + (4k+1) = k(2k+3)$

    $P(k+1)$: $5 + 9 + 13 + \cdots + [4(k+1)+1] = (k+1)[2(k+1)+3]$

    $5 + 9 + 13 + \cdots + (4k+5) = (k+1)(2k+5)$

13. $P(k)$: $\quad 2^2 + 4^2 + 6^2 + \cdots + (2k)^2 = \dfrac{2k(k+1)(2k+1)}{3}$

$P(k+1)$: $\quad 2^2 + 4^2 + 6^2 + \cdots + [2(k+1)]^2 = \dfrac{2(k+1)[(k+1)+1][2(k+1)+1]}{3}$

$\qquad\qquad 2^2 + 4^2 + 6^2 + \cdots + (2k+2)^2 = \dfrac{(2k+2)(k+2)(2k+3)}{3}$

15. $P(k)$: $\quad \cos(\theta + k\pi) = (-1)^k \cos\theta$

$P(k+1)$: $\quad \cos[\theta + (k+1)\pi] = (-1)^{k+1} \cos\theta$

$\qquad\qquad \cos(\theta + k\pi + \pi) = (-1)^{k+1} \cos\theta$

17. $P(k)$: $\quad k^2 + k$ is even

$P(k+1)$: $\quad (k+1)^2 + (k+1)$ is even

19. $P(k)$: $\quad \left(\dfrac{2}{3}\right)^{k+1} < \left(\dfrac{2}{3}\right)^{k}$

$P(k+1)$: $\quad \left(\dfrac{2}{3}\right)^{(k+1)+1} < \left(\dfrac{2}{3}\right)^{k+1}$

$\qquad\qquad \left(\dfrac{2}{3}\right)^{k+2} < \left(\dfrac{2}{3}\right)^{k+1}$

21. Step 1: Prove $P(1)$: $\quad 1 = \dfrac{1(1+1)}{2}$

$\qquad\qquad\qquad\qquad 1 = 1; \ \text{TRUE}$

Step 2: Assume $P(k)$: Hypothesis $\quad 1 + 2 + 3 + \cdots + k = \dfrac{k(k+1)}{2}$

Step 3: Prove $P(k+1)$: $\quad 1 + 2 + 3 + \cdots + (k+1) = \dfrac{(k+1)(k+2)}{2}$

1.	$1 + 2 + 3 + \cdots + k = \dfrac{k(k+1)}{2}$	1. Hypothesis
2.	$1 + 2 + 3 + \cdots + k + (k+1) = \dfrac{k(k+1)}{2} + (k+1)$	2. Add $(k+1)$ to both sides
3.	$= \dfrac{k(k+1)}{2} + \dfrac{2(k+1)}{2}$	3. Add fractions
4.	$= \dfrac{k^2 + 3k + 2}{2}$	4. Combine like terms
5.	$= \dfrac{(k+1)(k+2)}{2}$	5. Factor

Step 4: The proposition $P(n)$ is true for all positive integers n by PMI.

23. Step 1: Prove $P(1)$: $5 = 1(2 \cdot 1 + 3)$

$\qquad\qquad\qquad\qquad 5 = 5$; TRUE

Step 2: Assume $P(k)$: Hypothesis $5 + 9 + 13 + \cdots + (4k + 1) = k(2k + 3)$

Step 3: Prove $P(k + 1)$: $5 + 9 + 13 + \cdots + [4(k + 1) + 1] = (k + 1)[2(k + 1) + 3]$

$\qquad\qquad\qquad\qquad\qquad 5 + 9 + 13 + \cdots + (4k + 5) = (k + 1)(2k + 5)$

1. $5 + 9 + 13 + \cdots + (4k + 1) = k(2k + 3)$ 1. Hypothesis

2. $5 + 9 + 13 + \cdots + (4k + 1) + (4k + 5)$ 2. Add $(4k + 5)$ to both sides

$\qquad\qquad = k(2k + 3) + (4k + 5)$

3. $\qquad\qquad = 2k^2 + 7k + 5$ 3. Combine like terms

4. $\qquad\qquad = (k + 1)(2k + 5)$ 4. Factor

Step 4: The proposition $P(n)$ is true for all positive integers n by PMI.

25. Step 1: Prove $P(1)$: $2 = \dfrac{1(5 \cdot 1 - 1)}{2}$

$\qquad\qquad\qquad\qquad 2 = 2$; TRUE

Step 2: Assume $P(k)$: Hypothesis $2 + 7 + 12 + \cdots + (5k - 3) = \dfrac{k(5k - 1)}{2}$

Step 3: Prove $P(k + 1)$: $2 + 7 + 12 + \cdots + [5(k + 1) - 3] = \dfrac{(k + 1)[5(k + 1) - 1]}{2}$

$\qquad\qquad\qquad\qquad\qquad 2 + 7 + 12 + \cdots + (5k + 2) = \dfrac{(k + 1)(5k + 4)}{2}$

1. $2 + 7 + 12 + \cdots + (5k - 3) = \dfrac{k(5k - 1)}{2}$ 1. Hypothesis

2. $2 + 7 + 12 + \cdots + (5k - 3) + (5k + 2)$ 2. Add $(5k + 2)$ to both sides

$\qquad\qquad = \dfrac{k(5k - 1)}{2} + (5k + 2)$

3. $\qquad\qquad = \dfrac{k(5k - 1)}{2} + \dfrac{2(5k + 2)}{2}$ 3. Add fractions

4. $\qquad\qquad = \dfrac{5k^2 + 9k + 4}{2}$ 4. Combine like terms

5. $\qquad\qquad = \dfrac{(k + 1)(5k + 4)}{2}$ 5. Factor

Step 4: The proposition $P(n)$ is true for all positive integers n by PMI.

27. Step 1: Prove $P(1)$: $1^2 = \dfrac{1(2 \cdot 1 - 1)(2 \cdot 1 + 1)}{3}$

$$1 = 1; \ \text{TRUE}$$

Step 2: Assume $P(k)$: Hypothesis $\quad 1^2 + 3^2 + 5^2 + \cdots + (2k - 1)^2 = \dfrac{k(2k - 1)(2k + 1)}{3}$

Step 3: Prove $P(k + 1)$: $1^2 + 3^2 + 5^2 + \cdots + [2(k + 1) - 1]^2$

$$= \dfrac{(k + 1)[2(k + 1) - 1][2(k+1) + 1]}{3}$$

$$1^2 + 3^2 + 5^2 + \cdots + (2k + 1)^2 = \dfrac{(k + 1)(2k + 1)(2k + 3)}{3}$$

1. $1^2 + 3^2 + 5^2 + \cdots + (2k - 1)^2 = \dfrac{k(2k - 1)(2k + 1)}{3}$ 1. Hypothesis

2. $1^2 + 3^2 + 5^2 + \cdots + (2k - 1)^2 + (2k + 1)^2$ 2. Add $(2k + 1)^2$ to both sides

$$= \dfrac{k(2k - 1)(2k + 1)}{3} + (2k + 1)^2$$

3. $= \dfrac{k(2k - 1)(2k + 1)}{3} + \dfrac{3(2k + 1)^2}{3}$ 3. Add fractions

4. $= \dfrac{(2k + 1)[k(2k - 1) + 3(2k + 1)]}{3}$ 4. Factor out $(2k + 1)$

5. $= \dfrac{(2k + 1)(2k^2 + 5k + 3)}{3}$ 5. Distribute and combine like terms

6. $= \dfrac{(2k + 1)(2k + 3)(k + 1)}{3}$ 6. Factor

Step 4: The proposition $P(n)$ is true for all positive integers n by PMI.

29. Step 1: Prove $P(1)$: $2^2 = \dfrac{2 \cdot 1(1 + 1)(2 \cdot 1 + 1)}{3}$

$$4 = 4; \ \text{TRUE}$$

Step 2: Assume $P(k)$: Hypothesis $\quad 2^2 + 4^2 + 6^2 + \cdots + (2k)^2 = \dfrac{2k(k + 1)(2k + 1)}{3}$

Step 3: Prove $P(k + 1)$: $2^2 + 4^2 + 6^2 + \cdots + [2(k + 1)]^2$

$$= \dfrac{2(k + 1)[(k + 1) + 1][2(k+1) + 1]}{3}$$

$$2^2 + 4^2 + 6^2 + \cdots + (2k + 2)^2 = \dfrac{(2k + 2)(k + 2)(2k + 3)}{3}$$

(Continued on next page)

1. $2^2 + 4^2 + 6^2 + \cdots + (2k)^2 = \dfrac{2k(k+1)(2k+1)}{3}$ 1. Hypothesis

2. $2^2 + 4^2 + 6^2 + \cdots + (2k)^2 + (2k+2)^2$ 2. Add $(2k+2)^2$ to both sides

$$= \dfrac{2k(k+1)(2k+1)}{3} + (2k+2)^2$$

3. $\quad = \dfrac{2k(k+1)(2k+1)}{3} + \dfrac{3(2k+2)^2}{3}$ 3. Add fractions

4. $\quad = \dfrac{k(2k+2)(2k+1) + 3(2k+2)^2}{3}$ 4. Distrnbute

5. $\quad = \dfrac{(2k+2)[k(2k+1) + 3(2k+2)]}{3}$ 5. Factor out $(2k+2)$

6. $\quad = \dfrac{(2k+2)(2k^2 + 7k + 6)}{3}$ 6. Distribute and combine like terms

7. $\quad = \dfrac{2(k+1)(k+2)(2k+3)}{3}$ 7. Factor

Step 4: The proposition $P(n)$ is true for all positive integers n by PMI.

31. **Step 1:** Prove $P(1)$: $1 \cdot 3 = \dfrac{1(1+1)(2 \cdot 1 + 7)}{6}$

$$3 = 3; \ \text{TRUE}$$

Step 2: Assume $P(k)$: Hypothesis $1 \cdot 3 + 2 \cdot 4 + 3 \cdot 5 + \cdots + k(k+2) = \dfrac{k(k+1)(2k+7)}{6}$

Step 3: Prove $P(k+1)$: $1 \cdot 3 + 2 \cdot 4 + 3 \cdot 5 + \cdots + (k+1)[(k+1)+2]$

$$= \dfrac{(k+1)[(k+1)+1][2(k+1)+7]}{6}$$

$$1 \cdot 3 + 2 \cdot 4 + 3 \cdot 5 + \cdots + (k+1)(k+3) = \dfrac{(k+1)(k+2)(2k+9)}{6}$$

1. $1 \cdot 3 + 2 \cdot 4 + 3 \cdot 5 + \cdots + k(k+2) = \dfrac{k(k+1)(2k+7)}{6}$ 1. Hypothesis

2. $1 \cdot 3 + 2 \cdot 4 + 3 \cdot 5 + \cdots + k(k+2) + (k+1)(k+3)$ 2. Add $(k+1)(k+3)$ to both sides

$$= \dfrac{k(k+1)(2k+7)}{6} + (k+1)(k+3)$$

3. $\quad = \dfrac{k(k+1)(2k+7)}{6} + \dfrac{6(k+1)(k+3)}{6}$ 3. Add fractions

4. $\quad = \dfrac{(k+1)[k(2k+7) + 6(k+3)]}{6}$ 5. Factor out $(k+1)$

6. $\quad = \dfrac{(k+1)(2k^2 + 13k + 18)}{6}$ 6. Distribute and combine like terms

7. $\quad = \dfrac{(k+1)(k+2)(2k+9)}{6}$ 7. Factor

Step 4: The proposition $P(n)$ is true for all positive integers n by PMI.

33. **Step 1:** Prove $P(1)$: $1 = \dfrac{5^1 - 1}{4}$

$1 = 1$; TRUE

Step 2:   Assume $P(k)$: Hypothesis $1 + 5 + 5^2 + \cdots + 5^{k-1} = \dfrac{5^k - 1}{4}$

Step 3:   Prove $P(k+1)$: $1 + 5 + 5^2 + \cdots + 5^{(k+1)-1} = \dfrac{5^{k+1} - 1}{4}$

$1 + 5 + 5^2 + \cdots + 5^k = \dfrac{5^{k+1} - 1}{4}$

1. $1 + 5 + 5^2 + \cdots + 5^k = \dfrac{5^{k-1} - 1}{4}$	1. Hypothesis
2. $1 + 5 + 5^2 + \cdots + 5^{k-1} + 5^k = \dfrac{5^k - 1}{4} + 5^k$	2. Add 5^k to both sides
3. $\qquad = \dfrac{5^k - 1}{4} + \dfrac{4 \cdot 5^k}{4}$	3. Add Fractions
4. $\qquad = \dfrac{5 \cdot 5^k - 1}{4}$	4. Combine like terms
5. $\qquad = \dfrac{5^{k+1} - 1}{4}$	5. Rules of Exponents

Step 4:   The proposition $P(n)$ is true for all positive integers n by PMI.

35. **Step 1:** Prove $P(1)$: $\log a_1 = \log a_1$; TRUE

Step 2:   Assume $P(k)$: Hypothesis $\log(a_1 a_2 \cdots a_k) = \log a_1 + \log a_2 + \cdots + \log a_k$, for $k \geq 2$

Step 3:   Prove $P(k+1)$: $\log(a_1 a_2 \cdots a_k a_{k+1})$

$= \log a_1 + \log a_2 + \cdots + \log a_k + \log a_{k+1}$, for $k \geq 2$

1. $\log(a_1 a_2 \cdots a_k) = \log a_1 + \log a_2 + \cdots + \log a_k$	1. Hypothesis
2. $\log(a_1 a_2 \cdots a_k) + \log_{k+1}$	2. Add $\log a_{k+1}$ to both sides
$\qquad = \log a_1 + \log a_2 + \cdots + \log a_k + \log_{k+1}$	
3. $\log(a_1 a_2 \cdots a_k a_{k+1})$	3. $\log x + \log y = \log xy$
$\qquad = \log a_1 + \log a_2 + \cdots + \log a_k + \log_{k+1}$	

Step 4:   The proposition $P(n)$ is true for all positive integers n by PMI.

37. Step 1: Prove $P(1)$: $\qquad (b^m)^1 = b^{m \cdot 1}$

$\qquad\qquad\qquad\qquad\qquad\qquad b^m = b^m; \ \text{TRUE}$

Step 2: Assume $P(k)$: Hypothesis $\qquad (b^m)^k = b^{mk}$

Step 3: Prove $P(k+1)$: $\qquad\qquad (b^m)^{k+1} = b^{m(k+1)}$

1.	$(b^m)^k = b^{mk}$	1. Hypothesis
2.	$(b^m)^k \cdot b^m = b^{mk} \cdot b^m$	2. Multiply both sides by b^m
3.	$(b^m)^{k+1} = b^{mk+m}$	3. Definition, Problem 36
4.	$(b^m)^{k+1} = b^{m(k+1)}$	4. Factor

Step 4: The proposition $P(n)$ is true for all positive integers n by PMI.

39. Step 1: Prove $P(1)$: $\qquad \left(\dfrac{a}{b}\right)^1 = \dfrac{a^1}{b^1}$

$\qquad\qquad\qquad\qquad\qquad\qquad \dfrac{a}{b} = \dfrac{a}{b}; \ \text{TRUE}$

Step 2: Assume $P(k)$: Hypothesis $\qquad \left(\dfrac{a}{b}\right)^k = \dfrac{a^k}{b^k}$

Step 3: Prove $P(k+1)$: $\qquad\qquad \left(\dfrac{a}{b}\right)^{k+1} = \dfrac{a^{k+1}}{b^{k+1}}$

1.	$\left(\dfrac{a}{b}\right)^k = \dfrac{a^k}{b^k}$	1. Hypothesis
2.	$\left(\dfrac{a}{b}\right)^k \cdot \dfrac{a}{b} = \dfrac{a^k}{b^k} \cdot \dfrac{a}{b}$	2. Multiply both sides by $\dfrac{a}{b}$
3.	$\left(\dfrac{a}{b}\right)^k \cdot \dfrac{a}{b} = \dfrac{a^k \cdot a}{b^k \cdot b}$	3. Multiply Fractions
4.	$\left(\dfrac{a}{b}\right)^{k+1} = \dfrac{a^{k+1}}{b^{k+1}}$	4. Definition

Step 4: The proposition $P(n)$ is true for all positive integers n by PMI.

41. Step 1: Prove $P(1)$: $\quad \sin\left(\frac{\pi}{4}+1\cdot\pi\right)=(-1)^1\left(\frac{\sqrt{2}}{2}\right)$

$$\sin\frac{5\pi}{4}=-\frac{\sqrt{2}}{2};\ \text{TRUE}$$

Step 2: Assume $P(k)$: Hypothesis $\quad \sin\left(\frac{\pi}{4}+k\pi\right)=(-1)^k\left(\frac{\sqrt{2}}{2}\right)$

Step 3: Prove $P(k+1)$: $\quad \sin\left(\frac{\pi}{4}+(k+1)\pi\right)=(-1)^{k+1}\left(\frac{\sqrt{2}}{2}\right)$

$$\sin\left(\frac{\pi}{4}+k\pi+\pi\right)=(-1)^{k+1}\left(\frac{\sqrt{2}}{2}\right)$$

1. $\sin\left(\frac{\pi}{4}+k\pi\right)=(-1)^k\left(\frac{\sqrt{2}}{2}\right)$ 1. Hypothesis

2. $\sin\left[\left(\frac{\pi}{4}+k\pi\right)+\pi\right]$ 2. Addition law for sine

$$= \sin\left(\frac{\pi}{4}+k\pi\right)\cos\pi+\cos\left(\frac{\pi}{4}+k\pi\right)\sin\pi$$

3. $\quad = -1\cdot\sin\left(\frac{\pi}{4}+k\pi\right)+0$ 3. $\cos\pi=-1;\ \sin\pi=0$

4. $\quad = (-1)\cdot(-1)^k\left(\frac{\sqrt{2}}{2}\right)$ 4. Substitution

5. $\quad = (-1)^{k+1}\left(\frac{\sqrt{2}}{2}\right)$ 5. Law of exponents

Step 4: The proposition $P(n)$ is true for all positive integers n by PMI.

43. Step 1: Prove $P(1)$: $\quad 1^2+1=2$ is even; TRUE

Step 2: Assume $P(k)$: Hypothesis $\quad k^2+k$ is even

Step 3: Prove $P(k+1)$: $\quad (k+1)^2+(k+1)$ is even

1. k^2+k is even   1. Hypothesis

2. $(k+1)^2+(k+1)=k^2+2k+1+k+1$ 2. Multiply

3. $\qquad\qquad = k^2+k+2k+2$ 3. Commutative

4. $\qquad\qquad = (k^2+k)+2(k+1)$ 4. Factor

5. $2(k+1)$ is even   5. Divisible by 2

6. $(k^2+k)+2(k+1)$ is even   6. The sum of evens is even

Step 4: The proposition $P(n)$ is true for all positive integers n by PMI.

45. Step 1: Prove $P(1)$: $1(1+1)(1+2) = 6$ is divisible by 6;   TRUE

Step 2: Assume $P(k)$: Hypothesis $k(k+1)(k+2)$ is divisible by 6

Step 3: Prove $P(k+1)$: $(k+1)[(k+1)+1][(k+1)+2]$ is divisible by 6

$$(k+1)(k+2)(k+3) \text{ is divisible by 6}$$

1. $k(k+1)(k+2)$ is divisible by 6	1. Hypothesis
2. $(k+1)(k+2)(k+3) = k(k+1)(k+2) + 3(k+1)(k+2)$	2. Distribute last factor
3. Either $(k+1)$ or $(k+2)$ is divisible by 2	3. Consecutive integers
4. $3(k+1)(k+2)$ is divisible by 6	4. Divisible by 2 and 3
5. $(k+1)(k+2)(k+3) = k(k+1)(k+2) + 3(k+1)(k+2)$ is divisible by 6	5. Each term on the right is divisible by 6, so the sum is divisible by 6

Step 4: The proposition $P(n)$ is true for all positive integers n by PMI.

47. Step 1: Prove $P(1)$: $10^{1+1} + 3 \cdot 10^1 + 5 = 135$ is divisible by 9;   TRUE

Step 2: Assume $P(k)$: Hypothesis $10^{k+1} + 3 \cdot 10^k + 5$ is divisible by 9

Step 3: Prove $P(k+1)$: $10^{(k+1)+1} + 3 \cdot 10^{(k+1)} + 5$ is divisible by 9

$$10^{k+2} + 3 \cdot 10^{k+1} + 5 \text{ is divisible by 9}$$

1. $10^{k+1} + 3 \cdot 10^k + 5$ is divisible by 9	1. Hypothesis
2. $10^{k+2} + 3 \cdot 10^{k+1} + 5 = 10(10^{k+1} + 3 \cdot 10^k) + 5$	2. Factor 10 from first 2 terms
3. $= 10(10^{k+1} + 3 \cdot 10^k + 5) + 5 - 50$	3. Add $10 \cdot 5$ and subtract 50
4. $= 10(10^{k+1} + 3 \cdot 10^k + 5) - 45$	4. $5 - 50 = 45$
5. $10(10^{k+1} + 3 \cdot 10^k + 5) - 45$ is divisible by 9	5. Each term on the right is divisible by 9, so the sum is divisible by 9

Step 4: The proposition $P(n)$ is true for all positive integers n by PMI.

49. Step 1: Prove $P(1)$: $2^1 > 1$; TRUE

Step 2: Assume $P(k)$: Hypothesis $2^k > k$

Step 3: Prove $P(k+1)$: $2^{k+1} > k+1$

1. $2^k > k$ 1. Hypothesis

2. $2^k \cdot 2 > k \cdot 2$ 2. Multiply both sides by 2

3. $2^{k+1} > 2k$ 3. Definition

4. Assume $k > 1$ 4. $P(1)$ is true, k is a positive integer

5. $2k = k + k > k+1$ 5. Add k to both sides

6. $2^{k+1} > k+1$ 6. Transitive property of inequalities

Step 4: The proposition $P(n)$ is true for all positive integers n by PMI.

51. Step 1: Prove $P(1)$: $1 + 2 \cdot 1 \leq 3^1$

$$3 \leq 3; \text{ TRUE}$$

Step 2: Assume $P(k)$: Hypothesis $1 + 2k \leq 3^k$

Step 3: Prove $P(k+1)$: $1 + 2(k+1) \leq 3^{k+1}$

1. $1 + 2k \leq 3^k$ 1. Hypothesis

2. $1 + 2k + 2 \leq 3^k + 2$ 2. Add 2 to both sides

3. $1 + 2(k+1) \leq 3^k + 2$ 3. Factor

4. $2 \leq 2 \cdot 3^k$ 4. Since $k > 1$, $3^k > 1$

5. $3^k + 2 \leq 3^k + 2 \cdot 3^k$ 5. Add 3^k to both sides

6. $3^k + 2 \leq 3 \cdot 3^k$ 6. Combine like terms

7. $3^k + 2 \leq 3^{k+1}$ 7. Definition

8. $1 + 2(k+1) \leq 3^{k+1}$ 8. Transitive property of inequalities

Step 4: The proposition $P(n)$ is true for all positive integers n by PMI.

53. Step 1: Prove $P(1)$: $\quad 1 < \dfrac{(2 \cdot 1 + 1)^2}{8}$

$\qquad\qquad\qquad\qquad 1 < \dfrac{9}{8};\quad$ TRUE

Step 2: Assume $P(k)$: Hypothesis $\quad 1 + 2 + 3 + \cdots + k < \dfrac{(2k+1)^2}{8}$

Step 3: Prove $P(k+1)$: $\quad 1 + 2 + 3 + \cdots + (k+1) < \dfrac{[2(k+1)+1]^2}{8}$

$\qquad\qquad\qquad\qquad 1 + 2 + 3 + \cdots + (k+1) < \dfrac{(2k+3)^2}{8}$

1.	$1 + 2 + 3 + \cdots + k < \dfrac{(2k+1)^2}{8}$	1. Hypothesis
2.	$1 + 2 + 3 + \cdots + k + (k+1) < \dfrac{(2k+1)^2}{8} + (k+1)$	2. Add $(k+1)$ to both sides
3.	$< \dfrac{(2k+1)^2}{8} + \dfrac{8(k+1)}{8}$	3. Add fractions
4.	$< \dfrac{4k^2 + 12k + 9}{8}$	4. Multiply and combine like terms
5.	$< \dfrac{(2k+3)^2}{8}$	5. Factor

Step 4: The proposition $P(n)$ is true for all positive integers n by PMI.

55. Conjecture: $\quad 1^3 + 2^3 + 3^3 + \cdots + n^3 = \dfrac{n^2(n+1)^2}{4}$

Step 1: Prove $P(1)$: $\quad 1^3 = \dfrac{1^2(1+1)^2}{4}$

$\qquad\qquad\qquad\qquad 1 = 1;\quad$ TRUE

Step 2: Assume $P(k)$: Hypothesis $\quad 1^3 + 2^3 + 3^3 + \cdots + k^3 = \dfrac{k^2(k+1)^2}{4}$

Step 3: Prove $P(k+1)$: $\quad 1^3 + 2^3 + 3^3 + \cdots + k^3 + (k+1)^3 = \dfrac{(k+1)^2[(k+1)+1]^2}{4}$

$\qquad\qquad\qquad\qquad 1^3 + 2^3 + 3^3 + \cdots + k^3 + (k+1)^3 = \dfrac{(k+1)^2(k+2)^2}{4}$

1.	$1^3 + 2^3 + 3^3 + \cdots + k^3 = \dfrac{k^2(k+1)^2}{4}$	1. Hypothesis
2.	$1^3 + 2^3 + 3^3 + \cdots + k^3 + (k+1)^3 = \dfrac{k^2(k+1)^2}{4} + (k+1)^3$	2. Add $(k+1)^3$ to both sides
3.	$= \dfrac{k^2(k+1)^2}{4} + \dfrac{4(k+1)^3}{4}$	3. Add fractions
4.	$= \dfrac{(k+1)^2[k^2 + 4(k+1)]}{4}$	4. Factor out $(k+1)^2$
5.	$= \dfrac{(k+1)^2(k^2 + 4k + 4)}{4}$	5. Distribute the 4
6.	$= \dfrac{(k+1)^2(k+2)^2}{4}$	6. Factor

Step 4: The proposition $P(n)$ is true for all positive integers n by PMI.

57. Conjecture: $2 + 6 + 18 + \cdots + 2 \cdot 3^{n-1} = 3^n - 1$

Step 1: Prove $P(1)$: $2 = 3^1 - 1$

$2 = 2$; TRUE

Step 2: Assume $P(k)$: Hypothesis $2 + 6 + 18 + \cdots + 2 \cdot 3^{k-1} = 3^k - 1$

Step 3: Prove $P(k+1)$: $2 + 6 + 18 + \cdots + 2 \cdot 3^{(k+1)-1} = 3^{k+1} - 1$

$2 + 6 + 18 + \cdots + 2 \cdot 3^k = 3^{k+1} - 1$

1.	$2 + 6 + 18 + \cdots + 2 \cdot 3^{k-1} = 3^k - 1$	1. Hypothesis
2.	$2 + 6 + 18 + \cdots + 2 \cdot 3^{k-1} + 2 \cdot 3^k = 3^k - 1 + 2 \cdot 3^k$	2. Add $2 \cdot 3^k$ to both sides
3.	$= 3 \cdot 3^k - 1$	3. Combine like terms
4.	$= 3^{k+1} - 1$	4. Definition

Step 4: The proposition $P(n)$ is true for all positive integers n by PMI.

59. Step 1: Prove $P(1)$: $\sin x = \dfrac{\sin^2 x - \sin x}{\sin x - 1}$

$\sin x = \dfrac{\sin x \, (\sin x - 1)}{\sin x - 1}$; TRUE

Step 2: Assume $P(k)$: Hypothesis $\sin x + \sin^2 x + \cdots + \sin^k x = \dfrac{\sin^{k+1} x - \sin x}{\sin x - 1}$

Step 3: Prove $P(k+1)$: $\sin x + \sin^2 x + \cdots + \sin^k x + \sin^{k+1} x = \dfrac{\sin^{k+2} x - \sin x}{\sin x - 1}$

1. $\sin x + \sin^2 x + \cdots + \sin^k x = \dfrac{\sin^{k+1} x - \sin x}{\sin x - 1}$ 1. Hypothesis

2. $\sin x + \sin^2 x + \cdots + \sin^k x + \sin^{k+1} x$ 2. Add $\sin^{k+1} x$ to both sides

$= \dfrac{\sin^{k+1} x - \sin x}{\sin x - 1} + \sin^{k+1} x$

3. $= \dfrac{\sin^{k+1} x - \sin x}{\sin x - 1} + \dfrac{\sin^{k+1} x \, (\sin x - 1)}{\sin x - 1}$ 3. Add fractions

4. $= \dfrac{\sin^{k+1} x - \sin x + \sin^{k+2} x - \sin^{k+1} x}{\sin x - 1}$ 4. Distributive property

5. $= \dfrac{\sin^{k+2} x - \sin x}{\sin x - 1}$ 5. Combine like terms

Step 4: The proposition $P(n)$ is true for all positive integers n by PMI.

61. Step 1: Prove $P(1)$: $(r \text{ cis } \theta)^1 = r^1 \text{ cis}(1 \cdot \theta)$

$r \text{ cis } \theta = r \text{ cis } \theta;$ TRUE

Step 2: Assume $P(k)$: Hypothesis $(r \text{ cis } \theta)^k = r^k \text{ cis}(k\theta)$

Step 3: Prove $P(k+1)$: $(r \text{ cis } \theta)^{k+1} = r^{k+1} \text{ cis}[(k+1)\theta]$

1.	$(r \text{ cis } \theta)^k = r^k \text{ cis}(k\theta)$	1. Hypothesis
2.	$(r \text{ cis } \theta)^k \cdot (r \text{ cis } \theta) = r^k \text{ cis}(k\theta) \cdot (r \text{ cis } \theta)$	2. Multiply both sides by $r \text{ cis } \theta$
3.	$(r \text{ cis } \theta)^{k+1} = r^k \text{ cis}(k\theta) \cdot (r \text{ cis } \theta)$	3. Rule of exponents
4.	$(r \text{ cis } \theta)^{k+1} = (r^k \cdot r) \text{ cis}(k\theta + \theta)$	4. Multiplication of imaginary nos.
5.	$(r \text{ cis } \theta)^{k+1} = r^{k+1} \text{ cis}[(k+1)\theta]$	5. Rule of exponents; factor

Step 4: The proposition $P(n)$ is true for all positive integers n by PMI.

PROBLEM SET 10.2

1. $4! - 2! = 24 - 2 = 22$

3. $(4-2)! = 2! = 2$

5. $\dfrac{9!}{7!} = \dfrac{9 \cdot 8 \cdot 7!}{7!} = 9 \cdot 8 = 72$

7. $\dfrac{12!}{10!} = \dfrac{12 \cdot 11 \cdot 10!}{10!} = 12 \cdot 11 = 132$

9. $\dfrac{12!}{3!(12-3)!} = \dfrac{12!}{3! \cdot 9!} = \dfrac{12 \cdot 11 \cdot 10 \cdot 9!}{3 \cdot 2 \cdot 1 \cdot 9!} = 220$

11. $\dfrac{20!}{3!(20-3)!} = \dfrac{20!}{3! \cdot 17!} = \dfrac{20 \cdot 19 \cdot 18 \cdot 17!}{3 \cdot 2 \cdot 1 \cdot 17!} = 1,140$

13. $\dbinom{8}{1} = \dfrac{8!}{1!(8-1)!} = \dfrac{8!}{1! \cdot 7!} = \dfrac{8 \cdot 7!}{1 \cdot 7!} = 8$

15. $\dbinom{8}{2} = \dfrac{8!}{2!(8-2)!} = \dfrac{8!}{2! \cdot 6!} = \dfrac{8 \cdot 7 \cdot 6!}{2 \cdot 1 \cdot 6!} = 28$

17. $\dbinom{8}{3} = \dfrac{8!}{3!(8-3)!} = \dfrac{8!}{3! \cdot 5!} = \dfrac{8 \cdot 7 \cdot 6 \cdot 5!}{3 \cdot 2 \cdot 1 \cdot 5!} = 56$

19. $\dbinom{8}{4} = \dfrac{8!}{4!(8-4)!} = \dfrac{8!}{4! \cdot 4!} = \dfrac{8 \cdot 7 \cdot 6 \cdot 5 \cdot 4!}{4 \cdot 3 \cdot 2 \cdot 1 \cdot 4!} = 70$

21. $\dbinom{52}{2} = \dfrac{52!}{2!(52-2)!} = \dfrac{52!}{2! \cdot 50!} = \dfrac{52 \cdot 51 \cdot 50!}{2 \cdot 1 \cdot 50!} = 1,326$

23. $\dbinom{1,000}{1} = \dfrac{1,000!}{1!(1,000-1)!} = \dfrac{1,000!}{1! \cdot 999!} = \dfrac{1,000 \cdot 999!}{1 \cdot 999!} = 1,000$

25. Using row 6 of Pascal's triangle:

$(a+b)^6 = a^6 + 6a^5b + 15a^4b^2 + 20a^3b^3 + 15a^2b^4 + 6ab^5 + b^6$

27. Using row 3 of Pascal's triangle:

$(2x+3)^3 = (2x)^3 + 3 \cdot (2x)^2(3) + 3 \cdot (2x)(3)^2 + (3)^3 = 8x^3 + 36x^2 + 54x + 27$

29. Using row 5 of Pascal's triangle:

$(x+y)^5 = x^5 + 5x^4y + 10x^3y^2 + 10x^2y^3 + 5xy^4 + y^5$

31. Using row 5 of Pascal's triangle:

$$(3x+2)^5 = (3x)^5 + 5 \cdot (3x)^4(2) + 10 \cdot (3x)^3(2)^2 + 10 \cdot (3x)^2(2)^3 + 5 \cdot (3x)(2)^4 + (2)^5$$

$$= 243x^5 + 810x^4 + 1,080x^3 + 720x^2 + 240x + 32$$

33. Using row 4 of Pascal's triangle:

$(x+y)^4 = x^4 + 4x^3y + 6x^2y^2 + 4xy^3 + y^4$

35. Using row 3 of Pascal's triangle:

$\left(\frac{1}{2}x + y^3\right)^3 = \left(\frac{1}{2}x\right)^3 + 3 \cdot \left(\frac{1}{2}x\right)^2(y^3) + 3 \cdot \left(\frac{1}{2}x\right)(y^3)^2 + (y^3)^3 = \frac{1}{8}x^3 + \frac{3}{4}x^2y^3 + \frac{3}{2}xy^6 + y^9$

37. Using row 4 of Pascal's triangle:

$$\left(x^{1/2} + y^{1/2}\right)^4 = (x^{1/2})^4 + 4 \cdot (x^{1/2})^3\left(y^{1/2}\right) + 6 \cdot (x^{1/2})^2\left(y^{1/2}\right)^2 + 4 \cdot (x^{1/2})\left(y^{1/2}\right)^3 + \left(y^{1/2}\right)^4$$

$$= x^2 + 4x^{3/2}y^{1/2} + 6xy + 4x^{1/2}y^{3/2} + y^2$$

39. Using row 8 of Pascal's triangle:

$$(1-x)^8 = (1)^8 + 8 \cdot (1)^7(-x)^1 + 28 \cdot (1)^6(-x)^2 + 56 \cdot (1)^5(-x)^3 + 70 \cdot (1)^4(-x)^4$$

$$+ 56 \cdot (1)^3(-x)^5 + 28 \cdot (1)^2(-x)^6 + 8 \cdot (1)^1(-x)^7 + (-x)^8$$

$$= 1 - 8x + 28x^2 - 56x^3 + 70x^4 - 56x^5 + 28x^6 - 8x^7 + x^8$$

41. First the first four terms of $(a+b)^{10}$

$$\binom{10}{0}a^{10} + \binom{10}{1}a^9b + \binom{10}{2}a^8b^2 + \binom{10}{3}a^7b^3$$

$$= \frac{10!}{0!10!}a^{10} + \frac{10!}{1!9!}a^9b + \frac{10!}{2!8!}a^8b^2 + \frac{10!}{3!7!}a^7b^3 = a^{10} + 10a^9b + 45a^8b^2 + 120a^7b^3$$

43. Find the first four terms of $(a+b)^{14}$

$$\binom{14}{0}a^{14} + \binom{14}{1}a^{13}b + \binom{14}{2}a^{12}b^2 + \binom{14}{3}a^{11}b^3$$

$$= \frac{14!}{0!14!}a^{14} + \frac{14!}{1!13!}a^{13}b + \frac{14!}{2!12!}a^{12}b^2 + \frac{14!}{3!11!}a^{11}b^3 = a^{14} + 14a^{13}b + 91a^{12}b^2 + 364a^{11}b^3$$

45. Find the first four terms of $(x+2y)^{16}$

$$\binom{16}{0}x^{16} + \binom{16}{1}x^{15}(2y)^1 + \binom{16}{2}x^{14}(2y)^2 + \binom{16}{3}x^{13}(2y)^3$$

$$= \frac{16!}{0!16!}x^{16} + \frac{16!}{1!15!}x^{15}(2y) + \frac{16!}{2!14!}x^{14}(4y^2) + \frac{16!}{3!13!}x^{13}(8y^3)$$

$$= x^{16} + 32x^{15}y + 480x^{14}y^2 + 4,480x^{13}y^3$$

47. Find the first four terms of $(x-2y)^{12}$

$$\binom{12}{0}x^{12} + \binom{12}{1}x^{11}(-2y)^1 + \binom{12}{2}x^{10}(-2y)^2 + \binom{12}{3}x^9(-2y)^3$$

$$= \frac{12!}{0!12!}x^{12} + \frac{12!}{1!11!}x^{11}(-2y) + \frac{12!}{2!10!}x^{10}(4y^2) + \frac{12!}{3!9!}x^9(-8y^3)$$

$$= x^{12} - 24x^{11}y + 264x^{10}y^2 - 1,760x^9y^3$$

49. Find the first four terms of $(1-.02)^{12}$

$$\binom{12}{0}(1)^{12} + \binom{12}{1}(1)^{11}(-.02)^1 + \binom{12}{2}(1)^{10}(-.02)^2 + \binom{12}{3}(1)^9(-.02)^3$$

$$= \frac{12!}{0!12!} + \frac{12!}{1!11!}(-.02) + \frac{12!}{2!10!}(.0004) + \frac{12!}{3!9!}(-.000008) \approx 1 - .24 + .0264 - .00176$$

51. $(b)^r = b$, so $r = 1$; $\binom{3}{1}(a^2)^2(-2b)^1 = \frac{3!}{1!2!}a^4(-2b) = -6a^4b$; Coefficient is -6.

53. $(\sqrt{y})^r = y^2$; so $r = 4$; $\binom{9}{4}(2x^2)^5(\sqrt{y})^4 = \frac{9!}{4!5!}(32x^{10})y^2 = 4,032x^{10}y^2$; Coefficient is 4,032.

55. The term will be constant when $(y^{-2})^{5-r}(y^3)^r = y^0$; so $-2(5-r) + 3r = 0$;

$$-10 + 2r + 3r = 0; \quad 5r = 10; \quad r = 2; \quad \binom{5}{2}(4y^{-2})^3\left(\frac{y^3}{2}\right)^2 = \frac{5!}{2!3!}64y^{-6}\left(\frac{y^6}{4}\right) = 160$$

272

57. $\displaystyle \binom{n-1}{r-1}+\binom{n-1}{r}=\frac{(n-1)!}{(r-1)![(n-1)-(r-1)]!}+\frac{(n-1)!}{r![(n-1)-r]!}$

$$=\frac{(n-1)!}{(r-1)!(n-r)!}+\frac{(n-1)!}{r!(n-r-1)!}$$

$$=\frac{(n-1)!\cdot r!(n-r-1)!+(n-1)!\cdot (r-1)!(n-r)!}{(r-1)!(n-r)!\cdot r!(n-r-1)!}$$

$$=\frac{(n-1)!(r-1)!(n-r-1)!\cdot [r+(n-r)]}{(r-1)!(n-r)!\cdot r!(n-r-1)!}$$

$$=\frac{(n-1)!\cdot n}{r!(n-r)!}$$

$$=\frac{n!}{r!(n-r)!}$$

$$=\binom{n}{r}$$

59. $\displaystyle \binom{k}{r}+\binom{k}{r-1}=\frac{k!}{r!(k-r)!}+\frac{k!}{(r-1)![k-(r-1)]!}$

$$=\frac{k!\cdot (r-1)!(k-r+1)!+k!\cdot r!(k-r)!}{r!(k-r)!(r-1)!(k-r+1)!}$$

$$=\frac{k!(r-1)!(k-r)!\cdot [(k-r+1)+r]}{r!(k-r)!(r-1)!(k-r+1)!}$$

$$=\frac{k!(k+1)}{r!(k+1-r)!}$$

$$=\frac{(k+1)!}{r![(k+1)-r]!}$$

$$=\binom{k+1}{r}$$

61. Hypothesis: $(a+b)^k=\binom{k}{0}a^k+\binom{k}{1}a^{k-1}b+\binom{k}{2}a^{k-2}b^2+\cdots+\binom{k}{k-1}ab^{k-1}+\binom{k}{k}b^k$

63. Step 4: Problems 60-62 show that the binomial theorem is true for all positive integers n by the principal of mathematical induction.

PROBLEM SET 10.3

1. a. Arithmetic b. $d = 3$ c. $14 + 3 = 17$

3. a. Geometric b. $r = 2$ c. $48 \cdot 2 = 96$

5. a. Neither b. The difference decreases by 1 between successive terms c. $90 - 5 = 85$

7. a. Geometric b. $r = q$ c. $pq^4 \cdot q = pq^5$

9. a. Neither b. The differences increases by 2 between successive terms c. $26 + 10 = 36$

11. a. Neither b. The number of 5's between 2's increases by 1 c. 2

13. a. Neither b. Listing of fractions c. $\frac{5}{6}$

15. a. Neither b. Listing of perfect cubes c. $6^3 = 216$

17. $s_1 = 4(1) - 3 = 1$, $s_2 = 4(2) - 3 = 5$, $s_3 = 4(3) - 3 = 9$; 1, 5, 9

19. $s_1 = ar^{1-1} = a$, $s_2 = ar^{2-1} = ar$, $s_3 = ar^{3-1} = ar^2$; a, ar, ar^2

21. $s_1 = (-1)^1 = -1$, $s_2 = (-1)^2 = 1$, $s_3 = (-1)^3 = -1$; -1, 1, -1

23. $s_1 = 1 + \frac{1}{1} = 2$, $s_2 = 1 + \frac{1}{2} = \frac{3}{2}$, $s_3 = 1 + \frac{1}{3} = \frac{4}{3}$; 2, $\frac{3}{2}$, $\frac{4}{3}$

25. $s_1 = 2$, $s_2 = 2$, $s_3 = 2$; 2, 2, 2

27. $s_1 = \cos(1 \cdot x) = \cos x$, $s_2 = \cos(2 \cdot x) = \cos 2x$, $s_3 = \cos(3 \cdot x) = \cos 3x$; $\cos x$, $\cos 2x$, $\cos 3x$

29. $\displaystyle\sum_{k=2}^{6} k = 2 + 3 + 4 + 5 + 6 = 20$

31. $\displaystyle\sum_{n=0}^{6} (2n + 1) = 1 + 3 + 5 + 7 + 9 + 11 + 13 = 49$

33. $\displaystyle\sum_{k=2}^{5} (10 - 2k) = 6 + 4 + 2 + 0 = 12$

35. $\displaystyle\sum_{k=1}^{5} (-2)^{k-1} = 1 - 2 + 4 - 8 + 16 = 11$

37. $\displaystyle\sum_{k=0}^{3} 2(3^k) = 2 + 6 + 18 + 54 = 80$

39. $\displaystyle\sum_{k=1}^{10}[1^k+(-1)^k]=(1-1)+(1+1)+(1-1)+(1+1)+(1-1)+(1+1)$
$$+(1-1)+(1+1)+(1-1)+(1+1)$$
$$=0+2+0+2+0+2+0+2+0+2=10$$

41. $s_{15}=4(15)-3=57$

43. $s_{10}=(-1)^{10}=1$

45. $s_3=(-1)^{3+1}5^{3+1}=(-1)^4 5^4=5^4=625$

47. $s_1=2,\ \ s_2=3s_1=3(2)=6,\ \ s_3=3s_2=3(6)=18,\ \ s_4=3s_3=3(18)=54,$
$s_5=3s_4=3(54)=162;\ \ 2,\ 6,\ 18,\ 54,\ 162$

49. $s_1=1,\ \ s_2=1,\ \ s_3=s_2+s_1=1+1=2,\ \ s_4=s_3+s_2=2+1=3,$
$s_5=s_4+s_3=3+2=5;\ \ 1,\ 1,\ 2,\ 3,\ 5$

51. $\frac{1}{2}+\frac{1}{4}+\frac{1}{8}+\cdots+\frac{1}{128}=\frac{1}{2^1}+\frac{1}{2^2}+\frac{1}{2^3}+\cdots+\frac{1}{2^7}=\displaystyle\sum_{k=1}^{7}\frac{1}{2^k}$

53. $1+6+36+216+1,296=6^0+6^1+6^2+6^3+6^4=\displaystyle\sum_{k=1}^{5}6^{k-1}$ or $\displaystyle\sum_{k=0}^{4}6^k$

55. Arithmetic   57. Geometric   59. Geometric

61. a. $\displaystyle\sum_{j=1}^{r}a_j b_j=a_1 b_1+a_2 b_2+a_3 b_3+\cdots+a_r b_r$

b. $\displaystyle\sum_{j=1}^{r}ma_j=ma_1+ma_2+ma_3+\cdots+ma_r$
$$=m(a_1+a_2+a_3+\cdots+a_r)$$
$$=m\sum_{j=1}^{r}a_j$$

c. $\displaystyle\sum_{j=1}^{r}m=(m+m+m+\cdots+m)$ (Note: r terms of m)
$$=mr$$

63. 47; add preceding two terms

65. 7; digits listed alphabetically

PROBLEM SET 10.4

1. $a_1 = 5$, $a_2 = 5 + 4 = 9$, $a_3 = 9 + 4 = 13$, $a_4 = 13 + 4 = 17$; 5, 9, 13, 17

3. $a_1 = 85$, $a_2 = 85 + 3 = 88$, $a_3 = 88 + 3 = 91$, $a_4 = 91 + 3 = 94$; 85, 88, 91, 94

5. $a_1 = 100$, $a_2 = 100 + (-5) = 95$, $a_3 = 95 + (-5) = 90$, $a_4 = 90 + (-5) = 85$;
   100, 95, 90, 85

7. $a_1 = -\frac{5}{2}$, $a_2 = -\frac{5}{2} + \frac{1}{2} = -2$, $a_3 = -2 + \frac{1}{2} = -\frac{3}{2}$, $a_4 = -\frac{3}{2} + \frac{1}{2} = -1$;
   $-\frac{5}{2}$, -2, $-\frac{3}{2}$, -1

9. $a_1 = \sqrt{12} = 2\sqrt{3}$, $a_2 = 2\sqrt{3} + \sqrt{3} = 3\sqrt{3}$, $a_3 = 3\sqrt{3} + \sqrt{3} = 4\sqrt{3}$, $a_4 = 4\sqrt{3} + \sqrt{3} = 5\sqrt{3}$;
   $2\sqrt{3}$, $3\sqrt{3}$, $4\sqrt{3}$, $5\sqrt{3}$

11. $a_1 = x$, $a_2 = x + y$, $a_3 = (x + y) + y = x + 2y$, $a_4 = (x + 2y) + y = x + 3y$;
    x, $x + y$, $x + 2y$, $x + 3y$

13. $a_1 = 5$, $d = 8 - 5 = 3$

15. $a_1 = 6$, $d = 11 - 6 = 5$

17. $a_1 = -8$, $d = -1 - (-8) = 7$

19. $a_1 = x$, $d = 2x - x = x$

21. $a_1 = x - 5b$, $d = (x - 3b) - (x - 5b) = 2b$

23. $a_n = 5$

25. $a_1 = 35$, $d = 46 - 35 = 11$; $a_n = 35 + (n-1)11 = 35 + 11n - 11 = 24 + 11n$

27. $a_1 = -1$, $d = 1 - (-1) = 2$; $a_n = -1 + (n-1)2 = -1 + 2n - 2 = -3 + 2n$

29. $a_1 = x + \sqrt{3}$, $d = (x + \sqrt{12}) - (x + \sqrt{3}) = \sqrt{12} - \sqrt{3} = 2\sqrt{3} - \sqrt{3} = \sqrt{3}$;
    $a_n = (x + \sqrt{3}) + (n-1)\sqrt{3} = x + \sqrt{3} + n\sqrt{3} - \sqrt{3} = x + n\sqrt{3}$

31. $a_{20} = 6 + (20 - 1)(5) = 6 + 19(5) = 101$

33. $a_{10} = -20 + (10 - 1)(5) = -20 + 9(5) = 25$

35. $A_{100} = \frac{100}{2}[2(-7) + (100 - 1)(-2)] = 50(-14 - 198) = -10,600$

37. $a_{30} = a_1 + (30 - 1)d$; $-63 = -5 + 29d$; $-58 = 29d$; $d = -2$

39. $a_{10} = a_1 + (10 - 1)d$; $5 = -13 + 9d$; $18 = 9d$; $d = 2$

41. $d = -2$ (problem 37); $A_{10} = \frac{10}{2}[2(-5) + (10 - 1)(-2)] = 5[-10 - 18] = -140$

43. $a_5 = a_1 + (5 - 1)d$; $27 = a_1 + 4d$; $a_{10} = a_1 + (10 - 1)d$; $47 = a_1 + 9d$

$$\begin{cases} 27 = a_1 + 4d \\ 47 = a_1 + 9d \end{cases} \Rightarrow + \begin{cases} -27 = -a_1 - 4d \\ 47 = a_1 + 9d \end{cases}$$

$$20 = 5d$$

$$d = 4$$

45. $a_3 = a_1 + (3 - 1)d$; $36 = a_1 + 2d$; $a_5 = a_1 + (5 - 1)d$; $60 = a_1 + 4d$;

$$\begin{cases} 36 = a_1 + 2d \\ 60 = a_1 + 4d \end{cases} \Rightarrow + \begin{cases} -36 = -a_1 - 2d \\ 60 = a_1 + 4d \end{cases}$$

$$24 = 2d$$

$$d = 12$$

47. $A_{20} = \frac{20}{2}[2(100) + (20 - 1)(50)] = 10[200 + 950] = 11{,}500$

49. $42 + 44 + 46 + \cdots + 98$; $98 = 42 + (n - 1)(2)$; $98 = 40 + 2n$; $2n = 58$; $n = 29$;

$A_{29} = \frac{29}{2}(42 + 98) = \frac{29}{2}(140) = 2{,}030$

51. $A_n = \frac{n}{2}[2(1) + (n - 1)(2)] = \frac{n}{2}[2 + 2n - 2] = \frac{n}{2}[2n] = n^2$

53. a. $1, 2, 3, 4, 5, \ldots$;   yes, $d = 1$

b. $2, 5, 8, 11, 14, \ldots$;   yes, $d = 3$

c. $\frac{1}{2}, \frac{3}{2}, \frac{5}{2}, \frac{7}{2}, \ldots$;   yes, $d = 1$

d. $5, -5, -15, -25, \ldots$;   yes, $d = -10$

e. $\frac{4}{3}, 2, 3, \frac{9}{2}, \ldots$;   no, no common difference

55. a. $x = \dfrac{4 + 20}{2} = 12$

b. $x = \dfrac{4 + 15}{2} = \dfrac{19}{2}$

c. $x = \dfrac{\frac{1}{2} + \frac{1}{3}}{2} = \dfrac{\frac{5}{6}}{2} = \dfrac{5}{12}$

d. $x = \dfrac{-10 + (-2)}{2} = -6$

e. $x = \dfrac{-\frac{2}{3} + \frac{4}{5}}{2} = \dfrac{\frac{2}{15}}{2} = \dfrac{1}{15}$

57. Option C: $n = 4$; $a_1 = \frac{1}{4}(21,000) = 5,250$; $d = 90$;

$A_4 = \frac{4}{2}[2(5,250) + (4 - 1)(90)] = \$21,540$

Option D: $n = 12$; $a_1 = \frac{1}{12}(21,000) = 1,750$; $d = 10$;

$A_{12} = \frac{12}{2}[2(1,750) + (12 - 1)(10)] = \$21,660$

Option D is the better choice.

59. Option A: $n = 2$; $a_1 = 21,000$; $d = 1,440$;

$A_2 = \frac{2}{2}[2(21,000) + (2 - 1)(1,440)] = \$43,440$

Option B: $n = 4$; $a_1 = \frac{1}{2}(21,000) = 10,500$; $d = 360$;

$A_4 = \frac{4}{2}[2(10,500) + (4 - 1)(360)] = \$44,160$; (\$720 more than Option A)

Option C: $n = 8$; $a_1 = \frac{1}{4}(21,000) = 5,250$; $d = 90$;

$A_8 = \frac{8}{2}[2(5,250) + (8 - 1)(90)] = \$44,520$;

(\$1,080 more than Option A, \$360 more than Option B)

Option D: $n = 24$; $a_1 = \frac{1}{12}(21,000) = 1,750$; $d = 10$;

$A_{24} = \frac{24}{2}[2(1,750) + (24 - 1)(10)] = \$44,760$

(\$1,320 more than Option A, \$600 more than Option B, and \$240 more than Option C)

PROBLEM SET 10.5

1. $g_1 = 5$, $g_2 = 5 \cdot 3 = 15$, $g_3 = 15 \cdot 3 = 45$; 5, 15, 45

3. $g_1 = 1$, $g_2 = 1 \cdot (-2) = -2$, $g_3 = (-2)(-2) = 4$; 1, -2, 4

5. $g_1 = -15$, $g_2 = -15 \cdot \frac{1}{5} = -3$, $g_3 = (-3) \cdot \frac{1}{5} = -\frac{3}{5}$; -15, -3, $-\frac{3}{5}$

7. $g_1 = 8$, $g_2 = 8x$, $g_3 = 8x \cdot x = 8x^2$; 8, $8x$, $8x^2$

9. $g_1 = 3$, $r = \frac{6}{3} = 2$

11. $g_1 = 1$, $r = \frac{\frac{1}{2}}{1} = \frac{1}{2}$

13. $g_1 = x$, $r = \frac{x^2}{x} = x$

15. $g_n = 3 \cdot 2^{n-1}$

17. $g_n = 1 \cdot \left(\frac{1}{2}\right)^{n-1} = \left(2^{-1}\right)^{n-1} = 2^{1-n}$

19. $g_n = x \cdot x^{n-1} = x^n$

21. $g_1 = 1, \quad r = \dfrac{\frac{1}{2}}{1} = \frac{1}{2}; \quad G = \dfrac{1}{1-\frac{1}{2}} = \dfrac{1}{\frac{1}{2}} = 2$

23. $g_1 = 100, \quad r = \frac{50}{100} = \frac{1}{2}; \quad G = \dfrac{100}{1-\frac{1}{2}} = \dfrac{100}{\frac{1}{2}} = 200$

25. $g_1 = -45, \quad r = \frac{-15}{-45} = \frac{1}{3}; \quad G = \dfrac{-45}{1-\frac{1}{3}} = \dfrac{-45}{\frac{2}{3}} = -\frac{135}{2}$

27. $g_5 = 6 \cdot 3^{5-1} = 6 \cdot 3^4 = 6 \cdot 81 = 486$

29. $G_5 = \dfrac{6(1-3^5)}{1-3} = \dfrac{6(1-243)}{-2} = \dfrac{6(-242)}{-2} = 726$

31. $G_{10} = \dfrac{1(1-10^{10})}{1-10} = \dfrac{1-10,000,000,000}{-9} = \dfrac{-9,999,999,999}{-9} = 1,111,111,111$

33. $G = \dfrac{\frac{1}{3}}{1-\frac{1}{3}} = \dfrac{\frac{1}{3}}{\frac{2}{3}} = \frac{1}{2}$

35. $G = \dfrac{1}{1-.08} = \dfrac{1}{.92} = \dfrac{100}{92} = \dfrac{25}{23}$

37. $\displaystyle\sum_{k=1}^{4} \left(\frac{1}{3}\right)^k = \frac{1}{3}+\frac{1}{9}+\frac{1}{27}+\frac{1}{81}; \quad g_1 = \frac{1}{3}, \quad r = \frac{1}{3};$

$G_4 = \dfrac{\frac{1}{3}\left[1-\left(\frac{1}{3}\right)^4\right]}{1-\frac{1}{3}} = \dfrac{\frac{1}{3}\left(1-\frac{1}{81}\right)}{\frac{2}{3}} = \dfrac{\frac{1}{3}\left(\frac{80}{81}\right)}{\frac{2}{3}} = \dfrac{\frac{80}{243}}{\frac{2}{3}} = \frac{40}{81}$

39. $\quad .\overline{4} = .4 + .04 + .004 + .0004 + \cdots$

$\quad\quad = .4 + .4(.1) + .4(.01) + .4(.001) + \cdots$

$\quad\quad = .4 + .4(.1) + .4(.1)^2 + .4(.1)^3 + \cdots; \quad g_1 = .4, \quad r = .1$

$\quad G = \dfrac{.4}{1-.1} = \dfrac{.4}{.9} = \dfrac{4}{9}$

41. $\quad .\overline{9} = .9 + .09 + .009 + .0009 + \cdots$

$\quad\quad = .9 + .9(.1) + .9(.01) + .9(.001) + \cdots$

$\quad\quad = .9 + .9(.1) + .9(.1)^2 + .9(.1)^3 + \cdots; \quad g_1 = .9, \quad r = .1$

$\quad G = \dfrac{.9}{1-.1} = \dfrac{.9}{.9} = 1$

43. $.\overline{18} = .18 + .0018 + .000018 + \cdots$

$\qquad = .18 + .18(.01) + .18(.0001) + \cdots$

$\qquad = .18 + .18(.01) + .18(.01)^2 + \cdots; \quad g_1 = .18, \quad r = .01$

$\qquad G = \dfrac{.18}{1 - .01} = \dfrac{.18}{.99} = \dfrac{18}{99} = \dfrac{2}{11}$

45. $.\overline{418} = .418 + .000418 + .000000418 + \cdots$

$\qquad = .418 + .418(.001) + .418(.000001) + \cdots$

$\qquad = .418 + .418(.001) + .418(.001)^2 + \cdots; \quad g_1 = .418, \quad r = .001$

$\qquad G = \dfrac{.418}{1 - .001} = \dfrac{.418}{.999} = \dfrac{418}{999}$

47. $.\overline{123} = .123 + .000123 + .000000123 + \cdots$

$\qquad = .123 + .123(.001) + .123(.000001) + \cdots$

$\qquad = .123 + .123(.001) + .123(.001)^2 + \cdots; \quad g_1 = .123, \quad r = .001$

$\qquad G = \dfrac{.123}{1 - .001} = \dfrac{.123}{.999} = \dfrac{123}{999} = \dfrac{41}{333}$

49. $5.03\overline{1} = 5.03 + .001 + .0001 + .00001 + .000001 + \cdots$

$\qquad = 5.03 + .001 + .001(.1) + .001(.01) + .001(.001) + \cdots$

$\qquad = 5.03 + .001 + .001(.1) + .001(.1)^2 + .001(.1)^3 + \cdots; \quad g_1 = .001, \quad r = .1$

$\qquad G = \dfrac{503}{100} + \dfrac{.001}{1 - .1} = \dfrac{503}{100} + \dfrac{.001}{.9} = \dfrac{503}{100} + \dfrac{1}{900} = \dfrac{4,527}{900} + \dfrac{1}{900} = \dfrac{4,528}{900} = \dfrac{1,132}{225}$

51. With the first mailing you reach 10 people, the second mailing reaches 100 people, the third mailing reaches 1000 people, etc. This forms a geometric sequence: 10, 100, 1000, ...
The total number of people involved is $1 + G_n$ (add 1 to include yourself). The number involved in five mailings is:

$$1 + G_5 = 1 + \frac{10(1 - 10^5)}{1 - 10}$$

$$1 + G_5 = 1 + \frac{10(1 - 100,000)}{-9}$$

$$1 + G_5 = 1 + \frac{10(-99,999)}{-9}$$

$$1 + G_5 = 1 + 111,111$$

$$G_5 = 111,111$$

280

53. The total distance the ball will travel is:

$$10 + \tfrac{9}{10}(10) + \tfrac{9}{10}(10) + \tfrac{9}{10}\cdot\tfrac{9}{10}(10) + \tfrac{9}{10}\cdot\tfrac{9}{10}(10) + \cdots \text{(Recall the upward and downward motion)}$$

$$= 10 + \left(9 + 9 + \tfrac{81}{10} + \tfrac{81}{10} + \cdots\right); \quad g_1 = 9, \quad r = \tfrac{9}{10}$$

$$= 10 + 2\left(\frac{9}{1 - \tfrac{9}{10}}\right)$$

$$= 10 + 2\left(\frac{9}{\tfrac{1}{10}}\right)$$

$$= 10 + 2(90)$$

$$= 190 \text{ ft}$$

55. The total depreciation is: $2{,}000 + 1{,}600 + 1{,}280 + \cdots$; $\quad g_1 = 2{,}000, \quad r = \frac{1{,}600}{2{,}000} = .8$

$$G = \frac{2{,}000}{1 - .8} = \frac{2{,}000}{.2} = \$10{,}000$$

57. a. $\quad x = \sqrt{1\cdot 8} = \sqrt{8} = 2\sqrt{2}$

    b. $\quad x = \sqrt{2\cdot 8} = \sqrt{16} = 4$

    c. $\quad x = -\sqrt{(-5)\cdot(-3)} = -\sqrt{15}$

    d. $\quad x = -\sqrt{(-10)\cdot(-2)} = -\sqrt{20} = -2\sqrt{5}$

    e. $\quad x = \sqrt{4\cdot 20} = \sqrt{80} = 4\sqrt{5}$

59. $g_1 = 3, \quad r = \frac{\sqrt{3}}{3}; \quad G = \frac{3}{1 - \frac{\sqrt{3}}{3}} = \frac{9}{3 - \sqrt{3}} = \frac{9}{3 - \sqrt{3}}\cdot\frac{3 + \sqrt{3}}{3 + \sqrt{3}} = \frac{9(3 + \sqrt{3})}{9 - 3} = \frac{9 + 3\sqrt{3}}{2}$

61. $g_1 = \sqrt{2} - 1, \quad r = \frac{\sqrt{2} + 1}{1} = \sqrt{2} + 1$; No sum since $|r| > 1$.

63. a. $\quad p = 1; \quad \lim\limits_{n\to\infty} \sum\limits_{k=1}^{n} \frac{1}{k} = 1 + \frac{1}{2} + \frac{1}{3} + \cdots$; No limit

    b. $\quad p = 2; \quad \lim\limits_{n\to\infty} \sum\limits_{k=1}^{n} \frac{1}{k^2} = 1 + \frac{1}{4} + \frac{1}{9} + \cdots$; Limit exists

    c. $\quad p = 3; \quad \lim\limits_{n\to\infty} \sum\limits_{k=1}^{n} \frac{1}{k^3} = 1 + \frac{1}{8} + \frac{1}{27} + \cdots$; Limit exists

    d. $\quad p = .5; \quad \lim\limits_{n\to\infty} \sum\limits_{k=1}^{n} \frac{1}{k^{.5}} = 1 + \frac{1}{\sqrt{2}} + \frac{1}{\sqrt{3}} + \cdots$; No limit

65.

$$\text{Area} = \frac{1}{8} + \frac{1}{2}\cdot\frac{1}{8} + \frac{1}{2}\cdot\left(\frac{1}{2}\cdot\frac{1}{8}\right) + \frac{1}{2}\cdot\left(\frac{1}{2}\cdot\frac{1}{2}\cdot\frac{1}{8}\right) + \cdots$$

$$= \frac{1}{8} + \frac{1}{8}\cdot\left(\frac{1}{2}\right) + \frac{1}{8}\cdot\left(\frac{1}{2}\right)^2 + \frac{1}{8}\cdot\left(\frac{1}{2}\right)^3 + \cdots; \quad g_1 = \frac{1}{8}, \quad r = \frac{1}{2}$$

$$\text{Area} = G = \frac{\frac{1}{8}}{1-\frac{1}{2}} = \frac{\frac{1}{8}}{\frac{1}{2}} = \frac{1}{4}$$

CHAPTER 10 SUMMARY

1. If a given proposition $P(n)$ is true for $P(1)$ and if the truth of $P(k)$ implies the truth of $P(k+1)$, then $P(n)$ is true for all positive integers.

3. $P(k):\ 4+8+12+\cdots+4k = 2k(k+1)$

$P(k+1):\ 4+8+12+\cdots+4(k+1) = 2(k+1)(k+2)$

5. Using row 5 of Pascal's triangle:

$(a+b)^5 = a^5 + 5a^4b + 10a^3b^2 + 10a^2b^3 + 5ab^4 + b^5$

7. Using row 5 of Pascal's triangle:

$$(2x+y)^5 = (2x)^5 + 5\cdot(2x)^4(y) + 10\cdot(2x)^3(y)^2 + 10\cdot(2x)^2(y)^3 + 5\cdot(2x)(y)^4 + (y)^5$$

$$= 32x^5 + 80x^4y + 80x^3y^2 + 40x^2y^3 + 10xy^4 + y^5$$

9. $\dfrac{52!}{5!47!} = \dfrac{52\cdot51\cdot50\cdot49\cdot48\cdot47!}{5\cdot4\cdot3\cdot2\cdot1\cdot47!} = 2{,}598{,}960$

11. $\dbinom{8}{4} = \dfrac{8!}{4!(8-4)!} = \dfrac{8!}{4!\cdot4!} = \dfrac{8\cdot7\cdot6\cdot5\cdot4!}{4\cdot3\cdot2\cdot1\cdot4!} = 70$

13. $(a+b)^n = \sum\limits_{k=0}^{n}\dbinom{n}{k}a^{n-k}b^k = \dbinom{n}{0}a^n + \dbinom{n}{1}a^{n-1}b + \dbinom{n}{2}a^{n-2}b^2 + \cdots + \dbinom{n}{n}b^n$

15. $\dbinom{15}{r}x^{15-r}(-y)^r$

17. Arithmetic; $d = 11-1 = 10$; $a_1 = 1$; $a_n = 1+(n-1)(10) = -9+10n$

19. Neither; 11111, 111111

21. $\sum\limits_{k=0}^{3} 3^k = 3^0 + 3^1 + 3^2 + 3^3 = 1+3+9+27 = 40$

23. $\displaystyle\sum_{k=1}^{10} 2(3)^{k-1} = 2 + 6 + 18 + \cdots$; a geometric series, $g_1 = 2$, $r = 3$

$$\sum_{k=1}^{10} 2(3)^{k-1} = G_{10} = \frac{2(1-3^{10})}{1-3} = 59{,}048$$

25. The sequence is arithmetic, $a_1 = 1$, $d = 10$;

$$A_{10} = \tfrac{10}{2}[2(1) + (10-1)(10)] = 5(2 + 90) = 460$$

27. The sequence is geometric, $g_1 = 54$, $r = \tfrac{1}{3}$;

$$G_{10} = \frac{54\left[1 - \left(\frac{1}{3}\right)^{10}\right]}{1 - \frac{1}{3}} = \frac{54\left[1 - \left(\frac{1}{3^{10}}\right)\right]}{1 - \frac{1}{3}} \cdot \frac{3}{3} = \frac{162\left[1 - \left(\frac{1}{3^{10}}\right)\right]}{3 - 1} = 81\left[1 - \left(\frac{1}{3^{10}}\right)\right]$$

$$= 81 - 3^{-6} \approx 80.99862826$$

29. $a_{10} = a_1 + (10-1)d$; $\quad 20 = 2 + 9d$; $\quad d = 2$; $\quad A_{10} = \tfrac{10}{2}[2 + 20] = 110$

31. $g_{10} = 5 \cdot 2^{10-1} = 2{,}560$; $\quad G_5 = \dfrac{5(1 - 2^5)}{1 - 2} = \dfrac{5(-31)}{-1} = 155$

33. $g_1 = 1{,}000$, $\quad r = \tfrac{1}{2}$; $\quad G = \dfrac{1{,}000}{1 - \frac{1}{2}} = \dfrac{1{,}000}{\frac{1}{2}} = 2{,}000$

35. $\quad 2.\overline{18} = 2 + .18 + .0018 + .000018 + \cdots$

$\qquad\qquad = 2 + .18 + .18(.01) + .18(.01)^2 + \cdots$; $\quad g_1 = .18$, $\quad r = .01$

$$G = 2 + \frac{.18}{1 - .01} = 2 + \frac{.18}{.99} = 2 + \frac{18}{99} = 2 + \frac{2}{11} = \frac{24}{11}$$

CHAPTER 11

1.

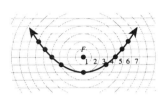

3.

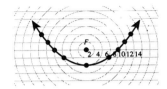

5.

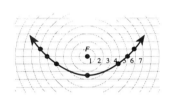

7

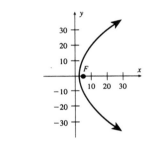

9.

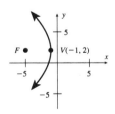

11.

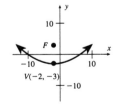

13.

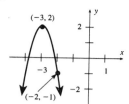

15.

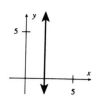

17. $y^2 = 8x$

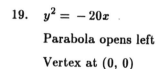

Parabola opens right

Vertex at $(0, 0)$

$4c = 8, \quad c = 2$

Focus at $(2, 0)$

Focal chord length $= 8$

19. $y^2 = -20x$

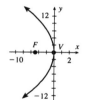

Parabola opens left

Vertex at $(0, 0)$

$4c = 20, \quad c = 5$

Focus at $(-5, 0)$

Focal chord length $= 20$

21. $3x^2 = -12y$

$x^2 = -4y$

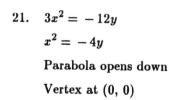

Parabola opens down

Vertex at $(0, 0)$

$4c = 4, \quad c = 1$

Focus at $(0, -1)$

Focal chord length $= 4$

23. $2x^2 + 5y = 0$

$x^2 = -\frac{5}{2}y$

Parabola opens down

Vertex at $(0, 0)$

$4c = \frac{5}{2}; \quad c = \frac{5}{8}$

Focus at $\left(0, -\frac{5}{8}\right)$

Focal chord length $= \frac{5}{2}$

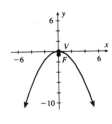

25. $3y^2 - 15x = 0$

$y^2 = 5x$

Parabola opens right

Vertex at $(0, 0)$

$4c = 5, \quad c = \frac{5}{4}$

Focus at $\left(\frac{5}{4}, 0\right)$

Focal chord length $= 5$

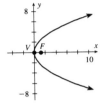

27. $5x^2 + 4y = 20; \quad 5x^2 = -4y + 20; \quad 5x^2 = -4(y - 5)$

$x^2 = -\frac{4}{5}(y - 5)$

Parabola opens down

Vertex at $(0, 5)$

$4c = \frac{4}{5}, \quad c = \frac{1}{5}$

Focus at $\left(0, 5 - \frac{1}{5}\right) = \left(0, \frac{24}{5}\right)$

Focal chord length $= \frac{4}{5}$

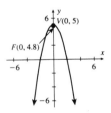

29. $(y - 1)^2 = 2(x + 2)$

Parabola opens right

Vertex at $(-2, 1)$

$4c = 2, \quad c = \frac{1}{2}$

Focus at $\left(-2 + \frac{1}{2}, 1\right) = \left(-\frac{3}{2}, 1\right)$

Focal chord length $= 2$

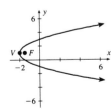

31. $(x + 2)^2 = 2(y - 1)$

   Parabola opens up

   Vertex at $(-2, 1)$

   $4c = 2, \quad c = \frac{1}{2}$

   Focus at $\left(-2, \; 1 + \frac{1}{2}\right) = \left(-2, \frac{3}{2}\right)$

   Focal chord length $= 2$

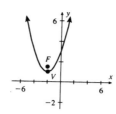

33. $(x - 1) = -2(y + 2)$

   The graph is a line.

   $x - 1 = -2y - 4$

   $2y = -x - 3$

   $y = -\frac{1}{2}x - \frac{3}{2}$

   Lines have no vertex or focus.

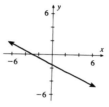

35. $y^2 + 4x - 3y + 1 = 0$

   $y^2 - 3y + \frac{9}{4} = -4x - 1 + \frac{9}{4}$

   $(y - \frac{3}{2})^2 = -4(x - \frac{5}{16})$

   Parabola opens left; Vertex at $\left(\frac{5}{16}, \frac{3}{2}\right)$

   $4c = 4; \quad c = 1$

   Focus at $\left(\frac{5}{16} - 1, \frac{3}{2}\right) = \left(-\frac{11}{16}, \frac{3}{2}\right)$

   Focal chord length $= 4$

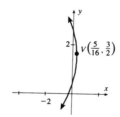

37. $2y^2 + 8y - 20x + 148 = 0$

   $y^2 + 4y + 4 = 10x - 74 + 4$

   $(y + 2)^2 = 10(x - 7)$

   Parabola opens right; Vertex at $(7, \; -2)$

   $4c = 10; \quad c = \frac{5}{2}$

   Focus at $\left(7 + \frac{5}{2}, \; -2\right) = \left(\frac{19}{2}, -2\right)$

   Focal chord length $= 10$

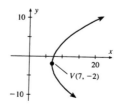

39. $9x^2 + 6x + 18y - 23 = 0$

$$9\left(x^2 + \tfrac{2}{3}x + \tfrac{1}{9}\right) = -18y + 23 + 1$$
$$9\left(x + \tfrac{1}{3}\right)^2 = -18\left(y - \tfrac{4}{3}\right)$$
$$\left(x + \tfrac{1}{3}\right)^2 = -2\left(y - \tfrac{4}{3}\right)$$

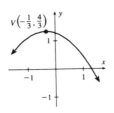

Parabola opens down; Vertex at $\left(-\tfrac{1}{3}, \tfrac{4}{3}\right)$

$4c = 2;\quad c = \tfrac{1}{2};\quad$ Focus at $\left(-\tfrac{1}{3}, \tfrac{4}{3} - \tfrac{1}{2}\right) = \left(-\tfrac{1}{3}, \tfrac{5}{6}\right)$

Focal chord length $= 2$

41. Sketch the curve as shown at the right. By inspection, the vertex is the point $\left(\tfrac{5}{2}, 0\right)$. Also $c = \tfrac{5}{2}$.

Since the parabola opens right, the equation has form $(y - k)^2 = 4c(x - h)$. Thus the equation is:

$$y^2 = 10\left(x - \tfrac{5}{2}\right)$$

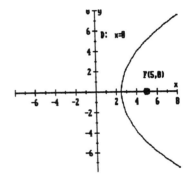

43. Sketch the curve as shown at the right. By inspection, $c = 4$. Since the parabola opens left, the equation has the form $(y - k)^2 = -4c(x - h)$. Thus the equation is $(y - 2)^2 = -16(x + 1)$.

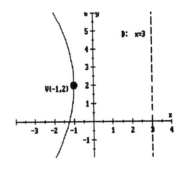

45. Sketch the curve as shown at the right. By inspection, $c = 6$. Since the parabola opens up, the equation has the form $(x - h)^2 = 4c(y - k)$. Thus the equation is $(x + 2)^2 = 24(y + 3)$

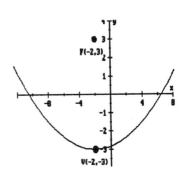

47. Since the axis is parallel to the y-axis, the equation has form $(x - h)^2 = 4c(y - k)$.

Using the vertex $(-3,\ 2)$ the equation becomes $(x + 3)^2 = 4c(y - 2)$. Substitute

the point $(-2,\ -1)$ for x and y in the equation: $(-2 + 3)^2 = 4c(-1 - 2)$;

$1 = -12c;\ c = -\frac{1}{12}$. So the equation is: $(x + 3)^2 = 4\left(-\frac{1}{12}\right)(y - 2)$ which

becomes $(x + 3)^2 = -\frac{1}{3}(y - 2)$

49. Let $(x,\ y)$ be any point on the curve. Then

the distance from $(4,\ 3)$ to $(x,\ y) =$ the distance from $(0,\ 3)$ to (x, y)

$$\sqrt{(x - 4)^2 + (y - 3)^2} = \sqrt{(x - 0)^2 + (y - 3)^2}$$
$$(x - 4)^2 + (y - 3)^2 = x^2 + (y - 3)^2$$
$$(x - 4)^2 = x^{2'}$$
$$x^2 - 8x + 16 = x^2$$
$$8x = 16$$
$$x = 2$$

51. From the sketch of the ball's path, it is clear that the
vertex is the point $(100, 50)$. So the equation is
$(x - 100)^2 = -4c(y - 50)$. The parabola passes
through the point $(0, 0)$. Hence, $(0 - 100)^2 = -4c(0 - 50)$;
$10000 = 200c;\ c = 50$. So the equation is:
$(x - 100)^2 = -200(y - 50);\quad D = [0,\ 200]$

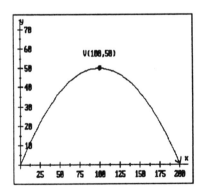

53. Sketch the parabola so the vertex is at the origin.
Thus the equation is $y^2 = 4cx$. Since the point $(4,\ 6)$ is
on the parabola, $(4,\ 6)$ satisfies the equation. Substituting
$(4,\ 6)$ into the equation gives: $6^2 = 4c(4);\quad 36 = 16c$;
$c = 2.25$. So, the focus is 2.25 m from the vertex on the
axis of the parabola.

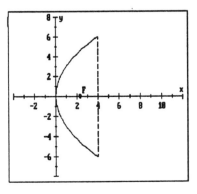

289

55.
$$\begin{cases} y = 2x + 10 \\ y = x^2 + 4x + 7 \end{cases}$$

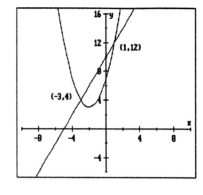

Using substitution:

$x^2 + 4x + 7 = 2x + 10$

$x^2 + 2x - 3 = 0$

$(x + 3)(x - 1) = 0$

$x = -3: \quad y = 2(-3) + 10 = 4$

$x = 1: \quad y = 2(1) + 10; \quad y = 12$

The solutions are $(-3, \ 4), \ (1, \ 12)$

57.
$$\begin{cases} 2x + y - 7 = 0 \\ y^2 - 6y - 4x + 17 = 0 \end{cases}$$

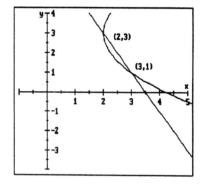

Solve the first equation for y: $\quad y = -2x + 7$

Substitute into the second equation:

$(-2x + 7)^2 - 6(-2x + 7) - 4x + 17 = 0$

$4x^2 - 28x + 49 + 12x - 42 - 4x + 17 = 0$

$\qquad\qquad\qquad 4x^2 - 20x + 24 = 0$

$\qquad\qquad\qquad\quad x^2 - 5x + 6 = 0$

$\qquad\qquad\qquad (x - 2)(x - 3) = 0$

$x = 2: \quad y = -2(2) + 7 = 3$

$x = 3: \quad y = -2(3) + 7 = 1$

The solutions are $(2, \ 3), \ (3, \ 1)$

59. Substitute the point $(20, \ 11.0)$ into the equation,

$11.0 = c_1(20 - 20)^2 + c_2; \quad c_2 = 11.0$

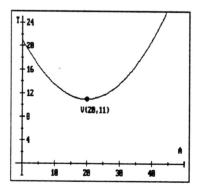

Subtitute the point $(16, \ 11.4)$ into the equation,

$11.4 = c_1(16 - 20)^2 + 11.0; \quad .4 = 16c_1; \quad c_1 = .025$

Therefore, the equation is: $\quad T = .025(A - 20)^2 + 11.0$

When $A = 40$: $\quad T = .025(40 - 20)^2 + 11.0;$

$T = 21.0$ seconds

290

61. Let (x, y) be a point P on the parabola. Then the distance from P to the line is $|c - y|$.
Using the definition of a parabola:
$$|PF| = |c - y|$$
$$\sqrt{(x-0)^2 + (y+c)^2} = |c - y|$$
$$x^2 + (y+c)^2 = (c - y)^2$$
$$x^2 + y^2 + 2cy + c^2 = c^2 - 2cy + y^2$$
$$x^2 = -4cy$$

63. Let (x, y) by a point P on the parabola. Then the distance from P to the line is $|c - x|$.
Using the definition of a parabola:
$$|PF| = |c - x|$$
$$\sqrt{(x+c)^2 + (y-0)^2} = |c - x|$$
$$(x+c)^2 + y^2 = (c - x)^2$$
$$x^2 + 2cx + c^2 + y^2 = c^2 - 2cx + x^2$$
$$y^2 = -4cx$$

65. a. Solving $Ax + By + C = 0$ for y yields: $y = -\frac{A}{B}x - \frac{C}{B}$. So the slope $m = -\frac{A}{B}$.

b. The slope of the line L' that is perpendicular to L is the negative reciprocal: $m' = \frac{B}{A}$.

c. Using the point slope form, $y - y_0 = \frac{B}{A}(x - x_0)$; $\quad Ay - Ay_0 = Bx - Bx_0$;
$$Bx - Ay - Bx_0 + Ay_0 = 0$$

d. Substitute $y = \frac{B}{A}(x - x_0) + y_0$ into the equation for L:
$$Ax + B\left[\frac{B}{A}(x - x_0) + y_0\right] + C = 0$$
$$Ax + \frac{B^2}{A}x - \frac{B^2}{A}x_0 + By_0 + C = 0$$
$$\left[A + \frac{B^2}{A}\right]x = \frac{B^2}{A}x_0 - By_0 - C$$
$$\left[\frac{A^2 + B^2}{A}\right]x = \frac{B^2}{A}x_0 - By_0 - C$$
$$x = \left[\frac{B^2x_0 - ABy_0 - AC}{A}\right] \cdot \left[\frac{A}{A^2 + B^2}\right]$$
$$x = \frac{B^2x_0 - ABy_0 - AC}{A^2 + B^2}$$

(Continued on next page)

$$y = \frac{B}{A} \cdot \left[\frac{B^2 x_0 - ABy_0 - AC}{A^2 + B^2} - x_0 \right] + y_0$$

$$= \frac{B(B^2 x_0 - ABy_0 - AC)}{A(A^2 + B^2)} - \frac{(A^2 + B^2)Bx_0}{A(A^2 + B^2)} + \frac{A(A^2 + B^2)y_0}{A(A^2 + B^2)}$$

$$= \frac{B^3 x_0 - AB^2 y_0 - ABC - A^2 Bx_0 - B^3 x_0 + A^3 y_0 + AB^2 y_0}{A(A^2 + B^2)}$$

$$= \frac{-A^2 Bx_0 + A^3 y_0 - ABC}{A(A^2 + B^2)}$$

$$= \frac{-ABx_0 + A^2 y_0 - BC}{A^2 + B^2}$$

e. $$d = \sqrt{\left[x_0 - \frac{B^2 x_0 - ABy_0 - AC}{A^2 + B^2} \right]^2 + \left[y_0 - \frac{-ABx_0 + A^2 y_0 - BC}{A^2 + B^2} \right]^2}$$

$$= \sqrt{\left[\frac{A^2 x_0 + ABy_0 + AC}{A^2 + B^2} \right]^2 + \left[\frac{ABx_0 + B^2 y_0 + BC}{A^2 + B^2} \right]^2}$$

$$= \sqrt{\frac{A^2(Ax_0 + By_0 + C)^2}{(A^2 + B^2)^2} + \frac{B^2(Ax_0 + By_0 + C)^2}{(A^2 + B^2)^2}}$$

$$= \sqrt{\frac{(Ax_0 + By_0 + C)^2(A^2 + B^2)}{(A^2 + B^2)^2}}$$

$$= \frac{|Ax_0 + By_0 + C|}{\sqrt{A^2 + B^2}}$$

67. For any point $P(x,\ y)$ on the parabola:

$$|PF| = \text{Distance from } P \text{ to the directrix}$$

$$\sqrt{(x-3)^2 + (y+5)^2} = \frac{|12x - 5y + 4|}{\sqrt{12^2 + 5^2}}$$

$$(x-3)^2 + (y+5)^2 = \frac{(12x - 5y + 4)^2}{13^2}$$

$$169(x^2 - 6x + 9 + y^2 + 10y + 25) = 144x^2 - 120xy + 96x + 25y^2 - 40y + 16$$

$$169x^2 - 1{,}014x + 169y^2 + 1{,}690y + 5{,}746 = 144x^2 - 120xy + 96x + 25y^2 - 40y + 16$$

$$25x^2 + 120xy + 144y^2 - 1{,}110x + 1{,}730y + 5{,}730 = 0$$

1.

3.

5.
$$25x^2 + 16y^2 = 400$$
$$\frac{x^2}{16} + \frac{y^2}{25} = 1$$

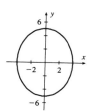

7.
$$3x^2 + 2y^2 = 6$$
$$\frac{x^2}{2} + \frac{y^2}{3} = 1$$

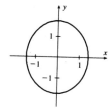

9.
$$5x^2 + 10y^2 = 7$$
$$\frac{5x^2}{7} + \frac{10y^2}{7} = 1$$
$$\frac{x^2}{7/5} + \frac{y^2}{7/10} = 1$$

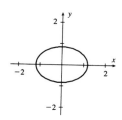

11. $(x+4)^2 + (y-2)^2 = 49$
$$\frac{(x+4)^2}{49} + \frac{(y-2)^2}{49} = 1$$

13.

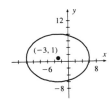

15.

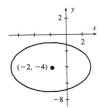

17.

$$10(x - 5)^2 + 6(y + 2)^2 = 60$$

$$\frac{(x - 5)^2}{6} + \frac{(y + 2)^2}{10} = 1$$

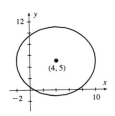

19.

21.

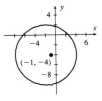

23. By inspection the ellipse is horizontal with center $(0, 0)$ and $c = 4$. Also $2a = 10$, so $a = 5$.

Since $c^2 = a^2 - b^2$, $16 = 25 - b^2$, $b^2 = 9$, so $b = 3$.

The intercepts are $(5, 0)$, $(-5, 0)$, $(0, 3)$, and $(0, -3)$.

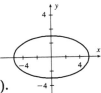

25. By inspection the ellipse is vertical with center $(0, 0)$, $c = 5$ and $a = 7$. Since $c^2 = a^2 - b^2$, $25 = 49 - b^2$, $b^2 = 24$, $b = \sqrt{24} = 2\sqrt{6}$. The pseudo-vertices are $(2\sqrt{6},\ 0)$ and $(-2\sqrt{6},\ 0)$.

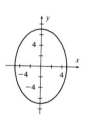

27. By inspection the ellipse is horizontal with center $(-1,\ 3)$, $c = 3$, and $a = 5$. Since $c^2 = a^2 - b^2$, $9 = 25 - b^2$, $b^2 = 16$, $b = 4$. The pseudo-vertices are $(-1,\ 7)$ and $(-1,\ -1)$.

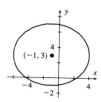

29. For any two points on the circle:

$(x_1 - h)^2 + (y_1 - k)^2 = (x_2 - h)^2 + (y_2 - k)^2$. So,

$(2 - h)^2 + (2 - k)^2 = (-2 - h)^2 + (-6 - k)^2$ and $(2 - h)^2 + (2 - k)^2 = (5 - h)^2 + (1 - k)^2$.

From the first equation:

$$4 - 4h + h^2 + 4 - 4k + k^2 = 4 + 4h + h^2 + 36 + 12k + k^2$$
$$8 - 4h - 4k = 40 + 4h + 12k$$
$$h + 2k = -4$$

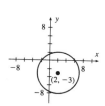

From the second equation:

$$4 - 4h + h^2 + 4 - 4k + k^2 = 25 - 10h + h^2 + 1 - 2k + k^2$$
$$8 - 4h - 4k = 26 - 10h - 2k$$
$$3h - k = 9$$

Solve the system: $\begin{cases} h + 2k = -4 \\ 3h - k = 9 \end{cases}$

(Continued on next page)

Solve the first equation for h, $h = -4 - 2k$

Substitute into the second equation: $3(-4-2k) - k = 9$; $-7k = 21$; $k = -3$

Substitute and solve for h, $h = -4 - 2(-3)$; $h = 2$

The circle has center $(2, -3)$. To find the radius, find the distance from the center to one of the given points: $r = \sqrt{(5-2)^2 + (1+3)^2} = \sqrt{9+16} = 5$

The equation of the circle is: $(x-2)^2 + (y+3)^2 = 25$

31. The graph is a circle with equation: $(x-4)^2 + (y-5)^2 = 36$

33. The graph is a circle with equation: $(x+1)^2 + (y+4)^2 = 36$

35. The graph is an ellipse with center $(0, 0)$, $c = 4$, and $2a = 10$, $a = 5$.

Since $c^2 = a^2 - b^2$, $16 = 25 - b^2$, $b^2 = 9$, $b = 3$.

The equation of the ellipse is: $\dfrac{x^2}{25} + \dfrac{y^2}{9} = 1$

37. By inspection the center of the ellipse is $(0, 0)$, $a = 7$, and $c = 5$.

Since $c^2 = a^2 - b^2$, $25 = 49 - b^2$, $b^2 = 24$, $b = 2\sqrt{6}$.

The equation of the ellipse is: $\dfrac{x^2}{25} + \dfrac{y^2}{49} = 1$

39. By inspection the center of the ellipse is $(-1, 3)$, $a = 5$, and $c = 3$.

Since $c^2 = a^2 - b^2$, $9 = 25 - b^2$, $b^2 = 16$, $b = 4$.

The equation of the ellipse is: $\dfrac{(x+1)^2}{24} + \dfrac{(y-3)^2}{16} = 1$

41. $(x-2)^2 + (y+3)^2 = 25$ (See problem 29)

43.
$$x^2 + 4x + y^2 + 6y - 12 = 0$$
$$(x^2 + 4x + 4) + (y^2 + 6y + 9) = 12 + 4 + 9$$
$$(x+2)^2 + (y+3)^2 = 25$$

Circle with center at $(-2, -3)$ and radius 5

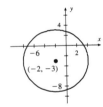

45.
$$16x^2 + 9y^2 + 96x - 36y + 36 = 0$$
$$16(x^2 + 6x + 9) + 9(y^2 - 4y + 4) = -36 + 144 + 36$$
$$16(x + 3)^2 + 9(y - 2)^2 = 144$$
$$\frac{(x + 3)^2}{9} + \frac{(y - 2)^2}{16} = 1$$

Ellipse with center at $(-3,\ 2)$, $a = 4$, and $b = 3$

47.
$$3x^2 + 4y^2 + 2x - 8y + 4 = 0$$
$$3(x^2 + \tfrac{2}{3}x + \tfrac{1}{9}) + 4(y^2 - 2y + 1) = -4 + \tfrac{1}{3} + 4$$
$$3(x + \tfrac{1}{3})^2 + 4(y - 1)^2 = \tfrac{1}{3}$$
$$\frac{(x + \tfrac{1}{3})^2}{1/9} + \frac{(y - 1)^2}{1/12} = 1$$

Ellipse with center $\left(-\tfrac{1}{3},\ 1\right)$, $a = \dfrac{1}{\sqrt{12}}$, and $b = \tfrac{1}{3}$

49.
$$y^2 + 4x^2 + 2y - 8x + 1 = 0$$
$$4(x^2 - 2x + 1) + (y^2 + 2y + 1) = -1 + 4 + 1$$
$$4(x - 1)^2 + (y + 1)^2 = 4$$
$$\frac{(x - 1)^2}{1} + \frac{(y + 1)^2}{4} = 1$$

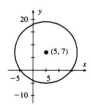

Ellipse with center $(1,\ -1)$, $a = 2$, and $b = 1$

51.
$$x^2 + y^2 - 10x - 14y - 70 = 0$$
$$(x^2 - 10x + 25) + (y^2 - 14y + 49) = 70 + 25 + 49$$
$$(x - 5)^2 + (y - 7)^2 = 144$$

Circle with center $(5,\ 7)$ and $r = 12$

297

53.

$$4x^2 + y^2 + 24x + 4y + 16 = 0$$

$$4(x^2 + 6x + 9) + (y^2 + 4y + 4) = -16 + 36 + 4$$

$$4(x + 3)^2 + (y + 2)^2 = 24$$

$$\frac{(x + 3)^2}{6} + \frac{(y + 2)^2}{24} = 1$$

Ellipse with center $(-3, -2)$, $a = 2\sqrt{6}$, and $b = \sqrt{6}$

55. Let $P(x, y)$ be a point on the ellipse. Then $|PF_1| = \sqrt{(x - 0)^2 + (y - c)^2}$ and $|PF_2| = \sqrt{(x - 0)^2 + (y + c)^2}$. Thus by definition of an ellipse,

$$\sqrt{x^2 + (y - c)^2} + \sqrt{x^2 + (y + c)^2} = 2a$$

$$\sqrt{x^2 + (y - c)^2} = 2a - \sqrt{x^2 + (y + c)^2}$$

$$x^2 + (y - c)^2 = 4a^2 - 4a\sqrt{x^2 + (y + c)^2} + x^2 + (y + c)^2$$

$$y^2 - 2cy + c^2 = 4a^2 - 4a\sqrt{x^2 + (y + c)^2} + y^2 + 2cy + c^2$$

$$4a\sqrt{x^2 + (y + c)^2} = 4a^2 + 4cy$$

$$a\sqrt{x^2 + (y + c)^2} = a^2 + cy$$

$$a^2\left[x^2 + (y + c)^2\right] = a^4 + 2a^2cy + c^2y^2$$

$$a^2x^2 + a^2y^2 + 2a^2cy + a^2c^2 = a^4 + 2a^2cy + c^2y^2$$

$$a^2x^2 + a^2y^2 - c^2y^2 = a^4 - a^2c^2$$

$$a^2x^2 + y^2(a^2 - c^2) = a^2(a^2 - c^2)$$

$$\frac{x^2}{a^2 - c^2} + \frac{y^2}{a^2} = 1$$

Let $b^2 = a^2 - c^2$
$$\frac{x^2}{b^2} + \frac{y^2}{a^2} = 1$$

57. Using $\epsilon = \frac{c}{a}$, $c = a\epsilon = \left(\frac{186,000,000}{2}\right)\left(\frac{1}{62}\right) = 1,500,000$

aphelion $= a + c = 93,000,000 + 1,500,000 = 94,500,000$ miles

perihelion $= a - c = 93,000,000 - 1,500,000 = 91,500,000$ miles

298

59. $a = \frac{1}{2}(378,000) = 189,000$.

apogee $= a + c$, so $c = $ apogee $- a = 199,000 - 189,000 = 10,000$

$\epsilon = \frac{c}{a} = \frac{10,000}{189,000} = \frac{10}{189} \approx .053$

61. Solve the equation of the line $4x - 3y - 23 = 0$ for y: $y = \frac{4x - 23}{3}$. Then substitute:

$$16(x - 2)^2 + 9\left(\frac{4x - 23}{3} + 1\right)^2 = 144$$

$$16(x - 2)^2 + 9\left(\frac{4x - 23 + 3}{3}\right)^2 = 144$$

$$16(x - 2)^2 + 9 \cdot \frac{(4x - 20)^2}{9} = 144$$

$$16(x - 2)^2 + (4x - 20)^2 = 144$$

$$16x^2 - 64x + 64 + 16x^2 - 160x + 400 = 144$$

$$32x^2 - 224x + 320 = 0$$

$$x^2 - 7x + 10 = 0$$

$$(x - 5)(x - 2) = 0$$

$$x = 5 \quad \text{or} \quad x = 2$$

When $x = 5$, $y = \frac{4(5) - 23}{3} = -1$. When $x = 2$, $y = \frac{4(2) - 23}{3} = -5$

The solution is: $(5, -1)$, $(2, -5)$

63. Let $P(x, y)$ be a point on the ellipse. Then, the distance from P to the line $x = \frac{a}{\epsilon}$ is $\left|\frac{a}{\epsilon} - x\right|$.

Also, $|PF| = \sqrt{(x - c)^2 + (y - 0)^2}$. We want to show that the equation

$\sqrt{(x - c)^2 + y^2} = \epsilon\left|\frac{a}{\epsilon} - x\right|$ is equivalent to $\frac{x^2}{a^2} + \frac{y^2}{b^2} = 1$, where $b^2 = a^2 - c^2$.

$$\sqrt{(x - c)^2 + y^2} = \epsilon\left|\frac{a}{\epsilon} - x\right|$$

$$(x - c)^2 + y^2 = \epsilon^2\left(\frac{a}{\epsilon} - x\right)^2$$

$$x^2 - 2cx + c^2 + y^2 = \epsilon^2\left(\frac{a - \epsilon x}{\epsilon}\right)^2$$

$$x^2 - 2cx + c^2 + y^2 = a^2 - 2a\epsilon x + \epsilon^2 x^2, \quad \text{substitute } \epsilon = \frac{c}{a}$$

$$x^2 - 2cx + c^2 + y^2 = a^2 - 2a\left(\frac{c}{a}\right)x + \left(\frac{c}{a}\right)^2 x^2$$

$$x^2 - 2cx + c^2 + y^2 = a^2 - 2cx + \frac{c^2 x^2}{a^2}$$

(Continued on next page)

299

$$a^2x^2 + a^2c^2 + a^2y^2 = a^4 + c^2x^2$$

$$a^2x^2 - c^2x^2 + a^2y^2 = a^4 - a^2c^2$$

$$x^2(a^2 - c^2) + a^2y^2 = a^2(a^2 - c^2)$$

$$\frac{x^2}{a^2} + \frac{y^2}{a^2 - c^2} = 1$$

$$\frac{x^2}{a^2} + \frac{y^2}{b^2} = 1$$

PROBLEM SET 11.3

1.

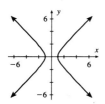

3.

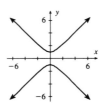

5.

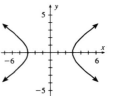

7.

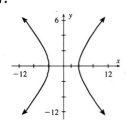

9.

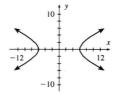

11. $\qquad 3x^2 - 4y^2 = 12$

$$\frac{x^2}{4} + \frac{y^2}{3} = 1$$

13. $3x^2 - 4y^2 = 5$

$\dfrac{x^2}{5/3} + \dfrac{y^2}{5/4} = 1$

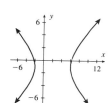

15. $4y^2 - x^2 = 9$

$\dfrac{y^2}{9/4} - \dfrac{x^2}{9} = 1$

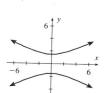

17.

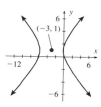

19.

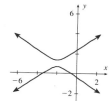

21.

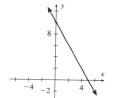

23. Line: $y = -2x + 10$

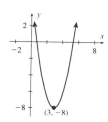

25. Parabola:

$$\dfrac{(x-3)^2}{4} - \dfrac{(y+2)}{6} = 1$$

$$6(x-3)^2 - 4(y+2) = 24$$

$$6(x-3)^2 - 4y - 8 = 24$$

$$6(x-3)^2 = 4y + 32$$

$$6(x-3)^2 = 4(y+8)$$

$$(x-3)^2 = \tfrac{2}{3}(y+8)$$

27. Line:
$$\frac{x-3}{9} + \frac{y-2}{25} = 1$$
$$25(x-3) + 9(y-2) = 225$$
$$25x - 75 + 9y - 18 = 225$$
$$9y = -25x + 318$$
$$y = -\frac{25}{9}x + \frac{106}{3}$$

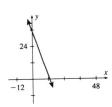

29. Parabola:
$$9(x+3)^2 + 4(y-2) = 0$$
$$9(x+3)^2 = -4(y-2)$$
$$(x+3)^2 = -\frac{4}{9}(y-2)$$

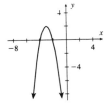

31. Ellipse:
$$x^2 + 8(y-12)^2 = 16$$
$$\frac{x^2}{16} + \frac{(y-12)^2}{2} = 1$$

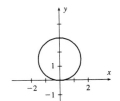

33. Circle:
$$x^2 + y^2 - 3y = 0$$
$$x^2 + \left(y^2 - 3y + \frac{9}{4}\right) = \frac{9}{4}$$
$$x^2 + \left(y - \frac{3}{2}\right)^2 = \frac{9}{4}$$

35. By inspection this is a horizontal hyperbola with center $(0, 0)$, $c = 6$ and $2a = 10$, $a = 5$. Using $c^2 = a^2 + b^2$, $36 = 25 + b^2$, $b^2 = 11$.

The equation of the hyperbola is: $\dfrac{x^2}{25} - \dfrac{y^2}{11} = 1$

37. By inspection this is a vertical hyperbola with center $(4, 6)$, $c = 3$ and $a = 2$. Using $c^2 = a^2 + b^2$, $9 = 4 + b^2$, $b^2 = 5$.

So the equation of the hyperbola is: $\dfrac{(y - 6)^2}{4} - \dfrac{(x - 4)^2}{5} = 1$

39. By inspection this is a horizontal hyperbola with center $(2, 0)$ and $a = 4$.

The equation has the form: $\dfrac{(x - 2)^2}{16} - \dfrac{y^2}{b^2} = 1$

Substitute the point $(10, 3)$: $\dfrac{(10 - 2)^2}{16} - \dfrac{3^2}{b^2} = 1$

$$\dfrac{64}{16} - \dfrac{9}{b^2} = 1$$

$$\dfrac{9}{b^2} = 3$$

$$b^2 = 3$$

The equation of the hyperbola is: $\dfrac{(x - 2)^2}{16} - \dfrac{y^2}{3} = 1$

41.

$$4(x + 4)^2 - 3(y + 3)^2 = -12$$

$$\dfrac{(x + 4)^2}{3} - \dfrac{(y + 3)^2}{4} = -1$$

$$\dfrac{(y + 3)^2}{4} - \dfrac{(x + 4)^2}{3} = 1$$

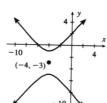

43.
$$9x^2 - 18x - 11 = 4y^2 + 16y$$
$$9(x^2 - 2x + 1) - 4(y^2 + 4y + 4) = 11 + 9 - 16$$
$$9(x - 1)^2 - 4(y + 2)^2 = 4$$
$$\frac{(x-1)^2}{4/9} - \frac{(y+2)^2}{1} = 1$$

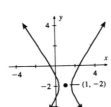

45.
$$x^2 - 4x + y^2 + 6y - 12 = 0$$
$$(x^2 - 4x + 4) + (y^2 + 6y + 9) = 12 + 4 + 9$$
$$(x - 2)^2 + (y + 3)^2 = 25$$
$$\text{Circle}$$

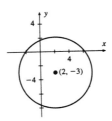

47.
$$3x^2 - 5y^2 + 18x + 10y - 8 = 0$$
$$3(x^2 + 6x + 9) - 5(y^2 - 2y + 1) = 8 + 27 - 5$$
$$3(x + 3)^2 - 5(y - 1)^2 = 30$$
$$\frac{(x+3)^2}{10} - \frac{(y-1)^2}{6} = 1$$

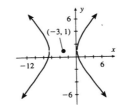

49. Parabola:
$$y^2 - 4x + 2y + 21 = 0$$
$$y^2 + 2y + 1 = 4x - 21 + 1$$
$$(y + 1)^2 = 4x - 20$$
$$(y + 1)^2 = 4(x - 5)$$

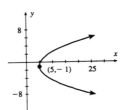

51. **Parabola:**
$$x^2 + 4x + 12y + 64 = 0$$
$$x^2 + 4x + 4 = -12y - 64 + 4$$
$$(x+2)^2 = -12y - 60$$
$$(x+2)^2 = -12(y+5)$$

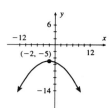

53. **Hyperbola:**
$$100x^2 - 7y^2 + 98y - 368 = 0$$
$$100x^2 - 7(y^2 - 14y + 49) = 368 - 343$$
$$100x^2 - 7(y-7)^2 = 25$$
$$\frac{x^2}{1/4} - \frac{(y-7)^2}{25/7} = 1$$

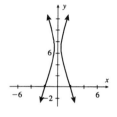

55. **Circle:**
$$4x^2 + 12x + 4y^2 + 4y + 1 = 0$$
$$4\left(x^2 + 3x + \tfrac{9}{4}\right) + 4\left(y^2 + y + \tfrac{1}{4}\right) = -1 + 9 + 1$$
$$4\left(x + \tfrac{3}{2}\right)^2 + 4\left(y + \tfrac{1}{2}\right)^2 = 9$$
$$\left(x + \tfrac{3}{2}\right)^2 + \left(y + \tfrac{1}{2}\right)^2 = \tfrac{9}{4}$$

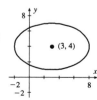

57. **Ellipse:**
$$9x^2 + 25y^2 - 54x - 200y + 256 = 0$$
$$9(x^2 - 6x + 9) + 25(y^2 - 8y + 16) = -256 + 81 + 400$$
$$9(x-3)^2 + 25(y-4)^2 = 225$$
$$\frac{(x-3)^2}{25} + \frac{(y-4)^2}{9} = 1$$

59. Let $P(x, y)$ be a point on the hyperbola, $F_1 = (-c, 0)$, and $F_2 = (c, 0)$.
Then $|PF_1| = \sqrt{(x+c)^2 + (y-0)^2}$ and $|PF_2| = \sqrt{(x-c)^2 + (y-0)^2}$. So,

$$\left\| PF_1| - |PF_2 \right\| = 2a$$

$$\left| \sqrt{(x+c)^2 + y^2} - \sqrt{(x-c)^2 + y^2} \right| = 2a$$

$$\sqrt{(x+c)^2 + y^2} = \sqrt{(x-c)^2 + y^2} \pm 2a$$

$$(x+c)^2 + y^2 = (x-c)^2 + y^2 \pm 4a\sqrt{(x-c)^2 + y^2} + 4a^2$$

$$x^2 + 2cx + c^2 + y^2 = x^2 - 2cx + c^2 + y^2 \pm 4a\sqrt{(x-c)^2 + y^2} + 4a^2$$

$$\pm 4a\sqrt{(x-c)^2 + y^2} = 4a^2 - 4cx$$

$$\pm a\sqrt{(x-c)^2 + y^2} = a^2 - cx$$

$$a^2\left[(x-c)^2 + y^2\right] = a^4 - 2a^2cx + c^2x^2$$

$$a^2x^2 - 2a^2xc + a^2c^2 + a^2y^2 = a^4 - 2a^2cx + c^2x^2$$

$$a^2x^2 - c^2x^2 + a^2y^2 = a^4 - a^2c^2$$

$$x^2(a^2 - c^2) + a^2y^2 = a^2(a^2 - c^2)$$

$$\frac{x^2}{a^2} + \frac{y^2}{a^2 - c^2} = 1 \qquad\qquad \text{Substitute } b^2 = a^2 - c^2$$

$$\frac{x^2}{a^2} + \frac{y^2}{b^2} = 1$$

61. For positive values of x and a, $x = \sqrt{x^2} > \sqrt{x^2 - a^2}$. In the first quadrant the vertical distance between the line and the hyperbola is:

$$d = \tfrac{b}{a}x - \tfrac{b}{a}\sqrt{x^2 - a^2} = \tfrac{b}{a}(x - \sqrt{x^2 - a^2}) = \tfrac{b}{a}\left[x - \sqrt{x^2\left(1 - \tfrac{a^2}{x^2}\right)}\right] = \tfrac{b}{a}x\left(1 - \sqrt{1 - \tfrac{a^2}{x^2}}\right).$$

Since a is a constant, $\dfrac{a^2}{x^2} \to 0$ as $x \to \infty$. So as $x \to \infty$, $d \to 0$.

63. The corners of the rectangle have coordinates (a, b), $(a, -b)$, $(-a, b)$, and $(-a, -b)$.

So the length of the diagonal is:

$$d = \sqrt{(a+a)^2 + (b+b)^2} = \sqrt{4a^2 + 4b^2} = 2\sqrt{a^2 + b^2} = 2c. \quad \text{(Note: } c^2 = a^2 + b^2,\ c > 0\text{)}$$

65. From problem 64, the directrix of this hyperbola is vertical.

Consider the focus $(c, 0)$. The y-coordinates when $x = c$ are:

$$\frac{c^2}{a^2} - \frac{y^2}{b^2} = 1; \quad \frac{y^2}{b^2} = \frac{c^2}{a^2} - 1; \quad y^2 = b^2\left(\frac{c^2}{a^2} - 1\right) = \frac{b^2(c^2 - a^2)}{a^2} = \frac{b^2 \cdot b^2}{a^2} = \frac{b^4}{a^2}$$

Therefore, $y^2 = \frac{b^4}{a^2}$, so $y = \pm\frac{b^2}{a}$.

The endpoints of the focal chord are $\left(c, \frac{b^2}{a}\right)$ and $\left(c, -\frac{b^2}{a}\right)$.

Thus the length of the focal chord $= \frac{b^2}{a} - \left(-\frac{b^2}{a}\right) = \frac{2b^2}{a}$

PROBLEM SET 11.4

1.
$$xy = 6$$
$$\frac{1}{\sqrt{2}}(x' - y') \cdot \frac{1}{\sqrt{2}}(x' + y') = 6$$
$$\frac{1}{2}(x' - y')(x' + y') = 6$$
$$x'^2 - y'^2 = 12$$

3.
$$x^2 - 4xy + 4y^2 + 5\sqrt{5}y - 10 = 0$$
$$\frac{1}{5}(2x' - y')^2 - 4\left(\frac{1}{5}\right)(2x' - y')(x' + 2y') + 4\left(\frac{1}{5}\right)(x' + 2y')^2 + 5\sqrt{5}\left(\frac{1}{\sqrt{5}}\right)(x' + 2y') - 10 = 0$$
$$(2x' - y')^2 - 4(2x' - y')(x' + 2y') + 4(x' + 2y')^2 + 25(x' + 2y') - 50 = 0$$
$$4x'^2 - 4x'y' + y'^2 - 8x'^2 - 12x'y' + 8y'^2 + 4x'^2 + 16x'y' + 16y'^2 + 25x' + 50y' - 50 = 0$$
$$25y'^2 + 25x' + 50y' - 50 = 0$$
$$25y'^2 + 50y' = -25x' + 50$$
$$y'^2 + 2y' + 1 = -x' + 2 + 1$$
$$(y' + 1)^2 = -(x' - 3)$$

5. $B^2 - 4AC = 1^2 - 4(0)(0) = 1 > 0$; hyperbola

7. $B^2 - 4AC = (-10)^2 - 4(13)(13) = 100 - 676 = -576 < 0$; ellipse

9. $B^2 - 4AC = 4^2 - 4(1)(4) = 0$; parabola

11. $B^2 - 4AC = (26\sqrt{3})^2 - 4(23)(-3) = 2{,}304 > 0$; hyperbola

13. $B^2 - 4AC = (16\sqrt{3})^2 - 4(24)(8) = 0$; parabola

15. $B^2 - 4AC = (6\sqrt{3})^2 - 4(13)(7) = -256 < 0$; ellipse

17. $B^2 - 4AC = (-3)^2 - 4(5)(1) = -11 < 0$; ellipse

19. $B^2 - 4AC = 3^2 - 4(4)(2) = -23 < 0$; ellipse

21. $B^2 - 4AC = (-\sqrt{56})^2 - 4(2)(7) = 0$; parabola

23.
$\cot 2\theta = \dfrac{0-0}{1}$

$\cot 2\theta = 0$

$2\theta = 90°$

$\theta = 45°$

$x = x'\cos\theta - y'\sin\theta$
$= x'\cos 45° - y'\sin 45°$
$= x'\left(\dfrac{1}{\sqrt{2}}\right) - y'\left(\dfrac{1}{\sqrt{2}}\right)$
$= \dfrac{1}{\sqrt{2}}(x' - y')$

$y = x'\sin\theta + y'\cos\theta$
$= x'\sin 45° + y'\cos 45°$
$= x'\left(\dfrac{1}{\sqrt{2}}\right) + y'\left(\dfrac{1}{\sqrt{2}}\right)$
$= \dfrac{1}{\sqrt{2}}(x' + y')$

25.
$\cot 2\theta = \dfrac{13-13}{-20}$

$\cot 2\theta = 0$

$2\theta = 90°$

$\theta = 45°$

$x = x'\cos\theta - y'\sin\theta$
$= x'\cos 45° - y'\sin 45°$
$= x'\left(\dfrac{1}{\sqrt{2}}\right) - y'\left(\dfrac{1}{\sqrt{2}}\right)$
$= \dfrac{1}{\sqrt{2}}(x' - y')$

$y = x'\sin\theta + y'\cos\theta$
$= x'\sin 45° + y'\cos 45°$
$= x'\left(\dfrac{1}{\sqrt{2}}\right) + y'\left(\dfrac{1}{\sqrt{2}}\right)$
$= \dfrac{1}{\sqrt{2}}(x' + y')$

27.
$\cot 2\theta = \dfrac{1-4}{4}$

$\cot 2\theta = -\dfrac{3}{4}$

$2\theta \approx 126.87°$

$\theta \approx 63.4°$

$\sec 2\theta = -\sqrt{\tan^2 2\theta + 1}$

$\sec 2\theta = -\sqrt{\tfrac{16}{9} + 1}$

$\sec 2\theta = -\dfrac{5}{3}$

$\cos 2\theta = -\dfrac{3}{5}$

$\cos\theta = \sqrt{\dfrac{1+\cos 2\theta}{2}}$
$= \sqrt{\dfrac{1-3/5}{2}} = \dfrac{1}{\sqrt{5}}$

$\sin\theta = \sqrt{\dfrac{1-\cos 2\theta}{2}}$
$= \sqrt{\dfrac{1+3/5}{2}} = \dfrac{2}{\sqrt{5}}$

Thus: $x = x'\cos\theta - y'\sin\theta = x'\left(\dfrac{1}{\sqrt{5}}\right) - y'\left(\dfrac{2}{\sqrt{5}}\right) = \dfrac{1}{\sqrt{5}}(x' - 2y')$

$y = x'\sin\theta + y'\cos\theta = x'\left(\dfrac{2}{\sqrt{5}}\right) - y'\left(\dfrac{1}{\sqrt{5}}\right) = \dfrac{1}{\sqrt{5}}(2x' + y')$

29.

$$\cot 2\theta = \frac{23 - (-3)}{26\sqrt{3}}$$

$$\cot 2\theta = \frac{1}{\sqrt{3}}$$

$$2\theta = 60°$$

$$\theta = 30°$$

$$x = x'\cos\theta - y'\sin\theta$$
$$= x'\cos 30° - y'\sin 30°$$
$$= x'\left(\frac{\sqrt{3}}{2}\right) - y'\left(\frac{1}{2}\right)$$
$$= \tfrac{1}{2}(\sqrt{3}\,x' - y')$$

$$y = x'\sin\theta + y'\cos\theta$$
$$= x'\sin 30° + y'\cos 30°$$
$$= x'\left(\frac{1}{2}\right) + y'\left(\frac{\sqrt{3}}{2}\right)$$
$$= \tfrac{1}{2}(x' + \sqrt{3}\,y')$$

31.

$$\cot 2\theta = \frac{24 - 8}{16\sqrt{3}}$$

$$\cot 2\theta = \frac{1}{\sqrt{3}}$$

$$2\theta = 60°$$

$$\theta = 30°$$

$$x = x'\cos\theta - y'\sin\theta$$
$$= x'\cos 30° - y'\sin 30°$$
$$= x'\left(\frac{\sqrt{3}}{2}\right) - y'\left(\frac{1}{2}\right)$$
$$= \tfrac{1}{2}(\sqrt{3}\,x' - y')$$

$$y = x'\sin\theta + y'\cos\theta$$
$$= x'\sin 30° + y'\cos 30°$$
$$= x'\left(\frac{1}{2}\right) + y'\left(\frac{\sqrt{3}}{2}\right)$$
$$= \tfrac{1}{2}(x' + \sqrt{3}\,y')$$

33.

$$\cot 2\theta = \frac{13 - 7}{-6\sqrt{3}}$$

$$\cot 2\theta = -\frac{1}{\sqrt{3}}$$

$$2\theta = 120°$$

$$\theta = 60°$$

$$x = x'\cos\theta - y'\sin\theta$$
$$= x'\cos 60° - y'\sin 60°$$
$$= x'\left(\frac{1}{2}\right) - y'\left(\frac{\sqrt{3}}{2}\right)$$
$$= \tfrac{1}{2}(x' - \sqrt{3}\,y')$$

$$y = x'\sin\theta + y'\cos\theta$$
$$= x'\sin 60° + y'\cos 60°$$
$$= x'\left(\frac{\sqrt{3}}{2}\right) + y'\left(\frac{1}{2}\right)$$
$$= \tfrac{1}{2}(\sqrt{3}\,x' + y')$$

35.

$$\cot 2\theta = \frac{5 - 1}{-3}$$

$$\cot 2\theta = -\frac{4}{3}$$

$$2\theta \approx 143.13°$$

$$\theta \approx 71.6°$$

$$\sec 2\theta = -\sqrt{\tan^2 2\theta + 1}$$
$$\sec 2\theta = -\sqrt{\frac{9}{16} + 1}$$
$$\sec 2\theta = -\frac{5}{4}$$
$$\cos 2\theta = -\frac{4}{5}$$

$$\cos\theta = \sqrt{\frac{1 + \cos 2\theta}{2}}$$
$$= \sqrt{\frac{1 - 4/5}{2}} = \frac{1}{\sqrt{10}}$$
$$\sin\theta = \sqrt{\frac{1 - \cos 2\theta}{2}}$$
$$= \sqrt{\frac{1 + 4/5}{2}} = \frac{3}{\sqrt{10}}$$

Thus:
$$x = x'\cos\theta - y'\sin\theta = x'\left(\frac{1}{\sqrt{10}}\right) - y'\left(\frac{3}{\sqrt{10}}\right) = \frac{1}{\sqrt{10}}(x' - 3y')$$

$$y = x'\sin\theta + y'\cos\theta = x'\left(\frac{3}{\sqrt{10}}\right) - y'\left(\frac{1}{\sqrt{10}}\right) = \frac{1}{\sqrt{10}}(3x' + y')$$

37. From problem 23: $x = \frac{1}{\sqrt{2}}(x' - y')$ and $y = \frac{1}{\sqrt{2}}(x' + y')$

$$xy = -1$$

$$\frac{1}{\sqrt{2}}(x' - y') \cdot \frac{1}{\sqrt{2}}(x' + y') = -1$$

$$\frac{1}{2}(x'^2 - y'^2) = -1$$

$$\frac{y'^2}{2} - \frac{x'^2}{2} = 1$$

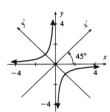

39. As in problem 23: $x = \frac{1}{\sqrt{2}}(x' - y')$ and $y = \frac{1}{\sqrt{2}}(x' + y')$

$$xy = 8$$

$$\frac{1}{\sqrt{2}}(x' - y') \cdot \frac{1}{\sqrt{2}}(x' + y') = 8$$

$$\frac{1}{2}(x'^2 - y'^2) = 8$$

$$\frac{x'^2}{16} - \frac{y'^2}{16} = 1$$

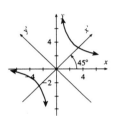

41. From problem 25: $x = \frac{1}{\sqrt{2}}(x' - y')$ and $y = \frac{1}{\sqrt{2}}(x' + y')$

$$13x^2 - 10xy + 13y^2 - 72 = 0$$

$$13 \cdot \frac{1}{2}(x' - y')^2 - 10 \cdot \frac{1}{2}(x' - y')(x' + y') + 13 \cdot \frac{1}{2}(x' + y')^2 - 72 = 0$$

$$\frac{13}{2}x'^2 - 13x'y' + \frac{13}{2}y'^2 - 5x'^2 + 5y'^2 + \frac{13}{2}x'^2 + 13x'y' + \frac{13}{2}y'^2 = 72$$

$$8x'^2 + 18y'^2 = 72$$

$$\frac{x'^2}{9} + \frac{y'^2}{4} = 1$$

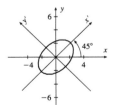

43. From problem 27: $x = \frac{1}{\sqrt{5}}(x' - 2y')$ and $y = \frac{1}{\sqrt{5}}(2x' + y')$

$$x^2 + 4xy + 4y^2 + 10\sqrt{5}\,x = 9$$

$$\tfrac{1}{5}(x' - 2y')^2 + \tfrac{4}{5}(x' - 2y')(2x' + y') + \tfrac{4}{5}(2x' + y')^2 + 10(x' - 2y') = 9$$

$$\tfrac{1}{5}x'^2 - \tfrac{4}{5}x'y' + \tfrac{4}{5}y'^2 + \tfrac{8}{5}x'^2 - \tfrac{12}{5}x'y' - \tfrac{8}{5}y'^2$$

$$+ \tfrac{16}{5}x'^2 + \tfrac{16}{5}x'y' + \tfrac{4}{5}y'^2 + 10x' - 20y' = 9$$

$$5x'^2 + 10x' - 20y' = 9$$

$$5(x'^2 + 2x' + 1) = 20y' + 9 + 5$$

$$5(x' + 1)^2 = 20(y' + \tfrac{7}{10})$$

$$(x' + 1)^2 = 4(y' + \tfrac{7}{10})$$

tan θ = 2

45. From problem 29: $x = \tfrac{1}{2}(\sqrt{3}\,x' - y')$ and $y = \tfrac{1}{2}(x' + \sqrt{3}\,y')$

$$23x^2 + 26\sqrt{3}xy - 3y^2 - 144 = 0$$

$$\tfrac{23}{4}(\sqrt{3}\,x' - y')^2 + \tfrac{26\sqrt{3}}{4}(\sqrt{3}\,x' - y')(x' + \sqrt{3}\,y') - \tfrac{3}{4}(x' + \sqrt{3}\,y')^2 - 144 = 0$$

$$\tfrac{69}{4}x'^2 - \tfrac{23}{2}\sqrt{3}x'y' + \tfrac{23}{4}y'^2 + \tfrac{78}{4}x'^2 + \tfrac{26}{2}\sqrt{3}x'y'$$

$$- \tfrac{78}{4}y'^2 - \tfrac{3}{4}x'^2 - \tfrac{3}{2}\sqrt{3}x'y' - \tfrac{9}{4}y'^2 = 144$$

$$36x'^2 - 16y'^2 = 144$$

$$\frac{x'^2}{4} - \frac{y'^2}{9} = 1$$

30°

47. From problem 31: $x = \tfrac{1}{2}(\sqrt{3}\,x' - y')$ and $y = \tfrac{1}{2}(x' + \sqrt{3}\,y')$

$$24x^2 + 16\sqrt{3}\,xy + 8y^2 - x + \sqrt{3}\,y - 8 = 0$$

$$\tfrac{24}{4}(\sqrt{3}\,x' - y')^2 + \tfrac{16\sqrt{3}}{4}(\sqrt{3}\,x' - y')(x' + \sqrt{3}\,y')$$

$$+ \tfrac{8}{4}(x' + \sqrt{3}\,y')^2 - \tfrac{1}{2}(\sqrt{3}\,x' - y') + \tfrac{\sqrt{3}}{2}(x' + \sqrt{3}\,y') - 8 = 0$$

$$18x'^2 - 12\sqrt{3}\,x'y' + 6y'^2 + 12x'^2 + 8\sqrt{3}x'y' - 12y'^2$$

$$+ 2x'^2 + 4\sqrt{3}\,x'y' + 6y'^2 - \tfrac{\sqrt{3}}{2}x' + \tfrac{1}{2}y' + \tfrac{\sqrt{3}}{2}x' + \tfrac{3}{2}y' = 8$$

$$32x'^2 + 2y' = 8$$

$$x'^2 = -\tfrac{1}{16}(y' - 4)$$

θ = 30°

49. $\cot 2\theta = \frac{3-3}{-10} = 0$ (as in problem 23), so

$x = \frac{1}{\sqrt{2}}(x' - y')$ and $y = \frac{1}{\sqrt{2}}(x' + y')$

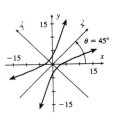

$$3x^2 - 10xy + 3y^2 - 32 = 0$$

$$\frac{3}{2}(x' - y')^2 - \frac{10}{2}(x' - y')(x' + y') + \frac{3}{2}(x' + y')^2 - 32 = 0$$

$$\frac{3}{2}x'^2 - 3x'y' + \frac{3}{2}y'^2 - 5x'^2 + 5y'^2 + \frac{3}{2}x'^2 + 3x'y' + \frac{3}{2}y'^2 = 32$$

$$-2x'^2 + 8y'^2 = 32$$

$$\frac{y'^2}{4} - \frac{x'^2}{16} = 1$$

51.

$\cot 2\theta = \frac{10-17}{24}$ $\sec 2\theta = -\sqrt{\tan^2 2\theta + 1}$ $\cos \theta = \sqrt{\frac{1 + \cos 2\theta}{2}}$

$\cot 2\theta = -\frac{7}{24}$ $\sec 2\theta = -\sqrt{\frac{576}{49} + 1}$ $\cos \theta = \sqrt{\frac{1 + (-7/25)}{2}} = \frac{3}{5}$

$2\theta \approx 106.26°$ $\sec 2\theta = -\frac{25}{7}$ $\sin \theta = \sqrt{\frac{1 - \cos 2\theta}{2}}$

$\theta \approx 53.1°$ $\cos 2\theta = -\frac{7}{25}$ $\sin \theta = \sqrt{\frac{1 - (-7/25)}{2}} = \frac{4}{5}$

Therefore: $x = x'\cos \theta - y'\sin \theta = x'\left(\frac{3}{5}\right) - y'\left(\frac{4}{5}\right) = \frac{1}{5}(3x' - 4y')$

$\qquad\qquad y = x'\sin \theta + y'\cos \theta = x'\left(\frac{4}{5}\right) + y'\left(\frac{3}{5}\right) = \frac{1}{5}(4x' + 3y')$

The equation is: $10x^2 + 24xy + 17y^2 - 9 = 0$

$$\frac{10}{25}(3x' - 4y')^2 + \frac{24}{25}(3x' - 4y')(4x' + 3y') + \frac{17}{25}(4x' + 3y')^2 = 9$$

$$\frac{90}{25}x'^2 - \frac{240}{25}x'y' + \frac{160}{25}y'^2 + \frac{288}{25}x'^2 - \frac{168}{25}x'y' - \frac{288}{25}y'^2 + \frac{272}{25}x'^2 + \frac{408}{25}x'y' + \frac{153}{25}y'^2 = 9$$

$$26x'^2 + y'^2 = 9$$

$$\frac{x'^2}{9/26} + \frac{y'}{9} = 1$$

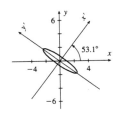

53.

$$\cot 2\theta = \frac{0-(-4)}{3} \qquad \sec 2\theta = \sqrt{\tan^2 2\theta + 1} \qquad \cos \theta = \sqrt{\frac{1+\cos 2\theta}{2}}$$

$$\cot 2\theta = \frac{4}{3} \qquad \sec 2\theta = \sqrt{\frac{9}{16}+1} \qquad \cos \theta = \sqrt{\frac{1+4/5}{2}} = \frac{3}{\sqrt{10}}$$

$$2\theta \approx 36.87° \qquad \sec 2\theta = \frac{5}{4} \qquad \sin \theta = \sqrt{\frac{1-\cos 2\theta}{2}}$$

$$\theta \approx 18.4° \qquad \cos 2\theta = \frac{4}{5} \qquad \sin \theta = \sqrt{\frac{1-4/5}{2}} = \frac{1}{\sqrt{10}}$$

Therefore: $\quad x = x'\cos \theta - y'\sin \theta = x'\left(\frac{3}{\sqrt{10}}\right) - y'\left(\frac{1}{\sqrt{10}}\right) = \frac{1}{\sqrt{10}}(3x' - y')$

$$y = x'\sin \theta + y'\cos \theta = x'\left(\frac{1}{\sqrt{10}}\right) + y'\left(\frac{3}{\sqrt{10}}\right) = \frac{1}{\sqrt{10}}(x' + 3y')$$

The equation is:

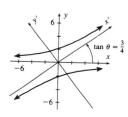

tan $\theta = \frac{3}{4}$

$$3xy - 4y^2 + 18 = 0$$

$$\frac{3}{10}(3x' - y')(x' + 3y') - \frac{4}{10}(x' + 3y')^2 + 18 = 0$$

$$\frac{9}{10}x'^2 + \frac{24}{10}x'y' - \frac{9}{10}y'^2 - \frac{4}{10}x'^2 - \frac{24}{10}x'y' - \frac{36}{10}y'^2 = -18$$

$$\frac{1}{2}x'^2 - \frac{9}{2}y'^2 = -18$$

$$\frac{y'^2}{4} - \frac{x'^2}{36} = 1$$

55. $\cot 2\theta = \frac{5-5}{-8} = 0$ (as in problem 23), so

$$x = \frac{1}{\sqrt{2}}(x' - y') \text{ and } y = \frac{1}{\sqrt{2}}(x' + y')$$

$$5x^2 - 8xy + 5y^2 - 9 = 0$$

$$\frac{5}{2}(x' - y')^2 - \frac{8}{2}(x' - y')(x' + y') + \frac{5}{2}(x' + y')^2 - 9 = 0$$

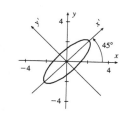

$$\frac{5}{2}x'^2 - 5x'y' + \frac{5}{2}y'^2 - 4x'^2 + 4y'^2 + \frac{5}{2}x'^2 + 5x'y' + \frac{5}{2}y'^2 = 9$$

$$x'^2 + 9y'^2 = 9$$

$$\frac{x'^2}{9} + \frac{y'}{1} = 1$$

57. From problem 33: $\quad x = \frac{1}{2}(x' - \sqrt{3}\,y') \ $ and $\ y = \frac{1}{2}(\sqrt{3}\,x' + y')$

$$13x^2 - 6\sqrt{3}\,xy + 7y^2 + (16\sqrt{3} - 8)x + (-16 - 8\sqrt{3})y + 16 = 0$$

$$\frac{13}{4}(x' - \sqrt{3}\,y')^2 - \frac{6\sqrt{3}}{4}(x' - \sqrt{3}\,y')(\sqrt{3}\,x' + y') + \frac{7}{4}(\sqrt{3}\,x' + y')^2$$

$$+ \frac{1}{2}(16\sqrt{3} - 8)(x' - \sqrt{3}\,y') + \frac{1}{2}(-16 - 8\sqrt{3})(\sqrt{3}\,x' + y') = -16$$

$$\frac{13}{4}x'^2 - \frac{26}{4}\sqrt{3}\,x'y' + \frac{39}{4}y'^2 - \frac{18}{4}x'^2 + \frac{12}{4}\sqrt{3}\,x'y' + \frac{18}{4}y'^2$$

$$+ \frac{21}{4}x'^2 + \frac{14}{4}\sqrt{3}\,x'y' + \frac{7}{4}y'^2 + 8\sqrt{3}x' - 4x' - 24y' + 4\sqrt{3}y'$$

$$- 8\sqrt{3}x' - 12x' - 8y' - 4\sqrt{3}\,y' = -16$$

$$4x'^2 + 16y'^2 - 16x' - 32y' = -16$$

$$4(x'^2 - 4x' + 4) + 16(y'^2 - 2y' + 1) = -16 + 16 + 16$$

$$\frac{(x' - 2)^2}{4} + \frac{(y' - 1)^2}{1} = 1$$

59. $\cot 2\theta = \frac{21 - 31}{10\sqrt{3}} = -\frac{1}{\sqrt{3}}\quad$ (as in problem 33), so

$$x = \frac{1}{2}(x' - \sqrt{3}\,y') \ \text{ and } \ y = \frac{1}{2}(\sqrt{3}\,x' + y')$$

$$21x^2 + 10\sqrt{3}\,xy + 31y^2 + 72x - 16\sqrt{3}\,x - 72\sqrt{3}\,y + 16y + 16 = 0$$

$$\frac{21}{4}(x' - \sqrt{3}\,y')^2 + \frac{10\sqrt{3}}{4}(x' - \sqrt{3}\,y')(\sqrt{3}\,x' + y')$$

$$+ \frac{31}{4}(\sqrt{3}\,x' + y')^2 - \frac{72\sqrt{3}}{2}(x' - \sqrt{3}\,y') - \frac{16\sqrt{3}}{2}(x' - \sqrt{3}\,y')$$

$$- \frac{72\sqrt{3}}{2}(\sqrt{3}\,x' + y') + \frac{16}{2}(\sqrt{3}\,x' + y') = -16$$

$$\frac{21}{4}x'^2 - \frac{21\sqrt{3}}{2}x'y' + \frac{63}{4}y'^2 + \frac{30}{4}x'^2 - 5\sqrt{3}\,x'y' - \frac{30}{4}y'^2$$

$$+ \frac{93}{4}x'^2 + \frac{31\sqrt{3}}{2}x'y' + \frac{31}{4}y'^2 - 36x' + 36\sqrt{3}\,y' - 8\sqrt{3}\,x' + 24y'$$

$$- 108x' - 36\sqrt{3}\,y' + 8\sqrt{3}\,x' + 8y' = -16$$

$$36x'^2 + 16y'^2 - 144x' + 32y' = -16$$

$$36(x'^2 - 4x' + 4) + 16(y'^2 + 2y' + 1) = -16 + 144 + 16$$

$$\frac{(x' - 2)^2}{4} + \frac{(y' + 1)^2}{9} = 1$$

PROBLEM SET 11.5

1. $\left|\vec{v}_x\right| = 12\cos 60° = 6$; $\left|\vec{v}_y\right| = 12\sin 60° = 6\sqrt{3}$; $\vec{v} = 6\vec{i} + 6\sqrt{3}\,\vec{j}$

3. $\left|\vec{v}_x\right| = \sqrt{2}\cos 45° = 1$; $\left|\vec{v}_y\right| = \sqrt{2}\sin 45° = 1$; $\vec{v} = \vec{i} + \vec{j}$

5. $\left|\vec{v}_x\right| = 7\cos 23° \approx 6.4435$; $\left|\vec{v}_y\right| = 7\sin 23° \approx 2.7351$; $\vec{v} = 6.4435\vec{i} + 2.7351\vec{j}$

7. $\left|\vec{v}_x\right| = 4\cos 112° \approx -1.4984$; $\left|\vec{v}_y\right| = 4\sin 112° \approx 3.7087$; $\vec{v} = -1.4984\vec{i} + 3.7087\vec{j}$

9. $\vec{v} = (2-4)\vec{i} + (3-1)\vec{j} = -2\vec{i} + 2\vec{j}$

11. $\vec{v} = (-5-1)\vec{i} + (-7+2)\vec{j} = -6\vec{i} - 5\vec{j}$

13. $\vec{v} = (5+3)\vec{i} + (-8-2)\vec{j} = 8\vec{i} - 10\vec{j}$

15. $\vec{v} = (0-7)\vec{i} + (0-1)\vec{j} = -7\vec{i} - \vec{j}$

17. $\vec{v} = (-3-6)\vec{i} + (-7+1)\vec{j} = -9\vec{i} - 6\vec{j}$

19. $\left|\vec{v}\right| = \sqrt{3^2 + 4^2} = \sqrt{25} = 5$

21. $\left|\vec{v}\right| = \sqrt{6^2 + (-7)^2} = \sqrt{85} \approx 9.2195$

23. $\left|\vec{v}\right| = \sqrt{(-2)^2 + 2^2} = \sqrt{8} = 2\sqrt{2} \approx 2.8284$

25. $\left|\vec{v}\right| = \sqrt{1^2 + (-3)^2} = \sqrt{10} \approx 3.1623$

27. $\left|\vec{v}\right| = \sqrt{4^2 + 5^2} = \sqrt{41} \approx 6.4031$

29. $\vec{v} \cdot \vec{w} = 2(6) + 3(-9) = -15$; no

31. $\vec{v} \cdot \vec{w} = 5(8) + 4(-10) = 0$; yes

33. $\vec{v} \cdot \vec{w} = 1(0) + 0(1) = 0$; yes

35. True course vector: $\vec{v} = -240\vec{i} - 43\vec{j}$

Ground speed $= \left|\vec{v}\right| = \sqrt{(-240)^2 + (-43)^2} \approx 244$ mph

$\tan\theta' = \frac{43}{240}$; $\theta' = \tan^{-1}\left(\frac{43}{240}\right) \approx 10°$

Bearing $=$ S80°W (heading 260°)

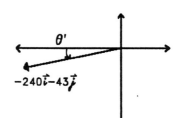

37. $\left|\vec{v}\right| =$ distance travelled $= 2 \cdot 248 = 496$ miles; $\theta = 90.0° - 43.0° = 47.0°$

The distance travelled east $= \vec{v}_x = 496\cos 47.0° \approx 338$ miles

39. $\vec{v} \cdot \vec{w} = 8(-5) + (-6)(12) = -112$; $\left|\vec{v}\right| = \sqrt{8^2 + (-6)^2} = \sqrt{100} = 10$;

$\left|\vec{w}\right| = \sqrt{(-5)^2 + 12^2} = \sqrt{169} = 13$; $\cos\theta = \frac{\vec{v} \cdot \vec{w}}{\left|\vec{v}\right|\left|\vec{w}\right|} = \frac{-112}{10 \cdot 13} = -\frac{56}{65}$

41. $\vec{v} \cdot \vec{w} = 7(2\sqrt{15}) + (-\sqrt{15})(14) = 0$; $|\vec{v}| = \sqrt{7^2 + (-\sqrt{15})^2} = \sqrt{64} = 8$;

$|\vec{w}| = \sqrt{(2\sqrt{15})^2 + 14^2} = \sqrt{256} = 16$; $\cos\theta = \dfrac{\vec{v} \cdot \vec{w}}{|\vec{v}||\vec{w}|} = \dfrac{0}{8 \cdot 16} = 0$

43. $\vec{v} \cdot \vec{w} = 3(2) + (9)(-5) = -39$; $|\vec{v}| = \sqrt{3^2 + 9^2} = \sqrt{90} = 3\sqrt{10}$;

$|\vec{w}| = \sqrt{2^2 + (-5)^2} = \sqrt{29}$; $\cos\theta = \dfrac{\vec{v} \cdot \vec{w}}{|\vec{v}||\vec{w}|} = \dfrac{-39}{3\sqrt{10} \cdot \sqrt{29}} = -\dfrac{13}{290}\sqrt{290}$

45. $\vec{v} \cdot \vec{w} = 0(0) + (1)(1) = 1$; $|\vec{v}| = \sqrt{0^2 + 1^2} = 1$;

$|\vec{w}| = \sqrt{0^2 + 1^2} = 1$; $\cos\theta = \dfrac{\vec{v} \cdot \vec{w}}{|\vec{v}||\vec{w}|} = \dfrac{1}{1 \cdot 1} = 1$

47. $\vec{v} \cdot \vec{w} = 5(2) + (-1)(3) = 7$; $|\vec{v}| = \sqrt{5^2 + (-1)^2} = \sqrt{26}$;

$|\vec{w}| = \sqrt{2^2 + 3^2} = \sqrt{13}$; $\cos\theta = \dfrac{\vec{v} \cdot \vec{w}}{|\vec{v}||\vec{w}|} = \dfrac{7}{\sqrt{26} \cdot \sqrt{13}} = \dfrac{7}{13\sqrt{2}} = \dfrac{7\sqrt{2}}{26}$

49. $\vec{v} \cdot \vec{w} = 1(1) + (1)(0) = 1$; $|\vec{v}| = \sqrt{1^2 + 1^2} = \sqrt{2}$;

$|\vec{w}| = \sqrt{1^2 + 0^2} = 1$; $\cos\theta = \dfrac{\vec{v} \cdot \vec{w}}{|\vec{v}||\vec{w}|} = \dfrac{1}{\sqrt{2} \cdot 1} = \dfrac{\sqrt{2}}{2}$

51. $\vec{v} \cdot \vec{w} = \sqrt{2}\left(\dfrac{\sqrt{3}}{2}\right) + (-\sqrt{2})(\tfrac{1}{2}) = \dfrac{\sqrt{6} - \sqrt{2}}{2}$; $|\vec{v}| = \sqrt{(\sqrt{2})^2 + (-\sqrt{2})^2} = \sqrt{4} = 2$;

$|\vec{w}| = \sqrt{\left(\dfrac{\sqrt{3}}{2}\right)^2 + \left(\dfrac{1}{2}\right)^2} = \sqrt{1} = 1$; $\cos\theta = \dfrac{\vec{v} \cdot \vec{w}}{|\vec{v}||\vec{w}|} = \dfrac{\frac{\sqrt{6} - \sqrt{2}}{2}}{2 \cdot 1} = \dfrac{\sqrt{6} - \sqrt{2}}{4}$

$\theta = \cos^{-1}\left(\dfrac{\sqrt{6} - \sqrt{2}}{4}\right) = 75°$

53. $\vec{v} \cdot \vec{w} = (-1)(-2\sqrt{2}) + (0)(2\sqrt{2}) = 2\sqrt{2}$; $|\vec{v}| = \sqrt{(-1)^2 + 0^2} = 1$;

$|\vec{w}| = \sqrt{(-2\sqrt{2})^2 + (2\sqrt{2})^2} = \sqrt{16} = 4$; $\cos\theta = \dfrac{\vec{v} \cdot \vec{w}}{|\vec{v}||\vec{w}|} = \dfrac{2\sqrt{2}}{1 \cdot 4} = \dfrac{\sqrt{2}}{2}$

$\theta = \cos^{-1}\left(\dfrac{\sqrt{2}}{2}\right) = 45°$

55. $\vec{v} \cdot \vec{w} = (-3)(6) + 2(9) = 0$; $\cos\theta = \dfrac{\vec{v} \cdot \vec{w}}{|\vec{v}||\vec{w}|} = \dfrac{0}{|\vec{v}||\vec{w}|} = 0$; $\theta = \cos^{-1}0 = 90°$

57. $\vec{v} \cdot \vec{w} = 4(-2) + (-a)(5) = 0$; $-8 - 5a = 0$; $-5a = 8$; $a = -\dfrac{8}{5}$

59. pilot's vector $= (241 \cos \theta)\vec{i} + (241 \sin \theta)\vec{j}$

wind vector $= -20.4\vec{i}$

plane's motion vector relative to the ground,

$$\vec{v} = (241 \cos \theta - 20.4)\vec{i} + (241 \sin \theta)\vec{j}$$

Since the plane is heading due north,

$$241 \cos \theta - 20.4 = 0$$
$$\cos \theta = \frac{20.4}{241}$$
$$\theta = \cos^{-1}\left(\frac{20.4}{241}\right) \approx 85.1°$$

So, the direction the pilot must head is $90° - \theta \approx 4.9°$ or N4.9°E

The ground speed $= |\vec{v}_y| = 241 \sin \theta \approx 240$ mph

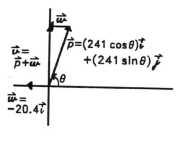

61. The weight of the astronaut is resolved into two
components, one parallel to the inclined plane with
length y and the other perpendicular to it with
length x. The weight of the astronaut is $|\vec{x}|$.

PROBLEM SET 11.6

1. $2\vec{i} - 3\vec{j}$

3. $\vec{i} - \vec{j}$

5. $3\vec{i} - 2\vec{j}$

7. $9\vec{i} + 7\vec{j}$

9. $4\vec{i} - \vec{j}$

11. $\frac{1}{2}\vec{i} + \vec{j} = \vec{i} + 2\vec{j}$

13. Find two points on the line, e.g. $A(-2, 0)$ and $B(1, 2)$. Then a vector determined by
the line is $\vec{v} = (1+2)\vec{i} + (2-0)\vec{j} = 3\vec{i} + 2\vec{j}$

15. As in problem 13: $A(-1, 2)$ and $B(0, 3)$; $\vec{v} = (0+1)\vec{i} + (3-2)\vec{j} = \vec{i} + \vec{j}$

17. As in problem 13: $A(1, 1)$ and $B(3, 4)$; $\vec{v} = (3-1)\vec{i} + (4-1)\vec{j} = 2\vec{i} + 3\vec{j}$

19. As in problem 13: $A(-1, 2)$ and $B(6, -7)$; $\vec{v} = (6+1)\vec{i} + (-7-2)\vec{j} = 7\vec{i} - 9\vec{j}$

21. As in problem 13: $A(3, 0)$ and $B(4, 4)$; $\vec{v} = (4-3)\vec{i} + (4-0)\vec{j} = \vec{i} + 4\vec{j}$

23. As in problem 13: $A(0, -10)$ and $B(2, -11)$; $\vec{v} = (2-0)\vec{i} + (-11+10)\vec{j} = 2\vec{i} - \vec{j}$

25. $\left| \dfrac{\vec{v} \cdot \vec{w}}{|\vec{w}|} \right| = \left| \dfrac{3(5) + 4(12)}{\sqrt{5^2 + 12^2}} \right| = \dfrac{63}{13}$

27. $\left| \dfrac{\vec{v} \cdot \vec{w}}{|\vec{w}|} \right| = \left| \dfrac{7(2\sqrt{15}) + (-\sqrt{15})(14)}{\sqrt{(2\sqrt{15})^2 + 14^2}} \right| = 0$

29. $\left| \dfrac{\vec{v} \cdot \vec{w}}{|\vec{w}|} \right| = \left| \dfrac{(-2)(6) + 3(5)}{\sqrt{6^2 + 5^2}} \right| = \dfrac{3}{\sqrt{61}} = \dfrac{3\sqrt{61}}{61}$

31. $\left(\dfrac{\vec{v} \cdot \vec{w}}{\vec{w} \cdot \vec{w}} \right) \vec{w} = \left(\dfrac{3(5) + 4(12)}{5^2 + 12^2} \right) (5\vec{i} + 12\vec{j}) = \dfrac{63}{169}(5\vec{i} + 12\vec{j}) = \dfrac{315}{169}\vec{i} + \dfrac{756}{169}\vec{j}$

33. $\left(\dfrac{\vec{v} \cdot \vec{w}}{\vec{w} \cdot \vec{w}} \right) \vec{w} = \left(\dfrac{7(2\sqrt{15}) + (-\sqrt{15})(14)}{(2\sqrt{15})^2 + 14^2} \right) (2\sqrt{15}\,\vec{i} + 14\vec{j}) = 0(2\sqrt{15}\,\vec{i} + 14\vec{j}) = \vec{0}$

35. $\left(\dfrac{\vec{v} \cdot \vec{w}}{\vec{w} \cdot \vec{w}} \right) \vec{w} = \left(\dfrac{(-2)(6) + 3(5)}{6^2 + 5^2} \right) (6\vec{i} + 5\vec{j}) = \dfrac{3}{61}(6\vec{i} + 5\vec{j}) = \dfrac{18}{61}\vec{i} + \dfrac{15}{61}\vec{j}$

37. $B(0, 2)$; $\overrightarrow{BP} = 4\vec{i} + 3\vec{j}$; $\vec{N} = 3\vec{i} - 4\vec{j}$; $d = \left| \dfrac{\overrightarrow{BP} \cdot \vec{N}}{|\vec{N}|} \right| = \left| \dfrac{4(3) + 3(-4)}{\sqrt{3^2 + (-4)^2}} \right| = 0$

39. $B(0, 2)$; $\overrightarrow{BP} = 9\vec{i} - 5\vec{j}$; $\vec{N} = 3\vec{i} - 4\vec{j}$; $d = \left| \dfrac{\overrightarrow{BP} \cdot \vec{N}}{|\vec{N}|} \right| = \left| \dfrac{9(3) + (-5)(-4)}{\sqrt{3^2 + (-4)^2}} \right| = \dfrac{47}{5}$

41. $B(-1, 3)$; $\overrightarrow{BP} = 0\vec{i} - 4\vec{j}$; $\vec{N} = 4\vec{i} + 3\vec{j}$; $d = \left| \dfrac{\overrightarrow{BP} \cdot \vec{N}}{|\vec{N}|} \right| = \left| \dfrac{0(4) + (-4)(3)}{\sqrt{4^2 + 3^2}} \right| = \dfrac{12}{5}$

43. $B(-1, 3)$; $\overrightarrow{BP} = 7\vec{i} - 2\vec{j}$; $\vec{N} = 4\vec{i} + 3\vec{j}$; $d = \left| \dfrac{\overrightarrow{BP} \cdot \vec{N}}{|\vec{N}|} \right| = \left| \dfrac{7(4) + (-2)(3)}{\sqrt{4^2 + 3^2}} \right| = \dfrac{22}{5}$

45. $B(0, 5)$; $\overrightarrow{BP} = \vec{i} - 11\vec{j}$; $\vec{N} = \vec{i} - 3\vec{j}$; $d = \left| \dfrac{\overrightarrow{BP} \cdot \vec{N}}{|\vec{N}|} \right| = \left| \dfrac{1(1) + (-11)(-3)}{\sqrt{1^2 + (-3)^2}} \right| = \dfrac{34}{\sqrt{10}} = \dfrac{17\sqrt{10}}{5}$

47. $B(0, 5)$; $\overrightarrow{BP} = 8\vec{i} + 9\vec{j}$; $\vec{N} = \vec{i} - 3\vec{j}$; $d = \left|\dfrac{\overrightarrow{BP} \cdot \vec{N}}{|\vec{N}|}\right| = \left|\dfrac{8(1) + 9(-3)}{\sqrt{1^2 + (-3)^2}}\right| = \dfrac{19}{\sqrt{10}} = \dfrac{19\sqrt{10}}{10}$

49. $B(5, 2)$; $\overrightarrow{BP} = -\vec{i} + 3\vec{j}$; $\vec{N} = 2\vec{i} - 5\vec{j}$; $d = \left|\dfrac{\overrightarrow{BP} \cdot \vec{N}}{|\vec{N}|}\right| = \left|\dfrac{(-1)(2) + 3(-5)}{\sqrt{2^2 + (-5)^2}}\right| = \dfrac{17}{\sqrt{29}} = \dfrac{17\sqrt{29}}{29}$

51. Area $= \frac{1}{2}bh$

$b = \sqrt{(4-1)^2 + (5-2)^2} = \sqrt{18} = 3\sqrt{2}$

h is the distance from P to the line $\overleftrightarrow{AB}$

Line $\overleftrightarrow{AB}$ has equation $y - 2 = \frac{5-2}{4-1} \cdot (x - 1)$

or $x - y = -1$. $\overrightarrow{BP} = -6\vec{i} + \vec{j}$ and $\vec{N} = \vec{i} - \vec{j}$.

$d = \left|\dfrac{\overrightarrow{BP} \cdot \vec{N}}{|\vec{N}|}\right| = \left|\dfrac{(-6)(1) + 1(-1)}{\sqrt{1^2 + (-1)^2}}\right| = \dfrac{7}{\sqrt{2}}$

Area $= \frac{1}{2}(3\sqrt{2})\left(\dfrac{7}{\sqrt{2}}\right) = \dfrac{21}{2}$

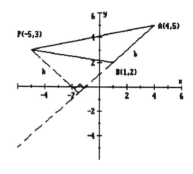

53. Area $= \frac{1}{2}bh$

$b = \sqrt{(5-0)^2 + (-3-0)^2} = \sqrt{34}$

h is the distance from P to the line $\overleftrightarrow{AB}$

Line $\overleftrightarrow{AB}$ has equation $y - 0 = \frac{-3-0}{5-0} \cdot (x - 0)$

or $3x + 5y = 0$. $\overrightarrow{BP} = -2\vec{i} - 7\vec{j}$ and $\vec{N} = 3\vec{i} + 5\vec{j}$.

$d = \left|\dfrac{\overrightarrow{BP} \cdot \vec{N}}{|\vec{N}|}\right| = \left|\dfrac{(-2)(3) + (-7)(5)}{\sqrt{3^2 + 5^2}}\right| = \dfrac{41}{\sqrt{34}}$

Area $= \frac{1}{2}(\sqrt{34})\left(\dfrac{41}{\sqrt{34}}\right) = \dfrac{41}{2}$

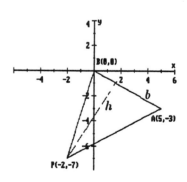

55. Area $= \frac{1}{2}bh$

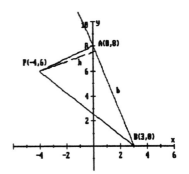

$b = \sqrt{(3-0)^2 + (0-8)^2} = \sqrt{73}$

h is the distance from P to the line $\overleftrightarrow{AB}$

Line $\overleftrightarrow{AB}$ has equation $y - 0 = \frac{8-0}{0-3} \cdot (x-3)$

or $8x + 3y = -24$. $\overrightarrow{BP} = -7\vec{i} + 6\vec{j}$ and $\vec{N} = 8\vec{i} + 3\vec{j}$.

$d = \left| \frac{\overrightarrow{BP} \cdot \vec{N}}{|\vec{N}|} \right| = \left| \frac{(-7)(8) + 6(3)}{\sqrt{8^2 + 3^2}} \right| = \frac{38}{\sqrt{73}}$

Area $= \frac{1}{2}(\sqrt{73})\left(\frac{38}{\sqrt{73}}\right) = \frac{38}{2} = 19$

57. $\vec{u} \cdot \vec{v} = (u_1)(v_1) + (u_2)(v_2)$
$= (v_1)(u_1) + (v_2)(u_2)$
$= \vec{v} \cdot \vec{u}$

59. $(\vec{v} \cdot \vec{u}) \cdot \vec{w} = \left[\vec{v} - \left(\frac{\vec{v} \cdot \vec{w}}{\vec{w} \cdot \vec{w}} \right) \vec{w} \right] \cdot \vec{w}$

$= \vec{v} \cdot \vec{w} - \left(\frac{\vec{v} \cdot \vec{w}}{\vec{w} \cdot \vec{w}} \right) \vec{w} \cdot \vec{w}$

$= \vec{v} \cdot \vec{w} - \vec{v} \cdot \vec{w}$

$= 0$

PROBLEM SET 11.7

1. $\frac{\pi}{4} + \pi = \frac{5\pi}{4}$;
$x = 4 \cos \frac{\pi}{4} = 4\left(\frac{\sqrt{2}}{2}\right) = 2\sqrt{2}, \quad y = 4 \sin \frac{\pi}{4} = 4\left(\frac{\sqrt{2}}{2}\right) = 2\sqrt{2}$;

P: $(4, \frac{\pi}{4})$, $\left(-4, \frac{5\pi}{4}\right)$
R: $(2\sqrt{2}, 2\sqrt{2})$

3. $\frac{2\pi}{3} + \pi = \frac{5\pi}{3}$;
$x = 5 \cos \frac{2\pi}{3} = 5\left(-\frac{1}{2}\right) = -\frac{5}{2}, \quad y = 5 \sin \frac{2\pi}{3} = 5\left(\frac{\sqrt{3}}{2}\right) = \frac{5\sqrt{3}}{2}$;

P: $(5, \frac{2\pi}{3})$, $(-5, \frac{5\pi}{3})$
R: $\left(-\frac{5}{2}, \frac{5}{2}\sqrt{3}\right)$

5. $-\frac{5\pi}{6} + 2\pi = \frac{7\pi}{6}$; $\quad -\frac{5\pi}{6} + \pi = \frac{\pi}{6}$;
$x = \frac{3}{2} \cos \frac{7\pi}{6} = \frac{3}{2}\left(-\frac{\sqrt{3}}{2}\right) = -\frac{3}{4}\sqrt{3}, \quad y = \frac{3}{2} \sin \frac{7\pi}{6} = \frac{3}{2}\left(-\frac{1}{2}\right) = -\frac{3}{4}$;

P: $(\frac{3}{2}, \frac{7\pi}{6})$, $(-\frac{3}{2}, \frac{\pi}{6})$
R: $\left(-\frac{3}{4}\sqrt{3}, -\frac{3}{4}\right)$

7. $4 - \pi \approx .86$;

    $x = -4\cos 4 \approx 2.61, \quad y = -4\sin 4 \approx 3.03$;

                                                     P: $(-4, 4), (4, .86)$

                                                     R: $(2.61, 3.03)$

9. $5\pi - 4\pi = \pi; \quad \pi - \pi = 0$;

    $x = -4\cos \pi = -4(-1) = 4, \quad y = -4\sin \pi = -4(0) = 0$;

                                                     P: $(-4, \pi), (4, 0)$

                                                     R: $(4, 0)$

11. $r = \sqrt{5^2 + 5^2} = 5\sqrt{2}; \quad \theta' = \tan^{-1}\left|\frac{5}{5}\right| = \tan^{-1} 1 = \frac{\pi}{4}$

    $\theta = \frac{\pi}{4}; \quad \frac{\pi}{4} + \pi = \frac{5\pi}{4}$ *P:* $\left(5\sqrt{2}, \frac{\pi}{4}\right), \left(-5\sqrt{2}, \frac{5\pi}{4}\right)$

13. $r = \sqrt{2^2 + (-2\sqrt{3})^2} = \sqrt{16} = 4; \quad \theta' = \tan^{-1}\left|\frac{-2\sqrt{3}}{2}\right|$

    $= \tan^{-1}\sqrt{3} = \frac{\pi}{3}; \quad \theta = 2\pi - \frac{\pi}{3} = \frac{5\pi}{3}; \quad \frac{5\pi}{3} - \pi = \frac{2\pi}{3}$ *P:* $\left(4, \frac{5\pi}{3}\right), \left(-4, \frac{2\pi}{3}\right)$

15. $r = \sqrt{3^2 + (-3)^2} = \sqrt{18} = 3\sqrt{2}; \quad \theta' = \tan^{-1}\left|\frac{-3}{3}\right|$

    $= \tan^{-1} 1 = \frac{\pi}{4}; \quad \theta = 2\pi - \frac{\pi}{4} = \frac{7\pi}{4}; \quad \frac{7\pi}{4} - \pi = \frac{3\pi}{4}$ *P:* $\left(3\sqrt{2}, \frac{7\pi}{4}\right), \left(-3\sqrt{2}, \frac{3\pi}{4}\right)$

17. $r = \sqrt{(-\sqrt{3})^2 + 1^2} = \sqrt{4} = 2; \quad \theta' = \tan^{-1}\left|\frac{1}{-\sqrt{3}}\right|$

    $= \tan^{-1}\frac{1}{\sqrt{3}} = \frac{\pi}{6}; \quad \pi - \frac{\pi}{6} = \frac{5\pi}{6}; \quad \frac{5\pi}{6} + \pi = \frac{11\pi}{6}$ *P:* $\left(2, \frac{5\pi}{6}\right), \left(-2, \frac{11\pi}{6}\right)$

19. $r = \sqrt{(-12)^2 + 5^2} = \sqrt{169} = 13; \quad \theta' = \tan^{-1}\left|\frac{5}{-12}\right| \approx .395$

    $\theta = \pi - \theta' \approx 2.75; \quad \theta + \pi \approx 5.89$ *P:* $(13, 2.75), (-13, 5.89)$

21. Lemniscate   **23.** 3-leaved rose   **25.** Cardiod   **27.** None

29. None   **31.** None   **33.** None   **35.** Cardiod

37.   **39.**   **41.**

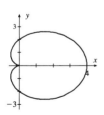   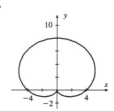   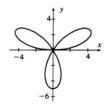

43.

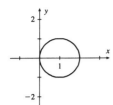

45.

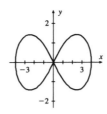

47.

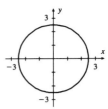

49.

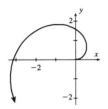

51. Given (r, θ) and (x, y). By definition of the trigonometric functions, $\cos \theta = \frac{x}{r}$ and $\sin \theta = \frac{y}{r}$. Therefore,

$$x = r \cos \theta \quad \text{and} \quad y = r \sin \theta$$

53. $\theta = \frac{\pi}{3} - \frac{\pi}{4} = \frac{\pi}{12}$

Using the law of cosines:

$$d = \sqrt{3^2 + 7^2 - 2(3)(7)\cos \frac{\pi}{12}} \approx 4.1751$$

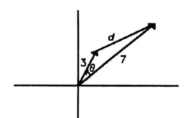

55. a.

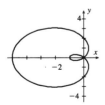

b.

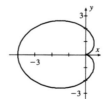

c.

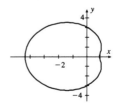

322

d.

57. **a.**

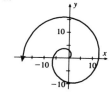

b.

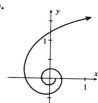

c.

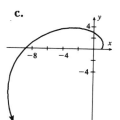

59.

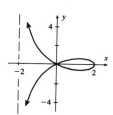

61.

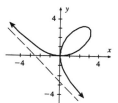

PROBLEM SET 11.8

1.

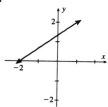

3.

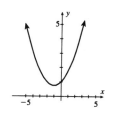

5.

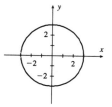

7.

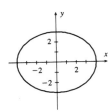

9.

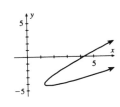

11.

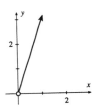

13.

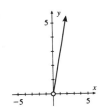

15. Since $x = t$, $y = 2 + \frac{2}{3}(x-1) = \frac{2}{3}x + \frac{4}{3}$ (same graph as #1)

17. Since $x = 2t$, $t = \frac{1}{2}x$; $y = (\frac{1}{2}x)^2 + \frac{1}{2}x + 1$; $y = \frac{1}{4}x^2 + \frac{1}{2}x + 1$; $4y = x^2 + 2x + 4$;

$4y - 4 + 1 = x^2 + 2x + 1$; $(x+1)^2 = 4(y - \frac{3}{4})$; (same graph as #3)

19. Since $x = 3\cos\theta$ and $y = 3\sin\theta$, then $\cos\theta = \frac{x}{3}$ and $\sin\theta = \frac{y}{3}$;

$\cos^2\theta + \sin^2\theta = 1$; $\left(\frac{x}{3}\right)^2 + \left(\frac{y}{3}\right)^2 = 1$; $\frac{x^2}{9} + \frac{y^2}{9} = 1$; $x^2 + y^2 = 9$; (same graph as #5)

21. Since $x = 4\cos\theta$ and $y = 3\sin\theta$, then $\cos\theta = \frac{x}{4}$ and $\sin\theta = \frac{y}{3}$;

$\cos^2\theta + \sin^2\theta = 1$; $\left(\frac{x}{4}\right)^2 + \left(\frac{y}{3}\right)^2 = 1$; $\frac{x^2}{16} + \frac{y^2}{9} = 1$ (same graph as #7)

23. $x - y = (t^2 + 2t + 3) - (t^2 + t - 4) = t + 7$, so $t = x - y - 7$. Since $y = t^2 + t - 4$,

$y = (x - y - 7)^2 + (x - y - 7) - 4 = x^2 - xy - 7x - xy + y^2 + 7y - 7x + 7y + 49 + x - y - 7 - 4$;

$x^2 - 2xy + y^2 - 13x + 12y + 38 = 0$. $B^2 - 4AC = (-2)^2 - 4(1)(1) = 0$, so the graph is

a parabola. $\cot 2\theta = \frac{A-C}{B} = \frac{1-1}{-2} = 0$. So $2\theta = 90°$, $\theta = 45°$ is the angle of rotation.

(same graph as #9)

25. $x = 3^t$, $y = 3^{t+1} = 3^t \cdot 3 = 3x$, so $y = 3x$, $x > 0$ (since $3^t > 0$ for all t) (same graph as #11)

27. $x = e^t$, $y = e^{t+1} = e^t \cdot e = ex$, so $y = ex$, $x > 0$ (since $e^t > 0$ for all t) (same graph as #13)

29. $m = \frac{9-0}{4-0} = \frac{9}{4}$; $y = \frac{9}{4}x$; let $x = 4t$, then $y = \frac{9}{4}(4t) = 9t$; $x = 4t$, $y = 9t$, $0 \le t \le 1$

31. $x = 3 - t$, $y = (3 - t)^2$, $0 \le t \le 3$

33. $\cos^2 t + \sin^2 t = 1$; $16 \cos^2 t + 16 \sin^2 t = 16$; $x = 4 \cos t$, $y = 4 \sin t$, $0 \le t \le \frac{\pi}{2}$

35. $\frac{x^2}{9} + \frac{y^2}{4} = 1$; $x = 3 \cos t$, $y = 2 \sin t$, $0 \le t \le 2\pi$

37. $x = t^2$, $y = \sqrt{t^2} = t$, $1 \le t \le 3$

39. $x = r \cos \theta$ and $y = r \sin \theta$, so $x = 2 \sin 5\theta \cos \theta$, $y = 2 \sin 5\theta \sin \theta$

41. Since $x = 60t$, $t = \frac{x}{60}$

So, $y = 80\left(\frac{x}{60}\right) - 16\left(\frac{x}{60}\right)^2$

$y = \frac{4}{3}x - \frac{1}{225}x^2$

$y = -\frac{1}{225}(x^2 - 300x)$

$y - 100 = -\frac{1}{225}(x^2 - 300x + 22,500)$

$y - 100 = \frac{1}{225}(x - 150)^2$

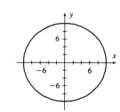

43. $x^2 + y^2 = (10 \cos t)^2 + (10 \sin t)^2$

$x^2 + y^2 = 100 \cos^2 t + 100 \sin^2 t$

$x^2 + y^2 = 100(\cos^2 t + \sin^2 t)$

$x^2 + y^2 = 100$

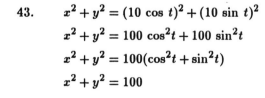

45. $x^2 + y^2 = (5 \cos \theta)^2 + (3 \sin \theta)^2$

$x^2 + y^2 = 25 \cos^2\theta + 9 \sin^2\theta$

$x^2 + y^2 = 16 \cos^2\theta + 9 \cos^2\theta + 9 \sin^2\theta$

$x^2 + y^2 = 16\left(\frac{x}{5}\right)^2 + 9(\cos^2\theta + \sin^2\theta)$

$x^2 + y^2 = \frac{16x^2}{25} + 9$

$\frac{9x^2}{25} + y^2 = 9$

$\frac{x^2}{25} + \frac{y^2}{9} = 1$

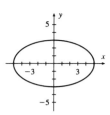

47. Plot points:

t	x	y
-2	4	-8
-1	1	-1
0	0	0
1	1	1
2	4	8

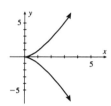

49. $x = e^t$

$y = e^{t+2} = e^t e^2 = x \cdot e^2 = e^2 x, \; x > 0$

(since $e^t > 0$ for all t)

$e = c^2$

51. Note that the value of y with respect to θ is periodic with a period of 2π. As θ increases in value from θ to $\theta + 2\pi$ the change in value of x is $[(\theta + 2\pi) - \sin(\theta + 2\pi)] - (\theta - \sin \theta)$
$= \theta + 2\pi - \sin \theta - \theta + \sin \theta = 2\pi$. So, every time θ increases by 2π, x increases by 2π and y repeats in value. Therefore, the value of y is periodic with respect to the value of x with a period of 2π. Plotting points between $t = 0$ and $t = 2\pi$ results in one period. This curve repeats every 2π units.

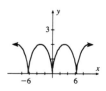

326

53. $y^2 - x^2 = (3 \sec 2t)^2 - (4 \tan 2t)^2$

$y^2 - x^2 = 9 \sec^2 2t - 16 \tan^2 2t$

$y^2 - x^2 = 9 \sec^2 2t - 9 \tan^2 2t - 7 \tan^2 2t$

$y^2 - x^2 = 9(\sec^2 2t - \tan^2 2t) - 7 \tan^2 2t$

$y^2 - x^2 = 9 - 7\left(\frac{x}{4}\right)^2$

$y^2 - \frac{9}{16}x^2 = 9$

$\frac{y^2}{9} - \frac{x^2}{16} = 1$

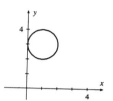

55. $\cos t = x - 1,\ \sin t = 3 - y$

$(x - 1)^2 + (3 - y)^2 = \cos^2 t + \sin^2 t$

$(x - 1)^2 + (y - 3)^2 = 1$

circle with center $(1, 3)$ and radius 1

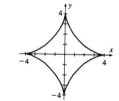

57. Plot points:

θ	x	y
0	4	0
$\frac{\pi}{6}$	2.6	0.5
$\frac{\pi}{2}$	0	4
π	-4	0

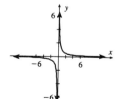

59. $y = \cos t = \frac{1}{\tan t} = \frac{1}{x}$

$y = \frac{1}{x}$

61. $x = |OA| = |OB| - |AB| = \text{arc } PB - |PQ| = a\theta - a \sin \theta$

$y = |PA| = |QB| = |CB| - |CQ| = a - a \cos \theta$

CHAPTER 11 SUMMARY

1.

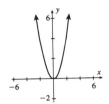

3. $8y^2 - x - 32y + 31 = 0$

$$8(y^2 - 4y + 4) = x - 31 + 32$$

$$8(y - 2)^2 = x + 1$$

$$(y - 2)^2 = \tfrac{1}{8}(x + 1)$$

5. By inspection the parabola opens right and $c = 5$. So $(y - 3)^2 = 20(x - 6)$

7. By inspection the parabola opens down and $c = 6$. So $(x + 3)^2 = -24(y - 5)$

9. $25x^2 + 16y^2 = 400$

$$\frac{x^2}{16} + \frac{y^2}{25} = 1$$

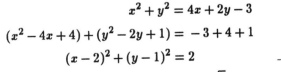

11. $x^2 + y^2 = 4x + 2y - 3$

$$(x^2 - 4x + 4) + (y^2 - 2y + 1) = -3 + 4 + 1$$

$$(x - 2)^2 + (y - 1)^2 = 2$$

Circle with center $(2, 1)$ and radius $\sqrt{2}$

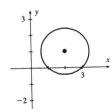

328

13. By inspection $c = 1$ and $a = 2$; $c^2 = a^2 - b^2$; $1 = 4 - b^2$; $b^2 = 3$.

So, $\dfrac{(x-4)^2}{4} + \dfrac{(y-1)^2}{3} = 1$]

15. This is a circle with center $(-1, -2)$ and radius 8; $(x+1)^2 + (y+2)^2 = 64$

17.
$$x^2 - y^2 + x - y = 3$$
$$(x^2 + x + \tfrac{1}{4}) - (y^2 + y + \tfrac{1}{4}) = 3 + \tfrac{1}{4} - \tfrac{1}{4}$$
$$(x + \tfrac{1}{2})^2 - (y + \tfrac{1}{2})^2 = 3$$
$$\dfrac{(x + \tfrac{1}{2})^2}{3} - \dfrac{(y + \tfrac{1}{2})^2}{3} = 1$$

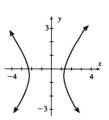

19.
$$5(x+2)^2 - 3(y+4)^2 = 60$$
$$\dfrac{(x+2)^2}{12} - \dfrac{(y+4)^2}{20} = 1$$

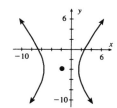

21. By inspection this is a horizontal hyperbola, center is $(-5, 4)$, and $c = 2$. Also, $2a = 2$ so $a = 1$. Using $c^2 = a^2 + b^2$, $4 = 1 + b^2$, so $b^2 = 3$. So, $\dfrac{(x+5)^2}{1} - \dfrac{(y-4)^2}{3} = 1$.

23. By inspection this is a vertical hyperbola, center is $(0, 0)$, and $a = 3$. Since $\epsilon = \dfrac{c}{a}$, $\dfrac{5}{3} = \dfrac{c}{3}$, so $c = 5$. Using $c^2 = a^2 + b^2$, $25 = 9 + b^2$, so $b^2 = 16$. So, $\dfrac{y^2}{9} - \dfrac{x^2}{16} = 1$.

25. $\dfrac{(x-h)^2}{a^2} + \dfrac{(y-k)^2}{b^2} = 1$

27. $(y-k)^2 = 4c(x-h)$, $c > 0$

329

29. Parabola:

$$3x - 2y^2 - 4y + 7 = 0$$
$$-2(y^2 + 2y + 1) = -3x - 7 - 2$$
$$-2(y+1)^2 = -3(y+3)$$
$$(y+1)^2 = \tfrac{3}{2}(y+3)$$

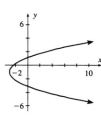

31. Ellipse:

$$25x^2 + 9y^2 = 225$$
$$\frac{x^2}{9} + \frac{y^2}{25} = 1$$

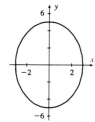

33. $\cot 2\theta = \dfrac{A-C}{B} = \dfrac{0-0}{1} = 0.$ So $2\theta = 90°$ and $\theta = 45°$.

35. $\cot 2\theta = \dfrac{A-C}{B} = \dfrac{4-1}{4} = \dfrac{3}{4}.$ So $2\theta \approx 53.13$ and $\theta \approx 26.6°$.

37. $B^2 - 4AC = 1 - 4(1)(0) = 1 > 0;$ Hyperbola

39. $B^2 - 4AC = 2^2 - 4(1)(1) = 0;$ Parabola

41. $\vec{v} + \vec{w} = (5-7)\vec{i} + (-2+3)\vec{j} = -2\vec{i} + \vec{j}$

43. $\alpha = 50°$ (alternate interior angles)

Using the law of cosines:

$$d = \sqrt{8^2 + 9^2 - 2(8)(9) \cos 70°} \approx 9.8$$
$$\frac{\sin(\gamma + 40°)}{8} = \frac{\sin 70°}{d}; \quad \gamma = \sin^{-1}\left(\frac{8 \sin 70°}{d}\right) - 40° \approx 10°$$

Direction N80°E

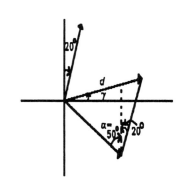

45. $\vec{v}_x = (4.5 \cos 51°)\vec{i} \approx 2.8\vec{i}; \quad \vec{v}_y = (4.5 \sin 51°)\vec{j} \approx 3.5\vec{j}$

47. $\overrightarrow{AB} = (5-0)\vec{i} + (12-0)\vec{j} = 5\vec{i} + 12\vec{j}$

49. $|\vec{v}| = \sqrt{(5)^2 + (-2)^2} = \sqrt{29}$

51. $|\vec{v}| = \sqrt{(6)^2 + (0)^2} = 6$

53. Let $\vec{v} = a\vec{i} + b\vec{j}$ and $\vec{w} = c\vec{i} + d\vec{j}$. Then the scalar product is $\vec{v} \cdot \vec{w} = ac + bd$.

55. $(3\vec{i} + 5\vec{j}) \cdot (-2\vec{i} + 3\vec{j}) = 3(-2) + 5(3) = 9$

57. $\dfrac{\vec{v} \cdot \vec{w}}{|\vec{v}||\vec{w}|} = \dfrac{3(5) + (-2)(12)}{\sqrt{(3)^2 + (-2)^2}\,\sqrt{(5)^2 + (12)^2}} = \dfrac{-9}{13\sqrt{13}} = -\dfrac{9}{169}\sqrt{13}$

59. $\dfrac{\vec{v} \cdot \vec{w}}{|\vec{v}||\vec{w}|} = \dfrac{\cos 15° \cos 20° + \sin 15° \sin 20°}{\sqrt{\cos^2 15° + \sin^2 15°}\,\sqrt{\cos^2 20° + \sin^2 20°}}$

$= \dfrac{\cos 15° \cos 20° + \sin 15° \sin 20°}{\sqrt{1}\,\sqrt{1}}$

$= \cos(15° - 20°)$

$= \cos(-5°)$

$= \cos 5°$

61. Normal vector: $\vec{v} = 5\vec{i} - 12\vec{j}$. Choose two points on the line, e.g. $A(-3, -1)$ and $B(9, 4)$.
A vector determined by the line is $\vec{w} = (9+3)\vec{i} + (4+1)j = 12\vec{i} + 5\vec{j}$ (or any scalar multiple).
$\vec{v} \cdot \vec{w} = 5(12) + (-12)(5) = 0$

63. Normal vector: $\vec{i} - 5\vec{j}$. Choose two points on the line, e.g. $A(-4, 0)$ and $B(1, 1)$.
A vector determined by the line is $\vec{w} = (1+4)\vec{i} + (1-0)\vec{j} = 5\vec{i} + \vec{j}$ (or any scalar multiple).
$\vec{v} \cdot \vec{w} = 1(5) + (-5)(1) = 0$

65. Vector projection: $\left(\dfrac{\vec{v} \cdot \vec{w}}{\vec{w} \cdot \vec{w}}\right)\vec{w} = \left(\dfrac{2(3\sqrt{5}) + \sqrt{5}(3)}{(3\sqrt{5})^2 + (3)^2}\right)(3\sqrt{5}\vec{i} + 3\vec{j})$

$= \dfrac{9\sqrt{5}}{54}(3\sqrt{5}\vec{i} + 3\vec{j}) = \tfrac{5}{2}\vec{i} + \dfrac{\sqrt{5}}{2}\vec{j}$

Scalar projection: $\left|\dfrac{\vec{v} \cdot \vec{w}}{|\vec{w}|}\right| = \dfrac{2(3\sqrt{5}) + \sqrt{5}(3)}{\sqrt{(3\sqrt{5})^2 + (3)^2}} = \dfrac{9\sqrt{5}}{\sqrt{54}} = \dfrac{9\sqrt{5}}{3\sqrt{6}} = \dfrac{3\sqrt{5}}{\sqrt{6}} \cdot \dfrac{\sqrt{6}}{\sqrt{6}} = \dfrac{\sqrt{30}}{2}$

67. Vector projection: $\left(\frac{\vec{v}\cdot\vec{w}}{\vec{w}\cdot\vec{w}}\right)\vec{w}=\left(\frac{9(1)+1(-9)}{(1)^2+(-9)^2}\right)(\vec{i}-9\vec{j})=\frac{0}{82}(\vec{i}-9\vec{j})=\vec{0}$

Scalar projection: $\left|\frac{\vec{v}\cdot\vec{w}}{|\vec{w}|}\right|=\frac{9(1)+1(-9)}{\sqrt{(1)^2+(-9)^2}}=\frac{0}{\sqrt{82}}=0$

69. Choose a point on the line, e.g. $B(8,-4)$; $\overrightarrow{BP}=(8-1)\vec{i}+(-4-5)\vec{j}=7\vec{i}-9\vec{j}$

Normal vector: $\vec{N}=5\vec{i}+12\vec{j}$

$d=\left|\frac{\overrightarrow{BP}\cdot\vec{N}}{|\vec{N}|}\right|=\left|\frac{7(5)+(-9)(12)}{\sqrt{(5)^2+(12)^2}}\right|=\left|\frac{-73}{13}\right|=\frac{73}{13}$

71. Choose a point on the line, e.g. $B(1,-1)$; $\overrightarrow{BP}=(-10-1)\vec{i}+(2+1)\vec{j}=-11\vec{i}+3\vec{j}$

Normal vector: $\vec{N}=2\vec{i}-3\vec{j}$

$d=\left|\frac{\overrightarrow{BP}\cdot\vec{N}}{|\vec{N}|}\right|=\left|\frac{(-11)(2)+3(-3)}{\sqrt{(2)^2+(-3)^2}}\right|=\left|\frac{-31}{\sqrt{13}}\right|=\frac{31\sqrt{13}}{13}$

73. $(5,\ \sqrt{75})\approx(5,\ 8.6603),\ \theta>2\pi;\ (5,\ \sqrt{75}-2\pi)\approx(5,\ 2.3771)$

$(-5,\ (\sqrt{75}-2\pi)+\pi)=(-5,\ 5.5187)$

75. $(-2,\ 2)$ is one primary representation. The other is $(2,\ 2+\pi)=(2,\ 5.1416)$

77. $x=r\cos\theta=3\cos\left(-\frac{2\pi}{3}\right)=3\left(-\frac{1}{2}\right)=-1.5000$

$y=r\sin\theta=3\sin\left(-\frac{2\pi}{3}\right)=3\left(-\frac{\sqrt{3}}{2}\right)=-2.5981$

The rectangular form is $(-1.5000,\ -2.5981)$

79. $r=\sqrt{x^2+y^2}=\sqrt{(3)^2+(-3)^2}=\sqrt{18}\approx4.2426;\ \theta'=\tan^{-1}\left|\frac{-3}{3}\right|=\tan^{-1}1=\frac{\pi}{4};$

Since $(3,\ -3)$ is in quadrant III, $\theta=2\pi-\theta'=\frac{7\pi}{4}\approx5.4978$

The polar form is $(4.2426,\ 5.4978)$

81.

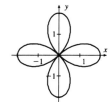

83.

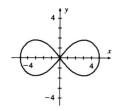

85. a. Lemniscate b. 4-leaved rose

87. a. None b. Cardioid

89. Plot points:

t	x	y
-2	-8	5
-1	-3	2
0	2	-1
1	7	-4
2	10	-7

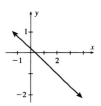

91. Plot points:

t	x	y
-2	e^{-8}	e^{-10}
-1	e^{-4}	e^{-6}
0	1	e^{-2}
1	e^{4}	e^{2}
2	e^{8}	e^{6}

93. Since $x = 2 + 5t$, then $t = \frac{x-2}{5}$; so $y = -1 - 3\left(\frac{x-2}{5}\right)$; $y = -1 - \frac{3}{5}x + \frac{6}{5}$;

$y = -\frac{3}{5}x + \frac{1}{5}$; $5y = -3x + 1$; $3x + 5y - 1 = 0$; same graph as problem 89.

95. $y = e^{4t-2} = e^{4t} \cdot e^{-2} = x \cdot e^{-2}$; $y = e^{-2}x$, $x > 0$; same graph as problem 91.

CUMULATIVE REVIEW III

1. a. $a_4 = 20\left(\frac{2}{3}\right) = \frac{40}{3}$; geometric sequence; $r = \frac{30}{45} = \frac{2}{3}$

   b. $a_4 = 15 + (-15) = 0$; arithmetic sequence; $d = 30 - 45 = -15$

   c. The next term is 30; neither;

   the odd terms form an arithmetic sequence, the even terms are 30.

2. a. $2 \begin{cases} 5x + 3y = 5 \\ -3 \begin{cases} 3x + 2y = 4 \end{cases} \end{cases}$

   $+ \begin{cases} 10x + 6y = 10 \\ -9x - 6y = -12 \end{cases}$

   $x \qquad = -2$

   $3(-2) + 2y = 4$

   $2y = 10$

   $y = 5$

   $(x, y) = (-2, 5)$

   b. $\begin{cases} x + y = 1 \\ y = x^2 + 4x + 5 \end{cases}$

   $x + (x^2 + 4x + 5) = 1$

   $x^2 + 5x + 4 = 0$

   $(x+4)(x+1) = 0$

   $x = -4$ or $x = -1$

   When $x = -4$, $(-4) + y = 1$, $y = 5$

   When $x = -1$, $(-1) + y = 1$, $y = 2$

   $(-1, 2)$, $(-4, 5)$

   c.

   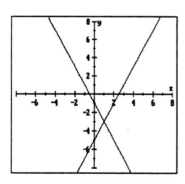

   $(x, y) = (1, -3)$

334

d. $\begin{bmatrix} 1 & 0 & 1 & \vdots & 1 \\ 1 & 1 & 1 & \vdots & 0 \\ 2 & 1 & 0 & \vdots & 3 \end{bmatrix} \xrightarrow[-2R1+R3]{-R1+R2} \begin{bmatrix} 1 & 0 & 1 & \vdots & 1 \\ 0 & 1 & 0 & \vdots & -1 \\ 0 & 1 & -2 & \vdots & 1 \end{bmatrix}$

$\xrightarrow{-R2+R3} \begin{bmatrix} 1 & 0 & 1 & \vdots & 1 \\ 0 & 1 & 0 & \vdots & -1 \\ 0 & 0 & -2 & \vdots & 2 \end{bmatrix} \xrightarrow{R3 \div 2} \begin{bmatrix} 1 & 0 & 1 & \vdots & 1 \\ 0 & 1 & 0 & \vdots & -1 \\ 0 & 0 & 1 & \vdots & -1 \end{bmatrix}$

$\xrightarrow{-R3+R1} \begin{bmatrix} 1 & 0 & 0 & \vdots & 2 \\ 0 & 1 & 0 & \vdots & -1 \\ 0 & 0 & 1 & \vdots & -1 \end{bmatrix}$ $\qquad (x,\ y,\ z) = (2,\ -1,\ -1)$

3. a. $A + BC = \begin{bmatrix} 1 & 0 \\ 2 & -1 \end{bmatrix} + \begin{bmatrix} 0 & 1 & 0 \\ 1 & 0 & -1 \end{bmatrix} \cdot \begin{bmatrix} 1 & 0 \\ -2 & 1 \\ 0 & 1 \end{bmatrix}$

$= \begin{bmatrix} 1 & 0 \\ 2 & -1 \end{bmatrix} + \begin{bmatrix} 0(1)+1(-2)+0(0) & 0(0)+1(1)+0(1) \\ 1(1)+0(-2)+(-1)(0) & 1(0)+0(1)+(-1)(1) \end{bmatrix}$

$= \begin{bmatrix} 1 & 0 \\ 2 & -1 \end{bmatrix} + \begin{bmatrix} -2 & 1 \\ 1 & -1 \end{bmatrix} = \begin{bmatrix} 1-2 & 0+1 \\ 2+1 & -1-1 \end{bmatrix} = \begin{bmatrix} -1 & 1 \\ 3 & -2 \end{bmatrix}$

b. Not conformable: # of columns of $A \neq$ # or rows of C

c. From part a: $BC = \begin{bmatrix} -2 & 1 \\ 1 & -1 \end{bmatrix}$. So,

$BCA = (BC)A$

$= \begin{bmatrix} -2 & 1 \\ 1 & -1 \end{bmatrix} \cdot \begin{bmatrix} 1 & 0 \\ 2 & -1 \end{bmatrix}$

$= \begin{bmatrix} (-2)(1)+1(2) & (-2)(0)+1(-1) \\ 1(1)+(-1)(2) & 1(0)+(-1)(-1) \end{bmatrix}$

$= \begin{bmatrix} 0 & -1 \\ -1 & 1 \end{bmatrix}$

4. a. Parabola   b. Hyperbola

   c. $B^2 - 4AC = 1^2 - 4(0)(1) = 1 > 0$; Hyperbola

   d. Circle   e. Line

   f. $xy + x - y - 8 = 0$; $B^2 - 4AC = 1 - 4(0)(0) = 1 > 0$;   Hyperbola

   g. 4-leaved rose   h. Lemniscate

   i. Cardiod   j. Circle

5. a. By inspection the ellipse is horizontal with $c = 1$ and $a = 2$. Since $c^2 = a^2 - b^2$,

   $1 = 4 - b^2$, so $b^2 = 3$. The equation is $\dfrac{(x-2)^2}{4} + \dfrac{(y-1)^2}{3} = 1$

   b. This is a hyperbola with center $(6, 0)$ and $c = 3$. Also, $2a = 4$, so $a = 2$. Since,

   $c^2 = a^2 + b^2$, $9 = 4 + b^2$, so $b^2 = 5$. The equation is: $\dfrac{(x-6)^2}{4} - \dfrac{y^2}{5} = 1$

   c. Since the vertex is vertical and left of the vertex, the parabola opens right.

   $c = 4 - 1 = 3$. The equation is $(y-3)^2 = 12(x-4)$

6. a.

$$3x - 2y^2 - 4y + 7 = 0$$
$$-2(y^2 + 2y + 1) = -3x - 7 - 2$$
$$-2(y+1)^2 = -3(x+3)$$
$$(y+1)^2 = \tfrac{3}{2}(x+3)$$

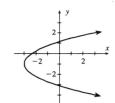

   b.

$$25x^2 - 16y^2 = 400$$
$$\frac{x^2}{16} - \frac{y^2}{25} = 1$$

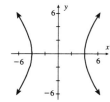

336

7.   a.
$$x^2 + y^2 = (4 \cos \theta)^2 + (3 \sin \theta)^2$$
$$x^2 + y^2 = 16 \cos^2\theta + 9 \sin^2\theta$$
$$x^2 + y^2 = 7 \cos^2\theta + 9 \cos^2\theta + 9 \sin^2\theta$$
$$x^2 + y^2 = 7\left(\frac{x}{4}\right)^2 + 9$$
$$\frac{9x^2}{16} + y^2 = 9$$
$$\frac{x^2}{16} + \frac{y^2}{9} = 1$$

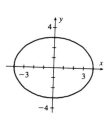

b.

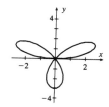

c.

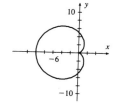

8.   a. $(a+b)^n = \sum_{k=0}^{n} \binom{n}{k} a^{n-k} b^k$

b. $(a+b)^n = \binom{n}{0} a^n + \binom{n}{1} a^{n-1} b + \cdots + \binom{n}{n-1} ab^{n-1} + \binom{n}{n} b^n$

c. $(x-3y)^4 = \binom{4}{0}x^4 + \binom{4}{1}x^3(-3y) + \binom{4}{2}x^2(-3y)^2 + \binom{4}{3}x(-3y)^3 + \binom{4}{4}(-3y)^4$

$$= \frac{4!}{0!\cdot 4!}x^4 + \frac{4!}{1!\cdot 3!}(-3x^3y) + \frac{4!}{2!\cdot 2!}(9x^2y^2) + \frac{4!}{3!\cdot 1!}(-27xy^3) + \frac{4!}{4!\cdot 0!}(81y^4)$$

$$= x^4 - 12x^3y + 54x^2y^2 - 108xy^3 + 81y^4$$

9. **Step 1:** Prove $P(1)$: $\qquad 1^3 = \dfrac{1^2(1+1)^2}{4}$

$$1 = 1; \ \ \text{TRUE}$$

Step 2: Assume $P(k)$: $\qquad 1^3 + 2^3 + 3^3 + \cdots + k^3 = \dfrac{k^2(k+1)^2}{4}$

Step 3: Prove $P(k+1)$: $\qquad 1^3 + 2^3 + 3^3 + \cdots + (k+1)^3 = \dfrac{(k+1)^2(k+2)^2}{4}$

1. $\qquad 1^3 + 2^3 + 3^3 + \cdots + k^3 = \dfrac{k^2(k+1)^2}{4}$ 1. **Hypothesis**

2. $\quad 1^3 + 2^3 + 3^3 + \cdots + k^3 + (k+1)^3 = \dfrac{k^2(k+1)^2}{4} + (k+1)^3$ 2. **Add $(k+1)^3$ to both sides**

3. $\quad 1^3 + 2^3 + 3^3 + \cdots + k^3 + (k+1)^3 = \dfrac{k^2(k+1)^2 + 4(k+1)^3}{4}$ 3. **Add fractions**

4. $\quad 1^3 + 2^3 + 3^3 + \cdots + k^3 + (k+1)^3 = \dfrac{(k+1)^2[k^2 + 4(k+1)]}{4}$ 4. **Factor out $(k+1)^2$**

5. $\quad 1^3 + 2^3 + 3^3 + \cdots + k^3 + (k+1)^3 = \dfrac{(k+1)^2[k^2 + 4k + 4]}{4}$ 5. **Simplify**

6. $\quad 1^3 + 2^3 + 3^3 + \cdots + k^3 + (k+1)^3 = \dfrac{(k+1)^2(k+2)^2}{4}$ 6. **Factor**

Step 4: The proposition is true for all positive integers n by PMI.

10. a. $\vec{v} - 2\vec{w} = (5\vec{i} - \sqrt{5}\,\vec{j}) - (-4\vec{i} + 6\sqrt{5})\vec{j} = (5+4)\vec{i} + (-\sqrt{5} - 6\sqrt{5})\vec{j} = 9\vec{i} - 7\sqrt{5}\,\vec{j}$

b. $\vec{v} \cdot \vec{w} = (5)(-2) + (-\sqrt{5})(3\sqrt{5}) = -10 - 15 = -25$

c. $|\vec{w}| = \sqrt{(-2)^2 + (3\sqrt{5})^2} = \sqrt{4+45} = \sqrt{49} = 7$

d. $\dfrac{\vec{v}}{|\vec{v}|} = \dfrac{5\vec{i} - \sqrt{5}\,\vec{j}}{\sqrt{(5)^2 + (\sqrt{5})^2}} = \dfrac{5\vec{i} - \sqrt{5}\,\vec{j}}{\sqrt{30}} = \dfrac{5}{\sqrt{30}}\vec{i} - \dfrac{\sqrt{5}}{\sqrt{30}}\vec{j} = \tfrac{1}{6}\sqrt{30}\,\vec{i} - \tfrac{1}{6}\sqrt{6}\,\vec{j}$

EXTENDED APPLICATION: PLANETARY ORBITS

1. $c = a\epsilon = (1.4 \times 10^8)(.093) \approx 1.3 \times 10^7$. The greatest distance is $a + c \approx 1.5 \times 10^8$ miles.
   The least distance is $a - c \approx 1.3 \times 10^8$ miles.

3. $c = a\epsilon = (.009)(3.66 \times 10^9) \approx 3.20 \times 10^7$. The greatest distance is $a + c \approx 3.69 \times 10^9$ miles.
   The least distance is $a + c \approx 3.63 \times 10^9$ miles.

5. $(\text{Semimajor axis of Phobos})^3$

$$= \frac{\text{mass of (Mars + Phobos)}}{\text{mass of (Sun + Mars + Phobos)}} \times \frac{(\text{period of Phobos})^2}{(\text{period of Mars})^2} \times (\text{semimajor axis of Mars})^3$$

$$= \frac{7.05 \times 10^{20} + 4.16 \times 10^{12}}{2.2 \times 10^{27} + 7.05 \times 10^{20} + 4.16 \times 10^{12}} \times \frac{(.3)^2}{(693.5)^2} \times (1.4 \times 10^8)^3 \approx 1.65 \times 10^{11}$$

The semimajor axis of Phobos' orbit is approximately $\sqrt[3]{1.65 \times 10^{11}} \approx 5,500$ miles

7. $a + c = 4.3 \times 10^7$ and $a - c = 2.9 \times 10^7$. So $(a + c) + (a - c) = 2a = 7.2 \times 10^7$. Thus,
   $a = 3.6 \times 10^7$ and $c = 4.3 \times 10^7 - a = .7 \times 10^2$. Also, $a^2 \approx 1.3 \times 10^{15}$ and $b^2 = a^2 - c^2$
   $\approx 1.2 \times 10^{15}$. The equation of orbit is: $\dfrac{x^2}{1.3 \times 10^{15}} + \dfrac{y^2}{1.2 \times 10^{15}} = 1$

   Multiplying both sides by $(1.2)(1.3) \times 10^{15}$ yields: $1.2x^2 + 1.3y^2 = 1.6 \times 10^{15}$

9. $a + c = 9.542 \times 10^7$ and $a - c = 9.225 \times 10^7$. So $(a + c) + (a - c) = 2a = 18.767 \times 10^7$.
   Thus, $a = 9.3855 \times 10^7$ and $c = 9.542 \times 10^7 - a = .1585 \times 10^7$. Also, $a^2 \approx 8.805 \times 10^{15}$
   and $b^2 = a^2 - c^2 \approx 8.802 \times 10^{15}$. The equation of orbit is: $\dfrac{x^2}{8.805 \times 10^{15}} + \dfrac{y^2}{8.802 \times 10^{15}} = 1$.

   Multiplying both sides by $(8.805)(8.802) \times 10^{15}$ yields: $8.802x^2 + 8.805y^2 = 7.751 \times 10^{16}$